Lecture Notes in Computer Science 14919

Founding Editors

Gerhard Goos
Juris Hartmanis

Editorial Board Members

Elisa Bertino, *Purdue University, West Lafayette, IN, USA*
Wen Gao, *Peking University, Beijing, China*
Bernhard Steffen , *TU Dortmund University, Dortmund, Germany*
Moti Yung , *Columbia University, New York, NY, USA*

The series Lecture Notes in Computer Science (LNCS), including its subseries Lecture Notes in Artificial Intelligence (LNAI) and Lecture Notes in Bioinformatics (LNBI), has established itself as a medium for the publication of new developments in computer science and information technology research, teaching, and education.

LNCS enjoys close cooperation with the computer science R & D community, the series counts many renowned academics among its volume editors and paper authors, and collaborates with prestigious societies. Its mission is to serve this international community by providing an invaluable service, mainly focused on the publication of conference and workshop proceedings and postproceedings. LNCS commenced publication in 1973.

Juliana Bowles · Harald Søndergaard
Editors

Logic-Based Program Synthesis and Transformation

34th International Symposium, LOPSTR 2024
Milan, Italy, September 9–10, 2024
Proceedings

Editors
Juliana Bowles ⓘ
University of St Andrews
Scotland, UK

Harald Søndergaard ⓘ
University of Melbourne
Melbourne, VIC, Australia

ISSN 0302-9743   ISSN 1611-3349 (electronic)
Lecture Notes in Computer Science
ISBN 978-3-031-71293-7   ISBN 978-3-031-71294-4 (eBook)
https://doi.org/10.1007/978-3-031-71294-4

© The Editor(s) (if applicable) and The Author(s), under exclusive license
to Springer Nature Switzerland AG 2024

This work is subject to copyright. All rights are solely and exclusively licensed by the Publisher, whether the whole or part of the material is concerned, specifically the rights of translation, reprinting, reuse of illustrations, recitation, broadcasting, reproduction on microfilms or in any other physical way, and transmission or information storage and retrieval, electronic adaptation, computer software, or by similar or dissimilar methodology now known or hereafter developed.
The use of general descriptive names, registered names, trademarks, service marks, etc. in this publication does not imply, even in the absence of a specific statement, that such names are exempt from the relevant protective laws and regulations and therefore free for general use.
The publisher, the authors and the editors are safe to assume that the advice and information in this book are believed to be true and accurate at the date of publication. Neither the publisher nor the authors or the editors give a warranty, expressed or implied, with respect to the material contained herein or for any errors or omissions that may have been made. The publisher remains neutral with regard to jurisdictional claims in published maps and institutional affiliations.

This Springer imprint is published by the registered company Springer Nature Switzerland AG
The registered company address is: Gewerbestrasse 11, 6330 Cham, Switzerland

If disposing of this product, please recycle the paper.

Preface

This volume contains the papers selected for presentation at LOPSTR 2024, the *34th International Symposium on Logic-Based Program Synthesis and Transformation*, held in Milan, Italy, September 9–10, 2024. The symposium was held in conjunction with PPDP 2024, the 26th International Symposium on Principles and Practice of Declarative Programming, as part of a suite of conferences under the umbrella of *Formal Methods 2024*. Details about the organisation of LOPSTR 2024 can be found at https://lopstr.git hub.io/2024/.

The LOPSTR series of symposia aim to stimulate and promote research and collaboration on logic-based program development, interpreted broadly and covering the entire software life cycle. LOPSTR has held a prominent place in this area for more than three decades, ever since the inaugural meeting in Manchester, in 1991. Information about the previous symposia can be found at https://lopstr.webs.upv.es/.

In response to the call for papers, 28 submissions were received. They were written by 80 authors from 14 different countries. After a screening for all conflicts of interest, the submissions went through a single-blind review process, each paper receiving three independent reviews. This was followed by an online discussion among the members of the program committee. Based on the final discussion, 13 papers were selected for presentation, namely 12 regular papers and one short paper.

The symposium further featured two distinguished invited speakers: Vítor Santos Costa, kindly sponsored by the Association for Logic Programming, and Ningning Xie, who was a shared invited speaker with PPDP 2024.

We would like to extend our thanks to all who submitted a paper, for their interest in LOPSTR, and to the invited speakers, for accepting to come and make highly valued contributions to the symposium. Sincere thanks are also due to the program committee and external reviewers for their careful work. It has been a pleasure to work with the committee, with Daniel Jurjo Rivas, the symposium's Publicity Chair, and with the PPDP 2024 chairs Alessandro Bruni and Alberto Momigliano. We also thank the Chair of the LOPSTR Steering Committee, Maurizio Proietti, and the organisers of FM 2024, Matteo Pradella and Matteo Rossi, for excellent stewardship. Finally, we thank the Springer staff for their guidance and collaboration in preparing the current volume.

August 2024

Juliana Bowles
Harald Søndergaard

Organization

Program Committee Chairs

Juliana Bowles	University of St Andrews, UK and SCCH, Austria
Harald Søndergaard	University of Melbourne, Australia

Steering Committee

Emanuele De Angelis	National Research Council, Italy
Maribel Fernández	King's College London, UK
Maurizio Gabbrielli	University of Bologna, Italy
Robert Glück	University of Copenhagen, Denmark
Bishoksan Kafle	IMDEA Software Institute, Spain
Fred Mesnard	University of Reunion, France
Maurizio Proietti (Chair)	National Research Council, Italy
Peter J. Stuckey	Monash University, Australia
Wim Vanhoof	University of Namur, Belgium
Germán Vidal	Universitat Politècnica de València, Spain
Alicia Villanueva	Universitat Politècnica de València, Spain

Program Committee

Elvira Albert	Complutense University of Madrid, Spain
Roberto Amadini	University of Bologna, Italy
Maribel Fernández	King's College London, UK
Fabio Fioravanti	University of Chieti-Pescara, Italy
Didier Galmiche	University of Lorraine, France
Robert Glück	University of Copenhagen, Denmark
Gopal Gupta	University of Texas at Dallas, TX, USA
Michael Hanus	Kiel University, Germany
Bishoksan Kafle	IMDEA Software Institute, Spain
Gabriele Keller	Utrecht University, Netherlands
Maja Hanne Kirkeby	Roskilde University, Denmark
Ekaterina Komendantskaya	University of Southampton, UK
Pedro Lopez-Garcia	Spanish Council for Scientific Research (CSIC) and IMDEA Software Institute, Spain

Fred Mesnard — University of Reunion, France
Koji Nakazawa — Nagoya University, Japan
Theresa Swift — Universidade Nova de Lisboa, Portugal
Laura Titolo — AMA/NASA Research, USA
Hans van Ditmarsch — CNRS Toulouse, France
Wim Vanhoof — University of Namur, Belgium
Germán Vidal — Universitat Politècnica de València, Spain

Publicity Chair

Daniel Jurjo Rivas — IMDEA Software Institute, Spain

Additional Reviewers

Kinjal Basu
Sopam Dasgupta
Emanuele De Angelis
Marco A. Feliu
Daniela Ferreiro
Miguel Isabel
Jérôme Maquoi
Enrique Martin-Martin
Manuel Montenegro

Naoki Nishida
Arend Rensink
Adrian Riesco
Julia Sapiña
Vincent van Oostrom
Sarat Chandra Varanasi
Huaduo Wang
Gonzague Yernaux

Invited Talks

Logic-Based Neural Networks

Vítor Santos Costa

University of Porto, Portugal
vscosta@fc.up.pt

Progress in computing thrives from progress in a diversity of fields. Matrices and Linear Algebra, an old stalwart of computing, are nowadays at the core of much work in Artificial Intelligence. Formalisms based on Mathematical Logic, such as propositional logic, datalog, or logic programming, enable principled programming where one can distinguish meaning from execution. Graphs arise naturally in programming, and are often used to understand properties of data.

The presentation starts by quickly mentioning some important properties connecting these areas. Graphs can encode programs, and they can be encoded as matrices, which can be encoded by programs themselves. This connection has led to interest in combining the approaches into a single system, and has led to developments such as Statistical Relational Learning (SRL) and the neuro-symbolic systems.

The SRL approach is about dealing with uncertainty on structured domains. In its simplest form, it can be seen as extending a graphical network with repeated information: say, students that perform well in Algebra I will also perform well in Algebra II. Logical rules can be used to express this common information, as exemplified in CLP(BN), an extension of Prolog that builds directed acyclic graphs with probabilistic annotations, that is, Bayesian networks. A different approach is used in Problog. This language is also based on Prolog, but it softens the language by introducing Probabilistic facts. These facts can be used to abduce on different values, but they can also be used to give probability to rules. One should also mention Markov Logic Networks. In these models, logic atoms are connected in clauses. These connections generate an undirected graphical model.

One problem with these models is that they may require heavy temporal and memory resources and can quickly become expensive. Deep neural networks have become successful by implementing a compound function $f = f_0 \cdot f_1 \ldots f_n$, where each f_i is a layer. Layers can be implemented through matrices, and parameter learning can use the derivative of f to optimize parameters through gradient-based search.

This development has led to rapid growth in the landscape of neuro-symbolic models. TensorLog is an influential example that constructs a deep neural network out of logical rules. This is achieved through restriction to chain-rules, where each goal has a single input variable, shared with the previous goal, and a single output variable, connected to the next goal. The model is suited for edge-detection problems, where it performs well, but is not ideal for pure classification. Several extensions have been proposed; we discuss how one, NeuralLog, tackles this problem. Notice these are but a few examples; very different approaches exist, such as Deep-Problog.

Meanwhile, there has been further progress in neural networks. Graph Neural Networks are close in many ways to logic-based proposals, but are based on graph theory, so they approach learning in a different way to logic-based systems. But focus has settled on Large Language Models. Although most developments are empirical, there exist already proposals for logic reasoning in LLMs, such as LOGIC-LLM and LOGICLM. We expect to see others soon.

Safe and Easy Compile-Time Generative Programming

Ningning Xie

University of Toronto, Canada
ningningxie@cs.toronto.com

Program generation is a powerful and expressive approach to eliminating abstraction overhead and improving program performance that has been studied and implemented in a variety of languages with different forms.

In this talk we overview MacoCaml, a new design and implementation of compile-time computation for OCaml. MacoCaml features a unifying and novel combination of phase separation and quotation-based staging. We review MacoCaml's recent developments, including a comprehensive formalism of a feature-rich macro calculus with key meta-theoretic properties, and an extension to module functors that leads to explicit phase distinction. We describe how the meta-theoretical results offer practical benefits for programmers, and conclude the talk with a few directions for future exploration.

Contents

Synthesis and Transformation

Parallel Assembly Synthesis 3
   Jingmei Hu, Stephen Chong, and Margo Seltzer

Improving Logic Programs by Adding Functions 27
   Michael Hanus

Decision Procedures

Deciding Knowledge Problems Modulo Classes of Permutative Theories 47
   *Serdar Erbatur, Andrew M. Marshall, Paliath Narendran,
   and Christophe Ringeissen*

Binary Implication Hypergraphs for the Representation and Simplification
of Propositional Formulae 64
   Jordina Francès de Mas

Combined Abstract Congruence Closure for Theories with Associativity
or Commutativity 82
   Christophe Ringeissen and Laurent Vigneron

A Certifying Algorithm for Linear (and Integer) Feasibility in Horn
Constraint Systems 99
   Piotr Wojciechowski and K. Subramani

Deployment

Pick a Flavour: Towards Sustainable Deployment of Cloud-Edge
Applications 117
   *Roberto Amadini, Simone Gazza, Jacopo Soldani, Monica Vitali,
   Antonio Brogi, Stefano Forti, Saverio Giallorenzo, Pierluigi Plebani,
   Francisco Ponce, and Gianluigi Zavattaro*

Specification, Refactoring and Testing

An Axiomatic Category-Based Access Control Model for Smart Homes 131
   Clara Bertolissi, Maribel Fernández, and Bhavani Thuraisingham

Towards Specification-Guarded Refactoring 149
    Adam D. Barwell, Christopher Brown, and Susmit Sarkar

Impact and Performance of Randomized Test-Generation Using Prolog 166
    Marcus Gelderie, Maximilian Luff, and Maximilian Peltzer

Term and Graph Rewriting

Proving Uniqueness of Normal Forms w.r.t Reduction of Term Rewriting
Systems 185
    Takahito Aoto

Rewriting Induction for Higher-Order Constrained Term Rewriting
Systems 202
    Kasper Hagens and Cynthia Kop

Introducing Quantification into a Hierarchical Graph Rewriting Language 220
    Haruto Mishina and Kazunori Ueda

Author Index 241

Synthesis and Transformation

Parallel Assembly Synthesis

Jingmei Hu[1](✉) [iD], Stephen Chong[1] [iD], and Margo Seltzer[2] [iD]

[1] Harvard University, Cambridge, MA 02138, USA
jingmei_hu@alumni.harvard.edu, chong@seas.harvard.edu
[2] University of British Columbia, Vancouver, BC, Canada
mseltzer@cs.ubc.ca

Abstract. Program synthesis offers an attractive alternative to the intricate and tedious process of writing assembly programs manually. Assembly program synthesis automatically generates implementations, given a high-level formal specification and a machine description. However, its limited scalability prevents widespread adoption. Automatic parallelization improves program synthesis in general, but parallelizing assembly synthesis is nontrivial as the realities that data are untyped and all state is global lead to an enormous search space and prevent straightforward decomposition into separable sub-problems that can be run in parallel. We present PASSES, a Parallel Assembly Synthesis System Exploiting Subspaces. PASSES uses five heuristics to transform an original assembly synthesis problem into a set of sub-problems; it runs multiple synthesis sub-problems in parallel and constructs the final result by combining them. We evaluate PASSES on 26 general bit manipulation assembly programming problems and 140 machine-dependent use cases from two operating systems. Compared to an existing assembly synthesis tool and a state-of-the-art parallel SMT solver, all five heuristics in PASSES significantly improve assembly synthesis scalability.

Keywords: program synthesis · assembly programming · parallel computing

1 Introduction

Assembly language is used in many mission-critical systems that require direct manipulation of hardware, access to performance-critical instructions, or access to special-purpose accelerators. Assembly language is also common in device drivers, real-time systems, low-level embedded systems, and machine-dependent operating system code. However, assembly language is fundamentally intricate and tedious to write. Writing and debugging assembly programs tends to take more time than debugging higher-level language programs, because each processor architecture has its own assembly (so one might need to debug the same functionality multiple times), and data in assembly programs are untyped and global.

Assembly program synthesis, as an alternative to manual implementation [17,22,45], is a promising approach for automated assembly programming. Existing assembly synthesis systems, in general, take a high-level formal specification, which describes *what* a program should do, and a machine description, which provides the executable model of an instruction set architecture (ISA) semantics, and uses *CounterExample Guided Inductive Synthesis (CEGIS)* [43] to generate a program, i.e., a sequence of assembly instructions that satisfies the specification. However, the limited scalability of synthesis, especially assembly language synthesis, prevents widespread adoption. Fundamentally, assembly synthesis is a search problem: it searches for an assembly program that satisfies the specification from the space of all instruction sequences. Compared to synthesis of higher-level language programs, the search space of possible programs in assembly synthesis is much larger, because data is untyped and global. The space is typically combinatorial in the number of machine state components (including dozens of registers and hundreds of memory locations) and exponential in the number of instructions [20,22,23]. To synthesize a single three-operand instruction OPCODE op1, op2, op3, the size of the search space is approximately $n \times x \times y \times z$ where n is the number of possible OPCODEs and x, y, and z represent the number of choices for op1, op2, and op3 respectively; for a program with N instructions, the space size is $(n \times x \times y \times z)^N$. Current tools and approaches are unable to synthesize assembly programs longer than a few instructions in reasonable time [4,22,45,46].

A natural step in improving the performance of program synthesis is parallelization. Parallel program synthesis has been proposed in recent work that enables efficient parallelization of challenging synthesis problems [10,25,26]. There are two common approaches for parallelization: One method is to search for a solution over a set of instances with different settings, e.g., running several instances of a sequential solver with different parameters or several different solvers in parallel. If any of the instances succeeds in finding a solution, all the instances are terminated. Another method is to find solutions to subsets of the original problem (i.e., *sub-problems*) and recombine them into a full correct solution, e.g., using divide-and-conquer techniques [2,3,13]. Small sub-problems typically have smaller search spaces to explore than does the original problem; we call these smaller search spaces *subspaces*. However, unlike the general parallelization problem, it is not obvious how to conduct parallel search in assembly synthesis. Although the specifications used in assembly synthesis are simple pre- and post-conditions, they lead to an enormous search space for SMT solvers [22,45], and the SMT expressions generated for assembly synthesis are difficult to decompose into separable conjunctions that can be solved in parallel, because data are untyped and machine state is global in assembly languages.

We present a novel parallel system for assembly synthesis, PASSES (Parallel Assembly Synthesis System Exploiting Subspaces), that uses domain knowledge of assembly language to parallelize synthesis. PASSES exploits this domain knowledge to identify multiple approaches to subspace creation. We first describe three general characteristics of subspaces in parallel synthesis problems: 1) whether the set of subspaces is exhaustive (i.e., *collective exhaustivity*), 2) whether the subspaces overlap (i.e., *mutual exclusivity*), and 3) whether they are of (approximately) equal size (i.e., *subspace size equality*). Based on those categories, PASSES identifies five complementary and reusable heuristics for creating subspaces, each of which represents a synthesis *sub-problem*. We deploy the techniques by running multiple parallel synthesis tasks and collecting the results from them. To evaluate the effectiveness of these heuristics and PASSES, we collected 26 general bit manipulation assembly programming problems, used in previous work [16], and 140 machine-dependent use cases from two pre-existing OSes with four machine architectures as benchmark examples, also used in previous work [20,22]. Compared with the state-of-the-art assembly synthesis approach [22,24], evaluation results show that all of PASSES's heuristics significantly improve the scalability of assembly synthesis, while preserving solution quality. In the best case, PASSES achieves a geometric mean speedup of more than 10× for programs that take the baseline more than 10 s to synthesize.

We summarize our contributions as follows:

- We develop a novel parallel assembly synthesis system, PASSES, to improve the scalability of assembly language synthesis.
- We design and implement five domain-specific heuristics, to transform an assembly language synthesis problem into a set of *sub-problems*, synthesize each sub-problem individually in parallel, and construct a solution from them.
- We evaluate the effectiveness of PASSES on general bit manipulation programs and use cases derived from the machine-dependent parts of two operating systems. Our evaluation demonstrates that PASSES greatly improves assembly synthesis scalability.

2 Background and Related Work

Program synthesis is used to automatically generate target executable programs that satisfy a given high-level formal specification, which is typically non-executable [5,8,14,30,34,51]. Modern program synthesis techniques fall into two main categories: deductive synthesis, which synthesizes programs by constructively proving a theorem, employing logical inference and constraint solving [27,30], and inductive synthesis, which finds programs matching a set of input-output examples and generalizes them to work for every input [11,33,49]. We focus more on the more relevant latter category of program synthesis.

In the area of inductive synthesis, there are two major directions: synthesis from informal and typically incomplete descriptions, e.g., Programming by Example [11,15,33,50] and synthesis from formal specifications, which began in 2006 with the introduction of SKETCH [42,44], and Syntax Guided Synthesis (SyGuS) [1]. Later, Solar-Lezama et al. introduced Counterexample Guided Inductive Synthesis (CEGIS) [43], which uses an iterative process to perform inductive generalization for all possible inputs. Our work lies in this area of synthesis, where we take formal specifications and use CEGIS to generate assembly programs satisfying the desired behaviors.

In the area of fully automated synthesis with parallelism, there are two main topics: using divide-and-conquer to decompose a synthesis problem into simpler ones, such as CYPRESS [40,41], FlashMeta [35], and EUSOLVER [3], and using parallel Boolean-satisfiability/satisfiability-modulo-theories (SAT/SMT) solvers to expedite synthesis solving [29], such as PaInleSS [28], PBoolector [37] (which is the parallel implementation of Boolector [7] splits a bit-vector formula and solves the subproblems in parallel; in Sect. 5, we compare it to our subspace decomposition approach) and the *parallel portfolio approach* such as ManySat [18,19,48]. The traditional divide-and-conquer approaches for program synthesis focus more on the *division of the specification*. However, assembly synthesis typically takes a relatively straightforward specification that describes program behavior and synthesizes instruction sequences with a complicated machine model. The machine model contains all possible machine states, including registers and memory locations, and all possible instructions that must be considered; it is one of the key performance bottlenecks for assembly synthesis: it leads to an enormous search space with these global states and untyped data. It is challenging to separate or decompose them for parallelization, especially at the specification level. Thus, applying divide-and-conquer only to specifications might not produce as significant a performance improvement as we might like. Rather than dividing the specification, we *divide the search space* into smaller sub-spaces, reducing the impact of the exponential state space explosion. In contrast, parallel SAT/SMT solvers directly incorporate parallel algorithms by modifying the state-of-the-art solvers accordingly; in contrast, we propose and evaluate parallelism in the synthesis procedure rather than in the solver implementation.

3 Preliminaries and Terminology

3.1 Baseline Assembly Synthesizer: Aquarium

We implement PASSES on top of a state-of-the-art assembly synthesis engine. To the best of our knowledge, there are two main assembly synthesis related tools: STOKE [39] and Aquarium [22,24]. STOKE [39] is a stochastic superoptimizer [31] that starts from an existing implementation, not an abstract specification, and optimizes the code to improve performance or reduce size. As such, it does not solve the problem we are trying to solve and is suitable neither as a starting point nor as a baseline comparison. Aquarium is a CEGIS-based assembly synthesis system designed to synthesize the machine-dependent parts of operating systems [22,24]. It takes as input a functional specification (pre- and post-condition) and a machine model description, and produces a sequence of satisfied assembly instructions. As such, it seems well-matched to our setting.

3.2 Assembly Instruction Types

Though different machine architectures have distinctive assembly syntax and semantics, we can categorize assembly instructions into types. The following six types are used throughout the rest of this paper; in the list below, we provide examples of the types from ARMv7[1]: 1) `ARITH`: arithmetic, such as addition (`add`); 2) `LOGIC`: logical, such as shifts(`lsl`); 3) `MEMOP`: memory handling, such as loads/stores (`ldr`/`str`); 4) `DATAOP`: register-to-register data transfer (`mov`); 5) `JMP`: conditional and unconditional branches (`b`); and 6) `COPROC`: coprocessor handling (`mcr`/`mrc`).

3.3 Subspace Creation

We introduce the following notation to describe subspaces. Given a synthesis problem Q, let $M = SearchSpace(Q)$ be the entire search space of Q, i.e., the set of all possible programs. Our goal is to produce a set of n subspaces $P_1 \ldots P_n$, each representing a sub-problem, $q_1 \ldots q_n$ of Q, i.e., $P_i = SearchSpace(q_i)$. Note that $\forall i, P_i \subseteq M$ and a solution to q_i is a solution to Q.

We describe three (generally desirable) characteristics of subspaces: *collective exhaustivity*, *mutual exclusivity*, and *subspace size equality*.

Collective Exhaustivity. A set of n sub-problems are *collectively exhaustive* if their subspaces P_i cover the entire search space M of the problem Q; that is, $M = \bigcup_{i=1}^{n} P_i$. We write $\sigma \vDash Q$ if a program σ satisfies Q, i.e., σ is a solution to Q. Then the following two statements hold for *collective exhaustivity*: $\exists \sigma \vDash Q \implies \exists i \in [1, n] \ \sigma \vDash q_i$ and $\forall \sigma \nvDash Q \implies \forall i \in [1, n] \ \sigma \nvDash q_i$. Otherwise, non-exhaustivity leaves *unsearched* portions in M that may contain possible solutions.

While collective exhaustivity is generally desirable, it is not required. A subspace creation technique that is not collectively exhaustive means that failure to find a solution in subspaces *does not* imply that the problem is unsatisfiable.

Mutual Exclusivity. A set of subspaces is *mutually exclusive*, or *disjoint*, if they are non-intersecting, i.e., $\forall i \in [1, n] \ \forall j \in [1, n] \quad P_i \cap P_j \neq \emptyset \implies i = j$.

Mutual exclusivity is generally desirable as it means that searching the subspaces does not duplicate work: any possible solution appears in at most one subspace. Indeed, it could be that the restrictions required to make subspaces disjoint (e.g., additional constraints) increases synthesis times, so it might be more efficient to allow subspaces to overlap.

Subspace Size Equality. In general, synthesis time is positively correlated to the size of the search space: the larger the search space, the longer it takes to find a solution. *Subspace size equality* means that all created subspaces have approximately the same size; in other words, $\forall i \in [1, n], |P_i| \approx |M|/n$ ($|P_i|$ is the number of possible programs in P_i). It equally divides M, which leads to approximately the same synthesis time for each sub-problem.

Subspace size equality is generally desirable, but not required. Under some conditions, such as some characteristics of the original synthesis problem, unequal division may achieve better performance; we discuss this in Sect. 4.2.

---
[1] The conditional execution feature in ARM does not affect the corresponding types.

4 PASSES: Parallel Assembly Synthesis System Exploiting Subspaces

Using insight specific to assembly synthesis, we develop five complementary and reusable heuristics to create subspaces. These heuristics are central to the design of PASSES, our novel parallel assembly synthesis system. PASSES uses these heuristics to create sets of subspaces, tries to synthesize a solution in each subspace in parallel, and collects a final solution from them. We refer to the subproblems that will be synthesized in parallel as *instances*. We built PASSES on top of an existing CEGIS-based assembly synthesizer, Aquarium [22]. We explain PASSES's algorithm in Appendix A.1. Table 1 includes the subspace creation categories for those five heuristics; we discuss more detail about each heuristic in the following sections.

Table 1. Five heuristics in PASSES. "(✓)" denotes that only some sub-problems belong to the corresponding category.

	Constraint-SC	RandSimpl-SC	TypeSimpl-SC	Inc-SC	PriorInc-SC
Subspace Creation	static	static	static	incremental	incremental
Collective Exhaustivity	✓	✓	✓		
Mutual Exclusivity	✓	✓	(✓)	(✓)	(✓)
Subspace Size Equality	✓	(✓)			

4.1 Constraint-Based Subspace Creation (Constraint-SC)

As the search space grows exponentially in the number of instructions in a program, so too does synthesis time. *Constraint-SC* creates subspaces by introducing *constraints* that require specific instructions to appear in specific locations in the target program; for example, if there is a pre-condition on a value stored in memory, perhaps it's reasonable to try programs where the first instruction loads a value from memory. The constraints applied to each instance determine not only which instructions should be considered but also, where they should appear, thus, there is no overlap among the subspaces and the sum of all subspaces covers the entire search space of the original problem. In practice, rather than select specific instructions, *Constraint-SC* constructs constraints using the six instruction types mentioned in Sect. 3 (Algorithm explained in Appendix A.2).

4.2 Model-Simplified Subspace Creation (Simpl-SC)

A complete machine model includes descriptions of all instructions; it contains more information than is strictly necessary to facilitate the synthesis of any single program (i.e., practically no program uses every instruction in an ISA).

Hence, instead of considering the entire ISA, synthesizing with only a partial instruction set should dramatically improve scalability. *Simpl-SC* simplifies the given machine model and creates multiple subspaces, each representing a subproblem with a different machine model containing fewer instructions (Algorithm explained in Appendix A.3). The main challenge is selecting appropriate partial instruction sets. Removing the instructions from the machine model might end up with something simple and convenient to consider, but the remaining machine model might not be capable of producing a correct target program.

Randomized Model-Simplified Subspace Creation (RandSimpl-SC) shuffles the whole instruction set and evenly divides them into groups as submodels, each containing the same number of instructions with various types. The subspaces are mutually exclusive, but may not cover the space. We add *RandSimpl-SC* to include the complete model to ensure that the entire space is covered.

Type-Based Model-Simplified Subspace Creation (TypeSimpl-SC) uses the six instruction types mentioned in Sect. 3 to construct sub-models. *TypeSimpl-SC* creates $\binom{6}{2} = 15$ sub-models with 2 different instruction types. Note that programs with a single instruction type are rare in real-world situations, thus, we do not create further detailed subspaces for brevity. Similarly, the complete-model instance (i.e., the possible program should contain at least three different types of instructions) explores the rest of the search space.

4.3 Incremental Subspace Creation (Inc-SC)

The previous heuristics create subspaces statically; they begin with a fixed number of divided subspaces that exhaustively cover the search space with the complete-model instance. In contrast, *Inc-SC* divides the search space dynamically without considering the complete model; it starts with a non-exhaustive division and incrementally expands the search to the entire space. We implement *Inc-SC* on top of *TypeSimpl-SC*, with algorithm explained in Appendix A.4.

4.4 Prioritized Incremental Subspace Creation (PriorInc-SC)

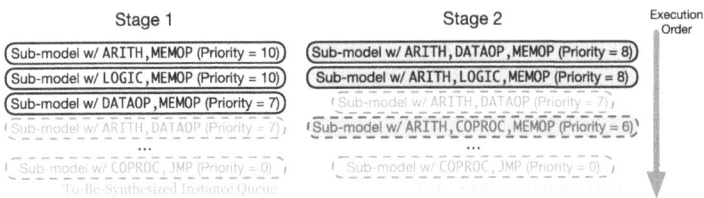

Fig. 1. Two stages in *PriorInc-SC*.

So far, all our heuristics select instructions with equal probability. However, sometimes the specification suggests that some instructions are more likely than others. Given this fact, *PriorInc-SC* extends *Inc-SC* with *Instruction Prioritization* using the following two-stage approach (Fig. 1).

First, it statically analyzes the specification, placing instructions into the following three classes: *must-class* instructions, i.e., there is a high possibility that the target program has them, *may-class* instructions, i.e., they may be used in the target program, and *not-class* instructions, i.e., the possibility of their usage is low. For example, if the postcondition uses data from memory locations and calculates on the value, MEMOP is *must-class*, ARITH and LOGIC are *may-class* (they can sometimes achieve equivalent behavior), and COPROC is *not-class*, since no coprocessor handling involved. We leave the exploration of other prioritization algorithms for future work. Figure 1 shows an instance initialization example with *PriorInc-SC*, where ARITH, LOGIC, and MEMOP get higher priority, and the top three instances get synthesized first (solid black lines).

In Stage 2, *PriorInc-SC* generates 20 cases with initial states that satisfy the precondition and presents the generated candidate in each CEGIS iteration with a *score*, i.e., the number of cases for which it satisfies the postcondition. For each failed instance, we record the highest score achieved so far and update the priority of each instruction type involved. We always choose to execute instances with the highest priority, i.e., the instructions involved are most likely to be present in the target program. In Fig. 1, after the instance with ARITH and MEMOP failed, newly added instances (gray boxes) might get higher priorities than the previous, not-executed ones (white boxes). Due to its dynamic enlargement, *PriorInc-SC* does not preserve *collective exhaustivity*, and for each newly added instance, we apply mutually exclusive constraints to avoid search space reconsideration.

5 Evaluation

We implemented PASSES in about 3500 lines of OCaml using *Aquarium* [20,22], implementing each of the five subspace creation heuristics. To demonstrate the effectiveness and efficacy of search space reduction and parallelization in PASSES, we evaluate its performance on 26 bit manipulation programming tasks and 140 machine-dependent operating system (OS) code examples, compared to *Aquarium* [20,22,24], as a baseline. We also consider two different SMT parallelization techniques. As mentioned in Sect. 2, we use *PBoolector* [37], the parallel SMT solver algorithm for Boolector [7]. We also implemented a synthesizer wrapper that runs multiple (sequential) Boolectors in parallel with different random seeds (hereinafter referred to as *RandomSeed*).

5.1 Benchmarks

We select the following two different categories of benchmark programs.

Bit Manipulation Benchmarks. Inspired by Gulwani et al. [16] and based on the scalability of current assembly synthesis approaches, we select 26 examples from the book *Hacker's Delight* [47], more detailed appears in Appendix B. We provided the specification of the desired behavior for each example in 32-bit MIPS, by specifying the pre- and post-condition. We first ran unmodified Aquarium [22], which is guaranteed to synthesize the shortest program satisfying the specification. All examples can be implemented with between 2 and 5 assembly instructions. Table 5 in Appendix B reports the number of instructions in each synthesized implementation, indicating the minimum length requirement for each benchmark.

Operating System Related Benchmarks. We obtained more complex examples using the 140 use cases from Aquarium [22], including 35 individual procedures from two pre-existing OSes, Barrelfish [6] and OS/161 [21], implementing machine-dependent OS functionality. They consist of machine-level system call trap and kernel entry code in Barrelfish, C standard library function `setjmp` and `longjmp`, user-level program startup code, system call stub, interrupt disable code, and kernel-level thread switch in OS/161. Each procedure is implemented with four different machine architectures: 32-bit ARM, 32-bit MIPS, 32-bit RISC-V, and 64-bit x86_64. Their length varies from 1 to 14 instructions.

5.2 Experimental Setup

We ran all experiments on an *m5.8xlarge* AWS EC2 instance with 32 virtual CPUs and 128 GB of memory; all benchmarks use eight processors for parallel synthesis. We also ran *PBoolector* with the maximum of eight sub-problems generated when splitting bit-vector formulas, and *RandomSeed* with eight different random seeds for synthesis. For bit manipulation examples, the synthesis target is MIPS programs that work on bitvectors of size 32 bits, while for OS-related examples, we synthesize for all four architectures listed above.

5.3 Performance Results

We measured synthesis performance for all of the aforementioned heuristics in PASSES. We ran each use case five times, varying the random number seed each time. We also randomized the variable names used for solver communication by prepending a random five-character alphabetical string, which produces significant variance in solver performance.

We run each synthesis task with a half-hour timeout; several use cases timed out under the baseline setting, denoted with "—" in the following tables.

Table 2. Performance of all 26 bit manipulation benchmarks, sorted by *Aquarium* baseline. For each benchmark, the table shows: (1) *Aquarium* baseline runtime, *PBoolector* runtime, and PASSES with five heuristics (in seconds, averaged across 5 trials), (2) number of instructions synthesized (in parentheses), and (3) the speedup, i.e., the ratio of the baseline time to the time with parallelism. "—" denotes cases where synthesis does not complete within the 1800-second timeout. A light gray background shows cases where a heuristic sped up synthesis (*Speedup* > 0); bold with a dark gray indicates large speedup (*Speedup* > 5).

Use Case	Aquarium Time (s)	Constraint-SC Time (s)	Speedup	RandSimpl-SC Time (s)	Speedup	TypeSimpl-SC Time (s)	Speedup	Inc-SC Time (s)	Speedup	PriorInc-SC Time (s)	Speedup	PBoolector Time (s)	Speedup
P1b	0.86 (2)	0.73 (2)	1.18	1.73 (2)	0.50	1.72 (2)	0.50	0.83 (2)	1.04	0.68 (2)	1.26	1.97 (2)	0.44
P5a	1.28 (2)	0.54 (2)	2.37	1.12 (2)	1.14	2.86 (2)	0.45	1.91 (2)	0.67	1.82 (2)	0.70	1.88 (2)	0.68
P7b	2.05 (2)	0.60 (2)	3.42	1.77 (2)	1.16	2.99 (2)	0.69	1.99 (2)	1.03	1.97 (2)	1.04	2.12 (2)	0.97
P1a	2.21 (2)	0.83 (2)	2.66	1.34 (3)	1.65	1.75 (2)	1.26	0.82 (2)	2.70	0.77 (2)	2.87	1.51 (2)	1.46
P4a	2.78 (2)	**0.52 (2)**	**5.35**	1.87 (2)	1.49	2.17 (2)	1.28	0.97 (2)	2.87	1.01 (2)	2.75	2.69 (2)	1.03
P2a	2.81 (2)	0.69 (2)	4.07	1.62 (2)	1.73	2.11 (2)	1.33	1.52 (2)	1.85	1.45 (2)	1.94	1.03 (2)	2.73
P2b	3.48 (2)	0.93 (2)	3.74	2.13 (2)	1.63	2.02 (2)	1.72	0.82 (2)	4.24	**0.66 (2)**	**5.27**	2.45 (2)	1.42
P7a	3.69 (2)	**0.69 (2)**	**5.35**	2.08 (2)	1.77	2.90 (2)	1.27	1.58 (2)	2.34	1.47 (2)	2.51	3.93 (2)	0.94
P13a	3.91 (3)	3.01 (3)	1.30	0.96 (3)	4.07	0.82 (3)	4.77	**0.75 (3)**	**5.21**	0.79 (3)	4.95	2.97 (3)	1.32
P8a	6.56 (3)	6.04 (3)	1.09	12.27 (3)	0.53	1.49 (3)	4.40	1.45 (3)	4.52	1.53 (3)	4.29	8.66 (3)	0.76
P3a	9.04 (3)	3.62 (3)	2.50	4.88 (3)	1.85	6.35 (3)	1.42	5.48 (3)	1.65	5.46 (3)	1.66	7.97 (3)	1.13
P5b	10.28 (3)	3.92 (3)	2.62	9.56 (3)	1.08	4.60 (3)	2.23	3.80 (3)	2.71	3.67 (3)	2.80	8.93 (3)	1.15
P9	10.86 (3)	8.58 (3)	1.27	7.26 (4)	1.50	2.46 (3)	4.41	2.51 (3)	4.33	2.67 (3)	4.07	7.49 (3)	1.45
P8b	12.64 (3)	5.46 (3)	2.32	5.28 (3)	2.39	**1.81 (3)**	**6.98**	**1.49 (3)**	**8.48**	**1.49 (3)**	**8.48**	8.09 (3)	1.56
P3b	12.73 (3)	5.71 (3)	2.23	12.43 (4)	1.02	8.58 (3)	1.48	7.10 (3)	1.79	7.09 (3)	1.80	12.77 (3)	1.00
P4b	12.91 (3)	5.63 (3)	2.29	13.83 (3)	0.93	11.88 (3)	1.09	10.00 (3)	1.29	9.01 (3)	1.43	12.19 (3)	1.06
P10	32.56 (3)	24.52 (3)	1.33	25.67 (3)	1.27	**5.26 (3)**	**6.19**	**5.19 (3)**	**6.27**	**5.27 (3)**	**6.18**	26.47 (3)	1.23
P11b	39.01 (3)	10.38 (3)	3.76	13.10 (3)	2.98	12.73 (3)	3.06	**5.67 (3)**	**6.88**	**5.65 (3)**	**6.90**	38.29 (3)	1.02
P11a	63.36 (3)	**11.27 (3)**	**5.62**	23.86 (3)	2.66	**8.03 (3)**	**7.89**	**4.79 (3)**	**13.23**	**4.79 (3)**	**13.23**	52.62 (3)	1.20
P6	70.27 (4)	27.75 (4)	2.53	56.22 (4)	1.25	46.23 (4)	1.52	49.24 (4)	1.43	49.31 (4)	1.43	58.10 (4)	1.20
P11c	73.88 (3)	16.21 (3)	4.56	97.46 (3)	0.76	14.89 (3)	4.96	**9.82 (3)**	**7.52**	**9.84 (3)**	**7.51**	71.19 (3)	1.04
P14b	97.64 (3)	109.61 (3)	0.89	102.05 (4)	0.96	99.05 (3)	0.99	95.48 (3)	1.02	94.07 (3)	1.04	86.49 (3)	1.13
P13b	134.02 (4)	145.82 (4)	0.92	163.17 (4)	0.82	29.06 (4)	4.61	28.75 (4)	4.66	**26.13 (4)**	**5.13**	129.48 (4)	1.04
P12b	297.69 (4)	229.05 (4)	1.30	409.52 (4)	0.73	291.61 (4)	1.02	313.56 (4)	0.95	310.26 (4)	0.96	156.65 (4)	1.90
P12a	666.35 (4)	**79.78 (4)**	**8.35**	370.10 (4)	1.80	286.06 (4)	2.33	307.98 (4)	2.16	305.15 (4)	2.18	651.56 (4)	1.02
P14a	—	—	—	—	—	1273.88 (5)	>1.41	1281.60 (5)	>1.40	1267.12 (5)	>1.42	—	—

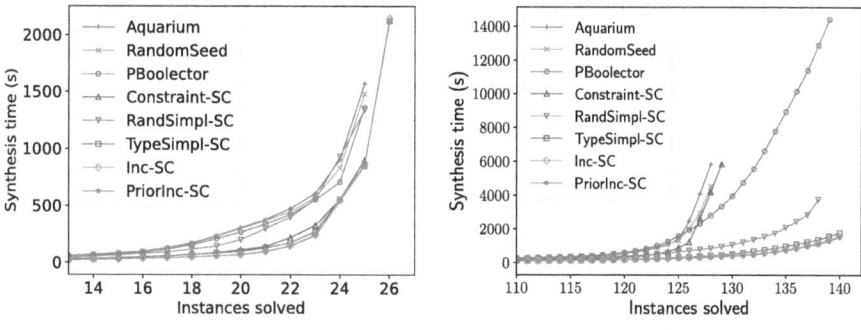

(a) Bit manipulation benchmarks.   (b) OS-related benchmarks.

Fig. 2. Heuristics performance relative to *Aquarium* baseline. Each plot shows each benchmark's average runtime across five trials that do not time out; we omit benchmarks in which all trials timed out. We omit the first 13 bit manipulation and 110 OS-related benchmark results to improve readability.

Performance of Bit Manipulation Benchmarks. Table 2 reports the synthesis performance comparison between *Aquarium* and five heuristics for all 26 bit manipulation benchmark examples, sorted by the *Aquarium* runtime; Fig. 2a adds comparison with the two other parallel implementations (*PBoolector* and *RandomSeed*). All PASSES heuristics, except *RandSimpl-SC*, generate the same length programs as *Aquarium* (reported in Table 5), although the actual synthesized programs are not identical; *RandSimpl-SC* introduces randomness, sometimes generating longer solution programs (with one more instruction) in 5 trials (in P1a, P3b, P6, P9, P14b). The synthesis times vary widely with different use cases; occasionally the heuristics take longer than *Aquarium*. We include the speedup information, i.e., the ratio of the baseline time to the synthesis time with each heuristic in Table 2, and we examine the performance result closely in the following.

Constraint-SC produces a geometric mean speedup of 2.43×. In the best case, it makes the synthesis up to 8.35× faster; it speeds up the synthesis in 23 out of 26 cases. However, it sometimes slows down synthesis due to the overhead caused by parallelism, such as in P14b and P13b. In the worst case, we observe a slowdown of 1.12×. Both *Aquarium* and *Constraint-SC* fail to synthesize P14a within the half-hour time limit. Figure 2a shows that *Constraint-SC* expedites assembly synthesis even though most programs are quite short.

When *Aquarium* is already quite fast, the overhead of PASSES can dominate performance. For example, when *Aquarium* takes less than 10 s, *RandSimpl-SC* frequently produces slowdown (up to 2.01×). However, for complicated programs, it makes the synthesis up to 4.07× faster, with a geometric mean speedup of 1.33×. *RandSimpl-SC* also fails to synthesize P14a within the half-hour time limit. Since *RandSimpl-SC* first randomly shuffles the instruction sets for later subspace creation, we also include the mean and standard deviation result across 5 trials in Table 3. This randomness produces large standard deviations on synthesis time for the complicated programs, especially for programs that take more than 100 s to synthesize.

Table 3. Means and standard deviation (SD) of *RandSimpl-SC* with all 26 bit manipulation benchmarks, sorted by *Aquarium* baseline. "—" denotes cases where synthesis does not complete within the 1800 s timeout.

Use Case		P1b	P5a	P7b	P1a	P4a	P2a	P2b	P7a	P13a	P8a	P3a	P5b	P9
Aquarium		0.86	1.28	2.05	2.21	2.78	2.81	3.48	3.69	3.91	6.56	9.04	10.28	10.86
RandSimpl-SC	Mean	1.73	1.12	1.77	1.34	1.87	1.62	2.13	2.08	0.96	12.27	4.88	9.56	7.26
	SD	0.12	0.22	0.25	0.81	0.48	0.44	0.49	0.05	0.16	2.18	1.65	4.67	4.27

Use Case		P8b	P3b	P4b	P10	P11b	P11a	P6	P11c	P14b	P13b	P12b	P12a	P14a
Aquarium		12.64	12.73	12.91	32.56	39.01	63.36	70.27	73.88	97.64	134.02	297.69	666.35	—
RandSimpl-SC	Mean	5.28	12.43	13.83	25.67	13.1	23.86	56.22	97.46	102.05	163.17	409.52	370.1	—
	SD	5.58	7.36	3.05	16.58	10.19	23.51	11.57	26.47	29.25	45.77	219.5	100.89	

TypeSimpl-SC produces more consistent results: 22 out of 26 cases achieve synthesis speedup. We observe a maximum speedup of 7.89×, with a geometric mean speedup of 1.99× (up to 2.23× slowdown due to the parallel overhead). For the programs that take *Aquarium* more than 3 s to synthesize, *TypeSimpl-SC* produces a significant and consistent speedup. It also synthesizes programs that are not accessible to *Aquarium*: it successfully synthesizes an assembly program for P14a in fewer than 1300 s; if we treat timeouts as taking 1800 s, this conservative speedup is 1.41×.

Inc-SC accelerates synthesis in most cases; it is beneficial even for small programs. It makes the synthesis up to 13.23× faster, with a geometric mean speedup of 2.69× (up to 1.49× slowdown). We also observe that *Inc-SC* induces an overhead due to the dynamic instance launches. Table 2 shows that, *Inc-SC* produces a slowdown for some complicated cases such as P12a and P14a, compared with the original *TypeSimpl-SC*, while it still expedites the synthesis compared with *Aquarium*. It slightly slows down synthesis for only two cases (P5a and P12b), compared to *Aquarium*.

PriorInc-SC is more effective. We notice that the previous heuristics sometimes generate unpromising instances that fail synthesis easily; due to our limited degree of parallelism (i.e., we benchmark with eight threads), those instances waste CPU resources during synthesis. *PriorInc-SC* prioritizes instances and executes those most likely to succeed first. Thus, it eliminates the overhead caused by unpromising instances. It produces a geometric mean speedup of 2.77× and a maximum speedup of 13.23× (up to 1.42× slowdown). It also successfully synthesizes the troublesome P14a. Compared to *Inc-SC*, it makes synthesis faster in some cases, especially for complicated programs. *Our high level observation is that Fig. 2a indicates that* TypeSimpl-SC, Inc-SC, *and* PriorInc-SC *outperform* Aquarium *and solve all benchmarks within the time limit.*

Comparison with Parallel SMT Solver and Different Random Seeds. Table 2 includes the detailed *PBoolector* synthesis runtime and speedup compared with *Aquarium* baseline, for each bit manipulation benchmark. As shown in Fig. 2a, for all 26 bit manipulation benchmarks, all five heuristics outperform both *PBoolector* and *RandomSeed*. In the best case, *PriorInc-SC* outperforms *PBoolector* and *RandomSeed* with a maximum speedup of 10.98× and 12.91× for the program that takes *Aquarium* about 60s to synthesize, respectively.

Search Space Comparison. We evaluate all bit manipulation benchmarks in 32-bit MIPS; the complete machine model includes 37 assembly instructions which covers all the basic operations. As a first-order approximation, the overall size of the search spaces for n-instruction programs is $2^{36.6n}$, while *TypeSimpl-SC*, for example, prunes the subspaces for 2-type machine models into $2^{26.7n}$ with ARITH and DATAOP, $2^{27.7n}$ with LOGIC and MEMOP, or even $2^{19.8n}$ with JMP and COPROC.

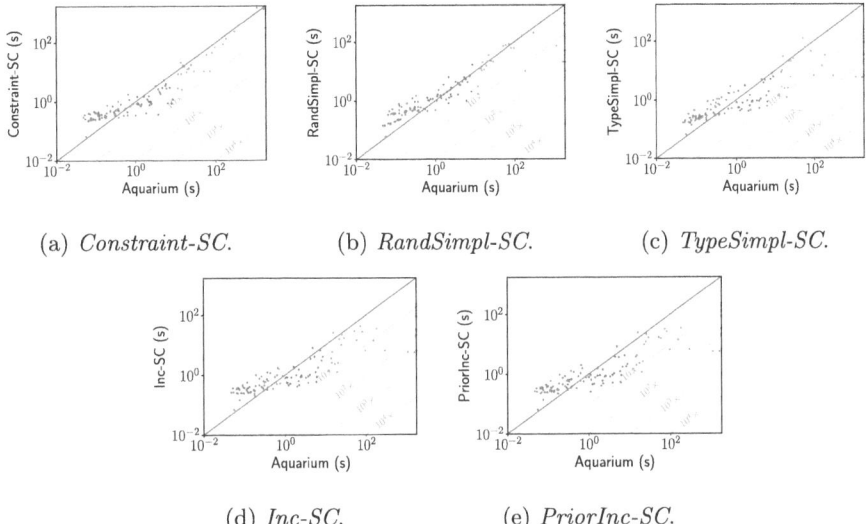

Fig. 3. Effect of PASSES with heuristics on synthesis runtimes, with all 140 OS-related benchmark examples. Each shows PASSES with heuristics against *Aquarium*. Each data point represents the runtime under both conditions for one OS-related benchmark, averaged over five trials (timeouts are counted as 1800 s). $1s-$ *programs* are colored in red, $10s-$ *programs* are colored in green, and $10s+$ *programs* are colored in pink. The blue diagonal line represents equal time under both conditions, so that points below/right of the diagonal line demonstrate better performance with the heuristic. Gray contours provide guidelines for visually estimating the speedup factor. The upper and right boundaries of the plot represent an 1800-s timeout. (Color figure online)

Performance of Operating System Related Benchmarks. Figure 2b and Fig. 3 show the synthesis performance comparison between *Aquarium* and PASSES with all five heuristics for the 140 OS-related benchmarks. We categorize those benchmarks into three groups: (1) $1s-$ *programs*: they take *Aquarium* no more than 1 s to complete, colored with red in Fig. 3; (2) $10s-$ *programs*: their *Aquarium* runtime is more than 1 s but no more than 10 s, colored with green in Fig. 3; (3) $10s+$ *programs*: they take *Aquarium* more than 10 s to synthesize, colored with pink in Fig. 3. In general, PASSES accelerates synthesis on the majority of the OS-related benchmark examples, especially as the synthesis time increases. In particular, for those $10s+$ programs, the heuristics and PASSES's ability to run them all in parallel reduces the synthesis time. However, the overhead of subspace creation and parallelism can cause a slowdown in synthesis for small programs, especially those $1s-$ programs. In Fig. 3, we observe a cluster of data points (red) above the diagonal line in the lower left-hand corner in all figures, indicating that for those $1s-$ programs, PASSES with heuristics slightly slows the synthesis. Similarly, all PASSES heuristics produce programs of the same length as *Aquarium*, except *RandSimpl-SC*, which may generate solutions with one additional instruction across 5 trials.

Table 4. Geometric mean speedups of all the heuristics compared to *Aquarium* with OS-related benchmarks.

	Constraint-SC	RandSimpl-SC	TypeSimpl-SC	Inc-SC	PriorInc-SC
$1s-$ Programs	0.43	0.41	0.46	0.42	0.42
$10s-$ Programs	2.15	1.16	2.27	2.46	2.75
$10s+$ Programs	1.49	3.42	9.21	10.71	11.30

Table 4 reports the geometric mean speedups of these three groups, evaluating with five heuristics against *Aquarium*. This indicates an entirely acceptable trade-off: a slowdown of 2.33× for $1s-$ programs on average is a small price to pay for a speedup of 2.16× for $10s-$ programs and 7.23× for $10s+$ programs on average.

Constraint-SC is often beneficial, although its benefit is not as outstanding as other heuristics. Figure 2b shows that though it slows synthesis for small programs with a maximum slowdown of 6.87×, if we assume a mix of large and small jobs, the speedup on the $10s+$ programs more than compensates for it. For those $10s+$ programs, it makes the synthesis faster with a geometric mean of 1.49×. In the best case, it produces a maximum speedup of 10.35×.

RandSimpl-SC and *TypeSimpl-SC* both outperform *Aquarium*, while, as discussed before, *TypeSimpl-SC* produces more consistent results than does *RandSimpl-SC*. They sometimes slow synthesis for small programs, but for those $10s+$ programs, *RandSimpl-SC* and *TypeSimpl-SC* produce a geometric mean speed up of 3.42× and 9.21×, respectively. In the best case, they synthesize programs that are not accessible to *Aquarium*, such as the SJ-1 and LJ-1 cases; counting timeouts as 1800 s, there is a speedup of 180.49× and 288.29×, respectively. Due to the subspace creation and the parallel overhead, they sometimes slow down synthesis up to 6.59× and 5.23× for the programs that take *Aquarium* less than 1 s to synthesize, respectively. For the $10s+$ programs, they produce a slowdown of 1.93× and 2.38×, respectively.

Inc-SC and *PriorInc-SC* are also effective, and they both perform better than *TypeSimpl-SC*. They produces a geometric mean speedup of 1.45× and 1.51× with all benchmarks, and 10.71× and 11.30× for the $10s+$ programs, respectively. Counting those timeouts as 1800 s, they make the synthesis for those programs not accessible to *Aquarium* up to 282.84× and 284.99×, respectively. In the worst case, they make our synthesis 6.88× and 6.44× slower for small programs, respectively. For the $10s+$ programs, they produces a slowdown of 1.64× and 1.56×, respectively.

Comparison with Parallel SMT Solver and Different Random Seeds. Compared to *PBoolector* and *RandomSeed*, all five heuristics produce better synthesis performance for the 140 OS-related benchmarks, especially in larger cases as shown in Fig. 2b. In the best case, *TypeSimpl-SC*, *Inc-SC* and *PriorInc-SC* all outperform *PBoolector* and *RandomSeed* with a maximum speedup of up to about 200× and 280×, respectively, for the programs that are not accessible to *Aquarium* (counting timeouts as 1800 s for comparison).

6 Discussion

6.1 Parallelism Trade-Off

PASSES generates multiple sub-problems and synthesizes them simultaneously, leveraging the fact that smaller search spaces make it easier, and therefore faster, to find a solution or determine that one does not exist. However, Amdahl's law [38] also clearly delineates the scenarios in which this analysis holds. That is, the overall performance speedup gained is limited by the fraction of time that the improved part is actually used [36]. Each parallel algorithm comes with its own overhead, particularly in terms of setup in apportioning the work to a set of sub-problems and tear-down in collecting the aggregated results from the sub-problems. Furthermore, multiple CPU and I/O resources are required for parallel execution, and synthesis itself is a memory-consuming work; coordinating multiple synthesis tasks, in general, leads to drastically high memory usage, which in turn slows down the entire procedure.

The evaluation clearly showed that there is a trade-off between creating sub-problems with small search spaces and reducing the parallel overhead for synthesis. Sub-problems with smaller search spaces, in general, can be solved more quickly, but to prevent missing some potential solutions, the number of sub-problems increases. In the evaluation, we initialize with eight processors for parallelism on the AWS instance to counteract memory consumption. Balancing between the number of parallel tasks and memory consumption for each task remains an important avenue of future investigation, to explore and identify the threshold where the benefit of parallelism for synthesis is maximized.

6.2 Performance Trade-Off

Another critical trade-off is between finding a solution quickly and synthesizing optimal solutions. The nature of synthesis indicates that there can exist multiple solutions that satisfy a given specification [12]. With multiple subspaces created, those possible solutions are distributed over potentially many sub-problems. The iterative procedure in CEGIS guarantees that for the original synthesis problem Q, the synthesized program always has the minimum program length. However, given a set of n sub-problems $q_1 \ldots q_n$, derived from Q, for example, q_i could produce an a-instruction program in t_i seconds, while q_j generates a b-instruction program in t_j seconds, where $a > b$ and $t_i < t_j$. Since PASSES always takes the first returned result from any sub-problem as the solution, it returns with an assembly program of a instructions instead of one with b instructions. It is possible that the length of the program produced by PASSES is not minimized.

Furthermore, assembly programs are frequently optimized for many metrics: code size, code density, execution speed, program latency, data throughput, or even energy consumption [9]. To address this concern, we assume that techniques such as superoptimization [31] can be performed after synthesis. Meanwhile, PASSES also has the ability to collect multiple solutions from sub-problems and evaluate them against different optimization or desirability metrics.

6.3 Generalizability of PASSES

Unlike prior work that used the divide-and-conquer methodology to expedite synthesis, in PASSES, none of the presented heuristics manipulate the specification: they directly decompose the search space by imposing instruction-level constraints or simplifying the machine model; the specification remains intact. Overall, assembly synthesis, as a special category in general program synthesis problems, has features that facilitate parallel processing. The assembly synthesizer, in general, uses symbolic execution to explore every execution path based on the program semantics, while with instruction-level *constraints* and machine model *simplification*, we instantiate symbolic values to prune impossible execution paths, achieving smaller search space for each sub-problems. Those symbolic values are highly related to the machine model and the target program; instantiating them based on assembly features does not require low-level operations, and it also benefits the following parallel computing. Morover, The incremental heuristics in PASSES share the idea with Lazy Task Creation [32] to explore parallel tasks dynamically at run-time. To generalize PASSES to other synthesis problems, it is important to identify those highly problem-related symbolic variables and concretize them with values. Though previous work on adaptive concretization demonstrates that randomly concretizing influential unknowns helps synthesis performance [25], PASSES still shows the significance of problem-related instantiation during symbolic execution in program synthesis.

7 Conclusion

To improve the scalability of assembly synthesis with parallelization, we present a novel parallel assembly synthesis system, PASSES. PASSES uses domain knowledge of assembly language to parallelize synthesis problems. It identifies multiple approaches to subspace creation, using this domain knowledge. We describe the PASSES subspace creation in terms of the three properties, *collective exhaustivity*, *mutual exclusivity*, and *subspace size equality*, and introduce five complementary and reusable heuristics to improve assembly synthesis performance using parallelism. We evaluate the performance of PASSES with general bit manipulation problems and machine-dependent code from pre-existing operating systems, showing that, compared to the state-of-the-art automated assembly synthesizer and SMT parallelization approaches, the heuristics in PASSES significantly improve assembly synthesis scalability for various realistic assembly programming problems.

A PASSES Heuristic Algorithms

A.1 PASSES Algorithm

Algorithm 1 shows pseudo code for PASSES. It first initializes the list of instances with different heuristics in line 4 and then parallelizes the synthesis for each instance in line 5, by invoking multiple individual synthesizers.

Given the specification, a machine model, and a list of constraints, in function *synth_instance* (lines 9–15), Synthesizer either successfully generates a target program *prog* in line 12 or gets a synthesis failure in line 13. Depending on the heuristic, PASSES handles the failure differently as shown in line 15. PASSES executes all running instances in parallel, and if one of them successfully produces a program, it returns the first result in line 7; otherwise, it waits until all instances finish with a synthesis failure in line 8.

Algorithm 1: The system, PASSES

Input : A specification *spec*, and a machine model *model*.
Output: The synthesis result { **SUCCESS**(*Program*), **FAILURE** }.

1 **begin**
2 $result \leftarrow$ **UNKNOWN**
3 $failed\_counts \leftarrow 0$
4 initialize the list of *instances* to be synthesized
   // instance: sub-problem with a machine model and a set of
      constraints
5 **foreach** *instance* $i \in$ *instances* **do in parallel** *synth_instance*(*i*)
6 **wait until**
7   **if** *result* = **SUCCESS** *(prog)* **then return SUCCESS**(*prog*)
8   **if** *failed_counts* = *the total amount of synthesized instances* **then
      return FAILURE**
9 **Function** *synth_instance (i: instance)* **is**
10   remove *i* from the list of *instances*
     // synthesize with specification, machine model and possible
        constraints
11   **switch** Synthesizer(*spec*, *i*) **do**
12   **case SUCCESS**(*prog*) **do** *result* $\leftarrow$ **SUCCESS**(*prog*)
13   **case FAILURE do**
14       $failed\_counts \leftarrow failed\_counts + 1$
15   (add new instance(s) to the list of *instances* with incremental subspace creation)

A.2   Constraint-SC Algorithm

Algorithm 2 shows the pseudo code for the instance initialization in *Constraint-SC*. In this code example, we initialize synthesis instances by forcing the first instruction in the target program to be different types (lines 3–4); one could imagine creating more detailed subspaces with constraints on other instruction locations, which further prunes the search space, expediting synthesis. In Sect. 5, we first apply instruction type constraints on the first instruction location. Since different instruction types can contain different numbers of instructions, and there are six categories of instruction types, to enable parallelism with more synthesis sub-problems and encourage *subspace size equality*, we take the cross product of the types of the first two instructions to produce $6^2 = 36$ groups, and merge parts of them to create groups with roughly equal sizes.

Algorithm 2: Instance Initialization in *Constraint-SC*

Input : A machine model *model*.
Output: A list of *instances*.

1 *instances* ← [ ]
2 **foreach** *instruction type* $t \in model$ **do**
3 $\quad$ *constraints* ← "the 1st instruction of the target program must belong to t"
4 $\quad$ append (*model*, *constraints*) to *instances*
5 **return** *instances*

A.3 Simpl-SC Basic Algorithm

Algorithm 3 shows the pseudo code for the instance initialization in *Simpl-SC*. We first divide the given complete machine model, *complete-model*, into a list of models with smaller instruction sets, *sub-models*, in line 2; each sub-model represents a subspace of the search space of the original synthesis problem;. We discuss model division in more detail later in this section. We refer to the synthesis problems for these sub-models as *sub-model instances* in line 4.

Algorithm 3: Instance Initialization in *Simpl-SC*

Input : A machine model *complete-model*.
Output: A list of *instances*.

1 *instances* ← [ ]; *constraints* ← [ ]
2 divide *complete-model* into a list of *sub-models* // with *RandSimpl-SC* or *TypeSimpl-SC*
3 **foreach** *sub-model* ∈ *sub-models* **do**
4 $\quad$ append (*sub-model*, [ ]) to the list of *instances* // sub-model instances
5 $\quad$ append "∃ instruction i in the target program, s.t. $i \in$ *complete-model* and $i \notin$ *sub-model*" to the list of *constraints* // avoid duplicate searches
6 append (*complete-model*, *constraints*) to the list of *instances* // complete-model instance
7 **return** *instances*

To preserve *collective exhaustivity*, *Simpl-SC* also adds a single synthesis instance with the complete-model, referred to as the *complete-model instance* (line 6), which allows the synthesizer to explore other available search spaces in the original problem. In other words, this guarantees that all subspaces explored by the complete-model and sub-model instances, exhaustively cover the entire search space. To avoid duplicate searches, we generate a set of mutually exclusive constraints (line 5) and apply them to the complete-model instance; those constraints force the synthesizer to produce a program with one or more instructions outside all possible sub-models. With those constraints, the complete-model instance precludes all subspaces searched by the sub-model instances and minimizes overlaps in the search spaces. Note that the complete-model instance preserves *mutual exclusivity* from all other sub-model instances, while those sub-model instances may not be mutually exclusive from each other.

A.4 Inc-SC Algorithm

Algorithm 4 shows the pseudo code for the function *synth_instance* in *Inc-SC* with synthesis failure handling shown in lines 5–10. *Inc-SC* first enlarges the machine model used in the failed instance into a list of new models (line 7) and then creates new instances with enlarged models and mutually exclusive constraints (lines 8–10). The model enlargement extends the given model with at least one more different instruction type from the complete machine model; in other words, whenever there is an instance with a x-type ($x \geq 2$) model failure, *Inc-SC* enlarges this x-type model into several ($x + 1$)-type models and launches the corresponding sub-model instances in parallel. The mutually exclusive constraints (line 9) guarantee that the target program cannot be synthesized with only x instruction types, i.e., the target program must include all ($x + 1$) types in its enlarged model; they utilize *mutual exclusivity* to assure that the search space explored by previously failed instances will not be reconsidered. Compared with the original complete-model, these enlarged models are still relatively smaller and their sub-model instances are easier to be evaluated by the synthesizer. Note that though *Inc-SC* does not preserve *collective exhaustivity* at the beginning, given that there may exist multiple programs that satisfy the specification, incremental model enlargement ensures that the solution will be eventually found.

Algorithm 4: *synth_instance(·)* in *Inc-SC*

---

1 **Function** *synth_instance (i: instance)* **is**
2   remove *i* from the list of *instances*
3   **switch** Synthesizer($spec, i$) **do**
4   **case** **SUCCESS**($prog$) **do** $result \leftarrow$ **SUCCESS**($prog$)
5   **case FAILURE do**
6             $failed\_counts \leftarrow failed\_counts + 1$
7   generate a list of *enlarged-models*, where each element *enlarged-models* = *model* (the machine model in instance *i*) + { at least one different instruction }
8   **foreach** *enlarged_model* $\in$ *enlarged_models* **do**
9                 $constraints \leftarrow$ "$\exists$ instruction x in the target program, s.t. $x \in$ *enlarged-model* and $x \notin$ *model*"
10   append (*enlarged-model, constraints*) to the list of *instances*

B Bit Manipulation Benchmark Examples

Figure 4 describes all the examples we used. The examples, numbered P1a – P14b, are grouped by similarity. Some examples include signed or unsigned comparisons, such as P11a and P14a, while others involve shifting with 0-fill (logical

P1a(x): Turn-off the right most 1-bit in a word.
$$x \,\&\, (x-1)$$
P1b(x): Turn-on the rightmost 0-bit in a word.
$$x \mid (x+1)$$
P2a(x): Turn-off the trailing 1's in a word.
$$x \,\&\, (x+1)$$
P2b(x): Turn-on the trailing 0's in a word.
$$x \mid (x-1)$$
P3a(x): Create a word with a single 1-bit at the position of the rightmost 0-bit in x.
$$\neg x \,\&\, (x+1)$$
P3b(x): Create a word with a single 0-bit at the position of the rightmost 1-bit in x.
$$\neg x \mid (x-1)$$
P4a(x): Create a word with 1's at the position of the trailing 0's in x.
$$\neg x \,\&\, (x-1) \text{ or } \neg(x \mid -x)$$
P4b(x): Create a word with 0's at the position of the trailing 1's in x.
$$\neg x \mid (x+1)$$
P5a(x): Isolate the rightmost 1-bit.
$$\neg x \,\&\, x$$
P5b(x): Isolate the rightmost 0-bit.
$$\neg(-x \,\&\, (x+1))$$

P6(x): Turn-off the rightmost contiguous string of 1's.
$$((x \mid (x-1))+1) \,\&\, x$$
P7a(x): Create a word with 1's at the position of the rightmost 1-bit and the trailing 0's in x.
$$x \oplus (x-1)$$
P7b(x): Create a word with 0's at the position of the rightmost 0-bit and the trailing 1's in x.
$$x \oplus (x+1)$$
P8a(x): Absolute Value Function.
$$y \leftarrow x \overset{\text{arithmetic}}{>>} 31$$
$$(x \oplus y) - y \text{ or } (x+y) \oplus y$$
P8b(x): Negative Absolute Value Function.
$$y \leftarrow x \overset{\text{arithmetic}}{>>} 31$$
$$y - (x \oplus y) \text{ or } (y-x) \oplus y$$
P9(x): Sign Function.
$$(x \overset{\text{arithmetic}}{>>} 31) \mid (-x \overset{\text{logical}}{>>} 31)$$
P10(x,y): Three-Valued Compare Function.
$$(x \overset{\text{signed}}{>} y) - (x \overset{\text{signed}}{<} y)$$
P11a(x,y): Max Function.
$$((x \oplus y) \,\&\, -(x \overset{\text{signed}}{\geq} y)) \oplus y$$
P11b(x,y): Min Function.
$$((x \oplus y) \,\&\, -(x \overset{\text{signed}}{\leq} y)) \oplus y$$
P11c(x,y): Doz Function (difference or zero).
$$(x-y) \,\&\, -(x \overset{\text{signed}}{\geq} y)$$

P12a(x,y): Floor of average of two integers without overflowing.
$$(x \,\&\, y) + ((x \oplus y) \overset{\text{logical}}{>>} 1)$$
P12b(x,y): Ceil of average of two integers without overflowing.
$$(x \mid y) - ((x \oplus y) \overset{\text{logical}}{>>} 1)$$
P13a(x,y): Exchange two registers without using a third.
$$x \leftarrow x \oplus y$$
$$y \leftarrow y \oplus x$$
$$x \leftarrow x \oplus y$$
P13b(x,m,k): Exchange two fields A and B of the same register x where m is a mask with 1's in field B and k is the shift distance (the number of bits from end of A to end of B).
$$t_1 = (x \,\&\, m) << k$$
$$t_2 = (x \overset{\text{logical}}{>>} k) \,\&\, m$$
$$x' = (x \,\&\, m') \mid t_1 \mid t_2$$
(m' is a mask that isolates fields other than A and B in register x.)
P14a(x,y): Test if $nlz(x) == nlz(y)$ where nlz is the number of leading zeroes.
$$(x \oplus y) \overset{\text{unsigned}}{\leq} (x \,\&\, y)$$
P14b(x,y): Test if $nlz(x) < nlz(y)$ where nlz is the number of leading zeroes.
$$(\neg y \,\&\, x) \overset{\text{unsigned}}{>} y$$

Fig. 4. Bit manipulation benchmark examples (26 in total).

shift) or sign-fill (arithmetic shift), such as P8a and P12a. We mark the type over each comparison and shift symbol.

Table 5 reports the number of instructions in each synthesized implementation by Aquarium, indicating the minimum length requirement for each benchmark.

Table 5. Size of the bit manipulation benchmarks with MIPS implementations (lines of code).

Benchmark	P1a	P1b	P2a	P2b	P3a	P3b	P4a	P4b	P5a	P5b	P6	P7a	P7b	P8a
Length (loc)	2	2	2	2	3	3	2	3	2	3	4	2	2	3

Benchmark	P8b	P9	P10	P11a	P11b	P11c	P12a	P12b	P13a	P13b	P14a	P14b
Length (loc)	3	3	3	3	3	3	4	4	3	4	5	3

References

1. Alur, R., et al.: Syntax-guided synthesis. In: Proceedings of the 2013 Formal Methods in Computer-Aided Design, FMCAD 2013, pp. 1–8. IEEE, Portland, OR, USA (10 2013). https://doi.org/10.1109/FMCAD.2013.6679385
2. Alur, R., Černý, P., Radhakrishna, A.: Synthesis through unification. In: Kroening, D., Păsăreanu, C.S. (eds.) CAV 2015. LNCS, vol. 9207, pp. 163–179. Springer, Cham (2015). https://doi.org/10.1007/978-3-319-21668-3_10
3. Alur, R., Radhakrishna, A., Udupa, A.: Scaling enumerative program synthesis via divide and conquer. In: Legay, A., Margaria, T. (eds.) TACAS 2017. LNCS, vol. 10205, pp. 319–336. Springer, Heidelberg (2017). https://doi.org/10.1007/978-3-662-54577-5_18
4. Barman, S., Bodik, R., Chandra, S., Torlak, E., Bhattacharya, A., Culler, D.: Toward tool support for interactive synthesis. In: 2015 ACM International Symposium on New Ideas, New Paradigms, and Reflections on Programming and Software (Onward!), Onward! 2015, pp. 121–136. Association for Computing Machinery, New York, NY, USA (2015). https://doi.org/10.1145/2814228.2814235
5. Basin, D., Deville, Y., Flener, P., Hamfelt, A., Fischer Nilsson, J.: Synthesis of programs in computational logic. In: Bruynooghe, M., Lau, K.-K. (eds.) Program Development in Computational Logic. LNCS, vol. 3049, pp. 30–65. Springer, Heidelberg (2004). https://doi.org/10.1007/978-3-540-25951-0_2
6. Baumann, A., et al.: The multikernel: a new OS architecture for scalable multicore systems. In: Proceedings of the ACM SIGOPS 22nd Symposium on Operating Systems Principles, SOSP 2009, pp. 29–44. Association for Computing Machinery, New York, NY, USA (2009). https://doi.org/10.1145/1629575.1629579
7. Brummayer, R., Biere, A.: Boolector: an efficient SMT solver for bit-vectors and arrays. In: Kowalewski, S., Philippou, A. (eds.) TACAS 2009. LNCS, vol. 5505, pp. 174–177. Springer, Heidelberg (2009). https://doi.org/10.1007/978-3-642-00768-2_16
8. Buchi, J.R., Landweber, L.H.: Solving sequential conditions by finite-state strategies. Trans. Am. Math. Soc. **138**, 295–311 (1969)
9. Cempron, J.P., Salinas, C.S., Uy, R.L.: Assembly program performance analysis metrics: instructions performed and program latency exemplified on loop unroll. Philippine J. Sci. **147**(3), 441–452 (2018)
10. Chennupati, G., Azad, R.M.A., Ryan, C., Eidenbenz, S., Santhi, N.: Synthesis of parallel programs on multi-cores. In: Ryan, C., O'Neill, M., Collins, J.J. (eds.) Handbook of Grammatical Evolution, pp. 289–315. Springer, Cham (2018). https://doi.org/10.1007/978-3-319-78717-6_12
11. Cypher, A.: Eager: programming repetitive tasks by example. In: Proceedings of the SIGCHI Conference on Human Factors in Computing Systems, CHI 1991, pp.

33–39. Association for Computing Machinery, New York, NY, USA (1991). https://doi.org/10.1145/108844.108850
12. David, C., Kroening, D.: Program synthesis: challenges and opportunities. Philos. Trans. Roy. Soc. A Math. Phys. Eng. Sci. **375**(2104), 20150403 (2017)
13. Farzan, A., Nicolet, V.: Phased synthesis of divide and conquer programs. In: Proceedings of the 42nd ACM SIGPLAN International Conference on Programming Language Design and Implementation, PLDI 2021, pp. 974–986. Association for Computing Machinery, New York, NY, USA (2021)
14. Flener, P., Partridge, D.: Inductive programming. Autom. Softw. Eng. **2**, 131–137 (2001). https://doi.org/10.1023/A:1008797606116
15. Gulwani, S.: Programming by examples. Dependable Softw. Syst. Eng. **45**(137), 3–15 (2016)
16. Gulwani, S., Jha, S., Tiwari, A., Venkatesan, R.: Synthesis of loop-free programs. In: Proceedings of the 32nd ACM SIGPLAN Conference on Programming Language Design and Implementation, PLDI 2011, pp. 62–73. Association for Computing Machinery, New York, NY, USA (2011). https://doi.org/10.1145/1993498.1993506
17. Gulwani, S., Polozov, O., Singh, R.: Program synthesis. In: Foundations and Trends in Programming Languages, vol. 4, pp. 1–119. NOW, Hanover, MA, USA, August 2017
18. Hamadi, Y., Jabbour, S., Sais, L.: ManySAT: solver description. Technical report, MSR-TR-2008-83, May 2008. https://www.microsoft.com/en-us/research/publication/manysat-solver-description/
19. Hamadi, Y., Jabbour, S., Sais, L.: ManySAT: a parallel SAT solver. J. Satisfiability Boolean Modeling Comput **6**(4), 245–262 (2010)
20. Holland, D.A., Hu, J., Kawaguchi, M., Lu, E., Chong, S., Seltzer, M.I.: Aquarium: Cassiopea and Alewife languages (2022). https://arxiv.org/abs/1908.00093
21. Holland, D.A., Lim, A.T., Seltzer, M.I.: A new instructional operating system. In: Proceedings of the 33rd SIGCSE Technical Symposium on Computer Science Education, SIGCSE 2002, pp. 111–115. Association for Computing Machinery, New York, NY, USA (2002). https://doi.org/10.1145/563340.563383
22. Hu, J., Lu, E., Holland, D.A., Kawaguchi, M., Chong, S., Seltzer, M.: Towards porting operating systems with program synthesis. ACM Trans. Program. Lang. Syst. **45**(1) (2023). https://doi.org/10.1145/3563943
23. Hu, J., Lu, E., Holland, D.A., Kawaguchi, M., Chong, S., Seltzer, M.I.: Trials and tribulations in synthesizing operating systems. In: Proceedings of the 10th Workshop on Programming Languages and Operating Systems, PLOS 2019, pp. 67–73. Association for Computing Machinery, New York, NY, USA (2019). https://doi.org/10.1145/3365137.3365401
24. Hu, J., Vaithilingam, P., Chong, S., Seltzer, M., Glassman, E.L.: Assuage: assembly synthesis using a guided exploration. In: The 34th Annual ACM Symposium on User Interface Software and Technology, pp. 134–148. Association for Computing Machinery, New York, NY, USA (2021)
25. Jeon, J., Qiu, X., Solar-Lezama, A., Foster, J.S.: Adaptive concretization for parallel program synthesis. In: Kroening, D., Păsăreanu, C.S. (eds.) CAV 2015. LNCS, vol. 9207, pp. 377–394. Springer, Cham (2015). https://doi.org/10.1007/978-3-319-21668-3_22
26. Jeon, J., Qiu, X., Solar-Lezama, A., Foster, J.S.: An empirical study of adaptive concretization for parallel program synthesis. Form. Methods Syst. Des. **50**(1), 75–95 (2017). https://doi.org/10.1007/s10703-017-0269-8

27. Jha, S., Seshia, S.A.: A theory of formal synthesis via inductive learning. Acta Inf. **54**(7), 693–726 (2017). https://doi.org/10.1007/s00236-017-0294-5
28. Le Frioux, L., Baarir, S., Sopena, J., Kordon, F.: PaInleSS: a framework for parallel SAT solving. In: Gaspers, S., Walsh, T. (eds.) SAT 2017. LNCS, vol. 10491, pp. 233–250. Springer, Cham (2017). https://doi.org/10.1007/978-3-319-66263-3_15
29. Le Frioux, L., Baarir, S., Sopena, J., Kordon, F.: Modular and efficient divide-and-conquer SAT solver on top of the painless framework. In: Vojnar, T., Zhang, L. (eds.) TACAS 2019. LNCS, vol. 11427, pp. 135–151. Springer, Cham (2019). https://doi.org/10.1007/978-3-030-17462-0_8
30. Manna, Z., Waldinger, R.: A deductive approach to program synthesis. ACM Trans. Program. Lang. Syst. **2**(1), 90–121 (1980). https://doi.org/10.1145/357084.357090
31. Massalin, H.: Superoptimizer: a look at the smallest program. In: Proceedings of the Second International Conference on Architectual Support for Programming Languages and Operating Systems, ASPLOS II, pp. 122–126. IEEE Computer Society Press, Washington, DC, USA (1987). https://doi.org/10.1145/36206.36194
32. Mohr, E., Kranz, D., Halstead, R.: Lazy task creation: a technique for increasing the granularity of parallel programs. IEEE Trans. Parallel Distrib. Syst. **2**(3), 264–280 (1991). https://doi.org/10.1109/71.86103
33. Myers, B.A.: Creating user interfaces using programming by example, visual programming, and constraints. ACM Trans. Program. Lang. Syst. **12**(2), 143–177 (1990). https://doi.org/10.1145/78942.78943
34. Partridge, D.: The case for inductive programming. Computer **30**(1), 36–41 (1997). https://doi.org/10.1109/2.562924
35. Polozov, O., Gulwani, S.: FlashMeta: a framework for inductive program synthesis. In: Proceedings of the 2015 ACM SIGPLAN International Conference on Object-Oriented Programming, Systems, Languages, and Applications, OOPSLA 2015, pp. 107–126. Association for Computing Machinery, New York, NY, USA (2015). https://doi.org/10.1145/2814270.2814310
36. Reddy, M.: Chapter 7 - performance. In: Reddy, M. (ed.) API Design for C++, pp. 209–240. Morgan Kaufmann, Boston (2011). https://doi.org/10.1016/B978-0-12-385003-4.00007-5
37. Reisenberger, C.: PBoolector: a parallel SMT solver for QF_BV by combining bit-blasting with look-ahead. Ph.D. thesis, Master's thesis, Johannes Kepler Univesität Linz, Linz, Austria (2014)
38. Rodgers, D.P.: Improvements in multiprocessor system design. In: Proceedings of the 12th Annual International Symposium on Computer Architecture, ISCA 1985, pp. 225–231. IEEE Computer Society Press, Washington, DC, USA (1985)
39. Schkufza, E., Sharma, R., Aiken, A.: Stochastic superoptimization. In: Proceedings of the Eighteenth International Conference on Architectural Support for Programming Languages and Operating Systems, ASPLOS 2013, pp. 305–316. Association for Computing Machinery, New York, NY, USA (2013). https://doi.org/10.1145/2451116.2451150
40. Smith, D.R.: The design of divide and conquer algorithms. Sci. Comput. Programm. **5**, 37–58 (1985). https://doi.org/10.1016/0167-6423(85)90003-6
41. Smith, D.R.: Top-down synthesis of divide-and-conquer algorithms. Artif. Intell. **27**(1), 43–96 (1985). https://doi.org/10.1016/0004-3702(85)90083-9
42. Solar-Lezama, A.: The sketching approach to program synthesis. In: Hu, Z. (ed.) APLAS 2009. LNCS, vol. 5904, pp. 4–13. Springer, Heidelberg (2009). https://doi.org/10.1007/978-3-642-10672-9_3

43. Solar-Lezama, A., Jones, C.G., Bodik, R.: Sketching concurrent data structures. In: Proceedings of the 29th ACM SIGPLAN Conference on Programming Language Design and Implementation, PLDI 2008, pp. 136–148. ACM, New York, NY, USA (2008). https://doi.org/10.1145/1375581.1375599
44. Solar-Lezama, A., Tancau, L., Bodik, R., Seshia, S., Saraswat, V.: Combinatorial sketching for finite programs. In: Proceedings of the 12th International Conference on Architectural Support for Programming Languages and Operating Systems, ASPLOS 2006, pp. 404–415. Association for Computing Machinery, New York, NY, USA (2006). https://doi.org/10.1145/1168857.1168907
45. Srinivasan, V., Reps, T.: Synthesis of machine code from semantics. In: Proceedings of the 36th ACM SIGPLAN Conference on Programming Language Design and Implementation, PLDI 2015, pp. 596–607. Association for Computing Machinery, New York, NY, USA (2015). https://doi.org/10.1145/2737924.2737960
46. Srinivasan, V., Sharma, T., Reps, T.: Speeding up machine-code synthesis. In: Proceedings of the 2016 ACM SIGPLAN International Conference on Object-Oriented Programming, Systems, Languages, and Applications, OOPSLA 2016, pp. 165–180. Association for Computing Machinery, New York, NY, USA (2016). https://doi.org/10.1145/2983990.2984006
47. Warren, H.S.: Hacker's Delight, 2nd edn. Addison-Wesley Professional, Boston, MA, USA (2012)
48. Wintersteiger, C.M., Hamadi, Y., de Moura, L.: A concurrent portfolio approach to SMT solving. In: Bouajjani, A., Maler, O. (eds.) CAV 2009. LNCS, vol. 5643, pp. 715–720. Springer, Heidelberg (2009). https://doi.org/10.1007/978-3-642-02658-4_60
49. Zhang, J., et al.: PyDex: repairing bugs in introductory python assignments using LLMs. Proc. ACM Program. Lang. **8**(OOPSLA1) (2024). https://doi.org/10.1145/3649850
50. Zhang, J., Li, D., Kolesar, J.C., Shi, H., Piskac, R.: Automated feedback generation for competition-level code. In: Proceedings of the 37th IEEE/ACM International Conference on Automated Software Engineering, ASE 2022. Association for Computing Machinery, New York, NY, USA (2023). https://doi.org/10.1145/3551349.3560425
51. Zhang, J., Piskac, R., Zhai, E., Xu, T.: Static detection of silent misconfigurations with deep interaction analysis. Proc. ACM Program. Lang. **5**(OOPSLA) (2021). https://doi.org/10.1145/3485517

Improving Logic Programs by Adding Functions

Michael Hanus[(✉)]

Institut für Informatik, Kiel University, Kiel, Germany
mh@informatik.uni-kiel.de

Abstract. Logic programming is based on defining relations. Functions are often considered as syntactic sugar which can be transformed into predicates so that their logic is not used for computational purposes. In this paper, we present a method to use functions to improve the operational behavior of logic programs without loosing the flexibility of logic programming. For this purpose, predicates and goals are transformed into functions and nested expressions. By evaluating these functions in a demand-driven manner wherever possible and taking potential failures into account, we ensure that the execution of the transformed programs will never require more steps than the original programs but can decrease the number of steps—in the best case reducing infinite search spaces to finite ones. Thus, we obtain a systematic method to improve the operational behavior of logic programs without changing their semantics.

1 Introduction

Logic programming supports flexible programming techniques by built-in non-determinism and free (logic) variables. Predicates can be called with unknown arguments so that there are no fixed input and output positions, in contrast to functional languages. Since functions can be represented as predicates by adding the result as a parameter, logic programming is often considered as the more expressive programming paradigm [36].

Due to these considerations, functions and nested function calls are considered as nice syntactic sugar which can be eliminated by translating them into predicates and flattening nested expressions [8,12,32]. A consequence of this traditional view is the fact that one does not gain any operational advantage by the presence of functions. Since it is well known in functional programming that demand-driven (lazy) evaluation supports new programming techniques, as computing with infinite data structures or modularity [25], there are also approaches to translate functions with a demand-driven evaluation strategy by exploiting coroutining in Prolog [12,32]. Since coroutining might influence completeness due to floundering, these approaches have an ad-hoc flavor—they are useful for particular examples, but general correctness results (soundness, completeness) are not provided.

To improve this situation, one could also take the opposite way. Instead of transforming functions into predicates, one could transform logic programs into programs of a language with another operational semantics. For instance, Van Roy and Haridi [42] proposed to translate Prolog programs into Oz programs where advanced features of Oz for concurrent and distributed computations are used. Although this is an interesting approach to move logic programs techniques into the distributed world, an operational improvement is not obtained by this transformation. This is different in [19] where various transformations of logic programs into functional logic programs are proposed. Functional logic languages, such as Curry [22], offer the same flexibility as logic languages (soundness and completeness w.r.t. computing with partial information). In addition, optimality properties are known for well-defined classes of programs (minimal number of computed solutions, minimal number of evaluation steps [2]). The latter properties are due to the demand-driven evaluation of functions, i.e., it is essential to keep nested functional expressions instead of flattening them.

Due to these results, one could transform logic programs into functional logic programs. However, there is still one obstacle. As shown in [19], this transformation yields equivalent computations (identical answers, same number of computation steps) if the functional logic program is evaluated with an eager (strict) strategy. When a lazy strategy is used, the transformed programs might be more efficient but there are also cases where they compute answers which are not justified by the logic program. This could be the case if some subgoal fails in the logic program but they are not evaluated (due to laziness) in the transformed program. Thus, the transformation might obtain more efficient programs but with logically different answers.

In this paper, we want to close this gap by incorporating the failure behavior into the transformation. Due to this improvement, we obtain a method to transform logic programs with the following properties.

- The transformed programs always compute the same or more general answers to a given goal.
- The evaluation of the transformed programs is guaranteed equal or better than the evaluation of the original programs: in the worst case, the same number of steps are performed, but there are also cases where less steps are required to compute the result.
- As a consequence, there are also cases where infinite search spaces are reduced to finite ones.

Sloppily speaking, the improvements introduced by our transformation are comparable to "green" cuts in logic programming.

To obtain this result, we extend the transformation of [19] by deciding the eager or lazy evaluation of operations based on potential failures. For this purpose, we carefully analyze the problem of different answers between an eager or lazy evaluation of transformed programs. Then we use an approximation of the totality property of the transformed functions to explicitly enforce strict evaluations at some points in the generated programs. If a subexpression is ensured to be totally defined, it can be evaluated in a demand-driven manner, otherwise its evaluation is enforced even if the result is not immediately demanded.

This paper is structured as follows. After sketching the basics of logic and functional logic programming, we review in Sect. 3 the existing method to transform logic into functional logic programs. Section 4 discusses the incorporation of failure information to obtain a transformation which is correct w.r.t. the logical consequences of the logic program. In order to improve the transformation, we show in Sect. 5 how the addition of types can help to deduce a more precise approximation of potentially failing computations. Section 6 sketches the implementation of our approach which is evaluated in Sect. 7. Section 8 discusses related work before we conclude.

2 Logic and Functional Logic Programming

In the following we briefly review some notions and features of logic and functional logic programming. More details can be found in [29] and in surveys on functional logic programming [5,17].

We use Prolog syntax to present logic programs. *Terms* in logic programs are constructed from variables $(X, Y, \ldots)$, numbers, atom constants $(c, d, \ldots)$, and functors or term constructors $(f, g, \ldots)$ applied to a sequence of terms, like $f(t_1, \ldots, t_n)$. A *literal* $p(t_1, \ldots, t_n)$ is a predicate p applied to a sequence of terms, and a *goal* $L_1, \ldots, L_k$ is a sequence of literals, where $\square$ denotes the empty goal $(k = 0)$. *Clauses* $L \text{ :- } B$ define predicates, where the *head* L is a literal and the *body* B is a goal (a *fact* is a clause with an empty body $\square$, otherwise it is a *rule*). A *logic program* is a sequence of clauses.

Logic programs are evaluated by SLD-resolution steps, where we consider the leftmost selection rule here. Thus, if $G = L_1, \ldots, L_k$ is a goal and $L \text{ :- } B$ is a variant of a program clause (with fresh variables) such that there exists a most general unifier[1] (*mgu*) σ of L_1 and L, then $G \vdash_\sigma \sigma(B, L_2, \ldots, L_k)$ is a *resolution step*. We denote by $G_1 \vdash^*_\sigma G_m$ a sequence $G_1 \vdash_{\sigma_1} G_2 \vdash_{\sigma_2} \ldots \vdash_{\sigma_{m-1}} G_m$ of resolution steps with $\sigma = \sigma_{m-1} \circ \cdots \circ \sigma_1$. A *computed answer* for a goal G is a substitution σ (restricted to the variables occurring in G) with $G \vdash^*_\sigma \square$.

Example 1. The following logic program defines a predicate `plus`, which relates two natural numbers in Peano representation, where o represents zero and s represents the successor of a natural [41] to its sum, a predicate `plus3`, which relates three natural numbers to its sum, and a predicate `thirdof` which is satisfied if the second argument is a third of the first one.

```
plus(o,Y,Y).
plus(s(X),Y,s(Z)) :- plus(X,Y,Z).
plus3(X,Y,Z,R) :- plus(X,Y,XY), plus(XY,Z,R).
thirdof(X,Y) :- plus3(Y,Y,Y,X).
```

For the goal `thirdof(o,T)`, the answer $\{T \mapsto o\}$ is computed but, after showing this answer, Prolog does not terminate due to an infinite search space (since it enumerates arbitrary large values for Y).

---

[1] Substitutions, variants, and unifiers are defined as usual [29].

Functional logic programming [5,17] integrates the most important features of functional and logic languages, such as higher-order functions and lazy (demand-driven) evaluation from functional programming and non-deterministic search and computing with partial information from logic programming. The declarative multi-paradigm language Curry [22], which we use in this paper, is a functional logic language with advanced programming concepts. Its syntax is close to Haskell [35], i.e., variables and names of defined operations start with lowercase letters and the names of data constructors start with an uppercase letter. The application of an operation f to e is denoted by juxtaposition ("$f\ e$").

In addition to Haskell, Curry allows *free (logic) variables* in program rules (equations) and initial expressions. Function calls with free variables are evaluated by a possibly non-deterministic instantiation of arguments.

Example 2. The following Curry program[2] defines the operations of Example 1 in a functional manner, where logic features (like the free variable y) are exploited to define thirdof:

```
plus 0     y = y
plus (S x) y = S (plus x y)
plus3 x y z = plus (plus x y) z
thirdof x | x =:= plus3 y y y
          = y
```

"|" introduces a condition, and "=:=" denotes semantic unification, i.e., the expressions on both sides are evaluated before unifying them.

Since plus can be called with arguments containing free variables, the condition in the definition of thirdof is solved by instantiating y to appropriate *values* (i.e., expressions without defined functions) before reducing a function call. This corresponds to narrowing [37,40]. $t \leadsto_\sigma t'$ is a *narrowing step* if there is some non-variable position p in t, an equation (program rule) $l = r$, and an mgu σ of $t|_p$ and l such that $t' = \sigma(t[r]_p)$,[3] i.e., t' is obtained from t by replacing the subterm $t|_p$ by the equation's right-hand side and applying the unifier. Conditional equations $l \mid c = r$ are considered as syntactic sugar for the unconditional equation $l = c$ &> r, where "&>" is defined by True &> x = x.

Curry is based on the *needed narrowing strategy* [2] which also uses non-most-general unifiers in narrowing steps to ensure the optimality of computations. Needed narrowing is a demand-driven evaluation strategy, i.e., it supports computations with infinite data structures [25] and can avoid superfluous computations so that it is optimal w.r.t. the number of computed solutions and the length of derivation (see [2] for precise statements). The latter property is our motivation to transform logic programs into Curry programs, since this can reduce infinite search spaces to finite ones. For instance, the evaluation of the expression thirdof 0 has a finite computation space: the generation of larger

---
[2] The concrete syntax is simplified by omitting the declaration of free variables, like y, which is required in Curry programs to enable consistency checks by the compiler.
[3] We use common notations from term rewriting [6].

numbers for the first argument of `plus3` is avoided since there is no demand for such numbers.

Curry has many more features which are useful to implement applications, like *set functions* [4] to encapsulate search, and standard features from functional programming, like modules and monadic I/O [43]. However, the kernel of Curry described so far should be sufficient to understand the remaining contents.

3 From Logic to Functional Logic Programs

Due to the fact that functional logic programming is an extension of pure logic programming, there is a simple way to transform logic into functional logic programs by mapping each predicate into a Boolean function and each clause into a (conditional) equation. This *conservative transformation* [19] does not change the structure of derivations since narrowing steps on Boolean functions correspond to resolution steps. Thus, there is no real advantage to perform this transformation.

In order to exploit the computational power of functional logic languages, one has to transform predicates into non-Boolean functions by selecting some arguments as results and generating function definitions according to this selection.

Example 3. Consider the predicate `plus` defined in Example 1. If the third argument is selected as a result argument (as often intended in logic programs), the clauses of `plus` can be transformed into the following functional logic program:

```
plus 0 y = y
plus (S x) y | z =:= plus x y = S z
```

In principle, any set of argument positions can be selected as results, as discussed in [19]. For instance, if the first two arguments of `plus` are selected as result arguments, the clauses of `plus` can be transformed into the following program:

```
plus y = (0, y)
plus (S z) | (x,y) =:= plus z = (S x, y)
```

Although this is not a function in a mathematical sense, it is a valid definition in Curry: it defines a *non-deterministic operation* which might deliver more than one result for a given argument. For a given natural number n, it returns all splittings into two numbers such that their sum is equal to n:

```
>   plus (S (S 0))
(S (S 0), 0)
(S 0, S 0)
(0, S (S 0))
```

Non-deterministic operations, which can formally be interpreted as mappings from values into sets of values [14], are an important feature of contemporary functional logic languages. The admissibility of non-deterministic operations provides more freedom when transforming logic programs into functional logic programs.

It is shown in [19] that, even if this *functional transformation* is used, there is a strong one-to-one correspondence, independent of the selection of result arguments, between resolution derivations w.r.t. the original logic program and narrowing derivations w.r.t. the transformed program. In order to get a real advantage, one has to replace the unification occurring in conditions by *let* bindings whenever possible[4] and inline these bindings if reasonable. For instance, one can transform the rule

```
plus (S x) y | z =:= plus x y = S z
```

into

```
plus (S x) y = let z = plus x y in S z
```

and inline the binding of z into

```
plus (S x) y = S (plus x y)
```

This *demand functional transformation* is described in detail in [19]. If the transformed program is eagerly evaluated, i.e., the arguments of a function call are evaluated before replacing the function call by its body ("call by value"), there is no operational difference between programs transformed by the functional and the demand functional transformation. This situation changes when the arguments are evaluated "by need," as in Haskell or Curry and discussed in [24,25]. For instance, consider the predicate signat defined by the clauses

```
signat(o,o).
signat(s(X),s(o)).
```

Its demand functional transformation is[5]

```
signat 0     = 0
signat (S x) = S 0
```

Now consider the evaluation of the expression signat (plus n_1 n_2), where n_1 is a big natural number. An eager evaluation requires $n_1 + 1$ rewrite steps, whereas a non-strict language needs only two steps.

Although it seems that the demand functional transformation is the way to go, there is one potential problem of this transformation: it might change the semantics, i.e., the set of computed solutions. This could be the case if the evaluation of some subexpression is not demanded and its evaluation would fail to yield a value. This failure would be propagated in the original logic program, but it might be "hidden" in the transformed program. For instance, consider a "decrement" predicate

```
dec(s(X),X).
```

and its application in the predicate

```
sig1(R) :- dec(o,X), plus(s(o),X,Y), signat(Y,R).
```

---

[4] This is possible when the variable in the left-hand side of the unification has no occurrences in result arguments of other goal literals, see [19] for a precise discussion.

[5] If it is not explicitly mentioned, the last argument of a predicate is considered as the result argument.

Due to the failure of the first subgoal, the goal "?- sig1(R)." fails. However, the demand functional transformation yields.

```
dec (S x) = x
sig1 = signat (plus (S 0) (dec 0))
```

so that a lazy evaluation of sig1 yields the value S 0.

One could argue that the latter result value is intended due to the mathematical principle of "replacing equals by equals." This point of view is taken in [19]. However, if we intend to use the transformation of logic programs into functional logic programs to obtain more efficient programs without changing the set of computed answers, this difference is not acceptable. Therefore, we develop in the next sections a slightly different transformation which does not change the answer behavior (except for computing more general answers) but improves the operational behavior in many cases.

4 Incorporating Failure Information

We have seen in the previous section that the demand functional transformation might yield a different answer behavior if functions generated by the transformation are partially defined, i.e., fail to compute a value for particular argument values. We call such functions *failing*. The reason could be a pattern matching failure as well as an infinite loop. Since failures due to infinite loops are seldom, we later concentrate on the approximation of pattern matching failures.

If all functions occurring in an evaluation are totally defined, i.e., always non-failing, then a demand-driven evaluation strategy, like needed narrowing, computes only more general answers compared to an eager strategy, as shown in the following result.

Theorem 1 (Correctness of the demand functional transformation).
Let P be a logic program and G be a goal. Let F be the functional logic program and e the expression obtained by applying the demand functional transformation to P and G, respectively.

1. *If there is a resolution derivation $G \vdash^*_\sigma \Box$ w.r.t. P, then there is a needed narrowing derivation $e \leadsto^*_{\sigma'}$ True w.r.t. F and a substitution φ with $\sigma = \varphi \circ \sigma'$.*
2. *If all functions in F are totally defined and there is a needed narrowing derivation $e \leadsto^*_\sigma$ True w.r.t. F, then there is a resolution derivation $G \vdash^*_{\sigma'} \Box$ w.r.t. P and a substitution φ with $\sigma' = \varphi \circ \sigma$.*

Proof. We sketch the proof which is a direct consequence of the soundness and completeness of needed narrowing [1,2].

If there is a resolution derivation $G \vdash^*_\sigma \Box$ w.r.t. P, then there is an innermost narrowing derivation $e \leadsto^*_\sigma$ True w.r.t. F computing the same answer [19]. By soundness of narrowing, σ is a solution of the Boolean expression e w.r.t. F. By completeness of needed narrowing, there is a needed narrowing derivation $e \leadsto^*_{\sigma'}$ True w.r.t. F and a substitution φ with $\sigma = \varphi \circ \sigma'$.

If there is a needed narrowing derivation $e \overset{*}{\leadsto}_\sigma$ True w.r.t. F, then, by soundness of needed narrowing, there is a rewriting of $\sigma(e)$ to True. This is not necessarily an innermost (eager) rewriting, i.e., there are some function calls in this rewrite sequence which might not be rewritten since their values are not needed. In an eager evaluation, all these function calls are evaluated. Due to the assumption that all functions are totally defined, they can always be evaluated to some value. An eager narrowing derivation might also bind some argument variables which were not bound in the needed narrowing evaluation. Thus, innermost narrowing might compute a more instantiated answer σ'. By [19], there is a resolution derivation $G \vdash^*_{\sigma'} \square$ w.r.t. P. □

For instance, consider the expression signat (plus x x) where x is a free variable. Needed narrowing computes only the answer substitutions $\{x \mapsto 0\}$ for the result 0 and $\{x \mapsto S\_\}$ for the result S 0, whereas innermost narrowing computes an infinite set of answer substitutions, i.e., $\{x \mapsto n\}$ for every natural number n. The same infinite set of answers is computed for the corresponding Prolog program.

In order to safely use the demand functional transformation w.r.t. the needed narrowing strategy, one has to ensure that all functions occurring in the evaluation steps are totally defined. However, one has not to give up in the presence of possibly failing operations since one only has to enforce the evaluation of such operations if they occur in the right-hand side of a rule, since the corresponding predicates are also evaluated in the logic program. This can be obtained by the following modification of the transformation. If a rule's right-hand side contains an application $f\ e$ and the evaluation of the expression e *might fail* to compute a value, then this application is replaced by

$f\ \$!\ e$

We call this modified transformation *fail-sensitive functional transformation*. The predefined infix operator "$!" denotes a function application with a strict evaluation of the argument. Thus, if the evaluation of e fails, the evaluation of $f\ \$!\ e$ fails.

For instance, the fail-sensitive functional transformation maps the predicate sig1 defined in the previous section into

sig1 = signat $! (plus (S 0) $! (dec 0))

so that the evaluation of sig1 leads to a failure due to the enforced evaluation of (dec 0). Note that the usage of "$!" is necessary at all places where a potentially failing expression occurs and not only where the failing expressions occurs first. For instance, the slightly modified expression

signat (S (plus (S 0) $! (dec 0)))

evaluates to S 0 when evaluated by needed narrowing, whereas

signat $! (S $! (plus (S 0) $! (dec 0))

fails, similarly to the corresponding logic program.

In order to perform our transformation, we need to know whether an operation is totally defined. Obviously, this is undecidable in general so that some approximation is required. We can split this property into two parts: termination and non-occurrence of failures. Termination of rewrite systems or functional programs is a well-studied topic so that various techniques are available to approximate this property, e.g., [13,28]. Actually, the Curry analysis framework CASS [21] also provides an analysis to approximate the termination of functions. Therefore, we concentrate the subsequent discussion to the problem to approximate fail-freeness.

An operation is called *fail-free* [18] if its evaluation will never run into an explicit failure (due to a pattern-match error). A recent technique to approximate and verify fail-freeness is a type-based approximation of the arguments describing values for which a function call does not fail [20]. For this purpose, sets of concrete argument values are approximated by some abstract type. A simple but practically useful approximation are depth-k abstractions [38]. For the case $k = 1$, depth-1 types are described by subsets of constructors representing all terms having one of these constructors at the root. Moreover, the special value $\top$ describes the set of all constructors occurring in the program (a refinement based on type information will be discussed later). A *call type* of an n-ary function f is sequence $\alpha_1, \ldots, \alpha_n$ of abstract types such that any evaluation of a call $f\,t_1 \ldots t_n$ does not fail if t_i belongs to the set represented by α_i ($i = 1, \ldots, n$). For instance, $\{\texttt{S}\}$ is a call type of the operation dec ($\varnothing$ would be another, less precise call type), and $\{\texttt{0}, \texttt{S}\}, \top$ is a call type of plus. Since $\{\texttt{0}, \texttt{S}\}$ are the only constructors in this program, we can also write this call type as $\top, \top$. A *trivial call type* is a sequence of $\top$s.

A method to infer call types is described in [20]. It is based on approximating the input/output behavior of operations and using this information to infer call types by considering the patterns and the operations occurring in the right-hand sides of the rules. Since these operations might demand for refined call types, the entire inference is a global fixpoint computation. For instance, the initial call type of

```
plusD x y = plus x (dec y)
```

is $\top, \top$ (since there are no restrictions in the left-hand side due to patterns). The call (dec y) demands that the second argument has type $\{\texttt{S}\}$ so that the initial call type is refined to $\top, \{\texttt{S}\}$. With this call type, plusD can be proved as fail-free.

An operation can have a trivial call type even if it calls potentially failing operations. For instance, consider the operation

```
dPlusS n = dec (plus (S n) n)
```

The tool described in [20] infers the trivial call type for dPlusS since the first argument of plus ensures that dec is always called with an S-rooted term.

The practical evaluation shows that only a few operations have non-trivial call types in larger programs since most operations are defined by complete pattern matchings on their input type. However, this demands for the consideration of the intended types of operations, as discussed next.

5 Type-Based Translation

Although Curry and other contemporary functional logic languages (e.g., TOY [31]) are strongly typed, type information is not considered in the fail-sensitive functional transformation. In principle, this is not necessary since types of functions are automatically inferred. Since data constructors need to be declared, the transformation adds a single definition of a type Term containing all constructors occurring in the program. For instance, the transformation of the previous examples using natural numbers in Peano representation adds the type declaration

```
data Term = O | S Term
```

If we extend the previous logic programs by a definition of a predicate to relate a binary tree with the number of its leaves, like

```
numleaves(leaf(_),s(o)).
numleaves(node(M1,M2),s(N)) :-
      numleaves(M1,N1), numleaves(M2,N2), plus(N1,N2,N).
```

then the following data type is generated:

```
data Term = O | S Term | Leaf Term | Node Term Term
```

Since it is not ensured that plus is called only with natural numbers as arguments, the call type inferred for plus is no longer trivial. This is correct since the call plus (Leaf O) O is allowed and fails.

In order to improve this situation, we extend the transformation by the inclusion of type declarations. Although there is some interest to add types to Prolog programs [39], there is no general agreement about its syntax and structure. For instance, CIAO-Prolog [23] allows the definition of regular and Hindley-Milner types, or [7] describes a tool to infer types from logic programs. For our purpose, we support the explicit definition of polymorphic algebraic data types in Prolog syntax, like

```
:- type nat = o ; s(nat).
:- type tree(A) = leaf(A) ; node(tree(A),tree(A)).
```

These will be translated into definitions in Curry syntax:

```
data Nat = O | S Nat
data Tree a = Leaf a | Node (Tree a) (Tree a)
```

With these definitions, the following types are inferred for the transformed operations plus and numleaves:

```
plus :: Nat → Nat → Nat
numleaves :: Tree a → Nat
```

Based on these types, both operations have trivial call types so that they are fail-free. This demonstrates that type information is useful to improve the transformation of logic programs into functional logic programs. In principle, it is not necessary. In the worst case, missing type information might have the effect that all operations are potentially failing so that the translated program is eagerly evaluated without any advantage compared to the original logic program.

6 Implementation

To evaluate our approach, we have extended the transformation tool presented in [19] by the inclusion of type and failure information.[6] Types can be declared in the source program as discussed in the previous section. They are directly mapped into corresponding data definitions in Curry. The remaining constructors occurring in the logic program but not mentioned in a type declaration are declared in a single `Term` type, as shown in the previous section.

Predicates are mapped into Curry functions without type annotations so that type signatures are inferred by the Curry system. The entire transformation process is performed in the following steps:

1. The given logic program is translated into a Curry program based on the demand functional transformation. The result argument positions are either inferred using a heuristic discussed in [19] or they can also be explicitly declared in the given logic program.
2. The generated functional logic program is analyzed with the tool described in [20] to check which operations are fail-free.[7]
3. Using the information computed in the previous step, the given logic programs is translated into a Curry program based on the fail-sensitive functional transformation. Thus, the same program structure is generated but the standard function application $f\ e$ is replaced by $f\ \$!\ e$ whenever the expressions e might fail, i.e., contains an operation which is not fail-free. Note that f might be an expression, e.g., another application. For instance, if f is a binary operation and e_1 as well as e_2 contain operations with non-trivial call types, then an application $f\ e_1\ e_2$ generated in step 1 is replaced by $(f\ \$!\ e_1)\ \$!\ e_2$.[8] If all operations in e_2 have trivial call types, the application $(f\ \$!\ e_1)\ e_2$ is generated.

7 Evaluation

The main motivation for this work is to show that functional logic programs have concrete operational advantages compared to pure logic programs. In principle, this has already been shown in [19] but that approach is based on changing the

---

[6] The tool, available at https://cpm.curry-lang.org/pkgs/prolog2curry-1.2.0.html, is implemented as a Curry package for easy installation. A script with all required tools is also available as a docker image at https://hub.docker.com/r/currylang/prolog2curry.
[7] The inclusion of automated termination checks is currently omitted since it is seldom that operations generated from Prolog are completely defined but non-terminating.
[8] Note that this might change the order of evaluation if both arguments are demanded by f, but this order is not relevant in a declarative language without side effects.

semantics of programs from strict to demand-driven evaluation. As discussed above, this could have the effect that the transformed program computes *more* answers than the original logic program. By incorporating failure information, our new transformation has not this effect but compute more general answers compared to the answers of the logic program. For instance, the Prolog goal

```
?- plus(X,o,N), signat(N,s(o)).
```

yields an infinite set of answers where X and N are bound to all non-zero natural numbers, whereas the equivalent Curry expression obtained by our transformation

```
> signat (plus x O) =:= S O
```

has a finite search space and yields the single answer substitution $\{x \mapsto S \_\}$.

In the following, we want to show the practical advantages of this transformation by various examples. Note that these examples are small programs since larger Prolog programs are seldom logic programs—they often use non-declarative features. Such non-declarative features are either not necessary in functional logic programs (e.g., cuts are replaced by exploiting functional dependencies) or can be reformulated in a declarative manner (e.g., declarative monadic I/O, state monads). Due to these reasons, functional logic programs have advantages from the view of software construction—which is less tangible to formalization. Therefore, we evaluate the advantages of functional logic programming only from the operational point of view.

Table 1. Execution times (in seconds) of Prolog and generated Curry programs

Language:	Prolog	Prolog	Curry
System:	SWI 9.0.4	SICStus 4.9.0	KiCS2 3.1.0
rev_4096	0.23	0.22	0.10
tak_27_16_8	6.97	3.23	0.74
ackermann_3_9	2.13	8.72	0.07
thirdof_0	∞	∞	0.01
signat_plus_0	∞	∞	0.01
numleaves_7	∞	∞	0.01
permsort_10	1.43	0.28	0.03
permsort_11	16.16	1.38	0.08
permsort_12	206.34	15.23	0.28

Table 1 contains the results of executing various Prolog programs with SWI-Prolog and SICStus-Prolog and the Curry programs obtained by applying the fail-sensitive functional transformation with the Curry system KiCS2 [9].[9] The direct comparison of original and transformed programs is not straightforward since the execution times depend on the implementation techniques used in the Prolog and Curry systems. KiCS2 compiles Curry programs into Haskell programs that are compiled to machine code by the Glasgow Haskell Compiler (GHC 9.4.5). GHC generates efficient code for functional computations. This is visible in the first three benchmarks which are purely deterministic computations. `rev_4096` is the naive list reversal applied to a list of 4096 elements. `tak_27_16_8` applies the highly recursive `tak` function, used in various benchmarks [34] for logic and functional languages, to the values (27,16,8) in Peano representation. Similarly, the Ackermann function is defined on Peano numbers and applied to the Peano representation of (3,9). The definition of these functions can be found in the source code of the tool. It is interesting to note that the definition of `tak` decrements some arguments based on the predicate `dec` defined in Sect. 3. Since the generated function `dec` might fail, our transformation inserts the strict application operator "$!" in several places in the body of `tak`. These operators cause some overhead[10] and are not really necessary since `tak` is strict in its first argument. With a strictness analysis, one could drop these operators. Although this is not implemented in our tool, a manual optimization of the Curry code results in an execution time of 0.38 s. Thus, there is potential to improve the generated Curry code.

For the first three benchmarks, the demand-driven evaluation strategy has no real advantage since the values of all subexpressions are required. Hence, the same steps, possibly in a different order, are performed in the Prolog and Curry programs. The situation is different in the next three benchmarks which use examples already discussed in this paper. `thirdof_0` is the evaluation of the literal `thirdof(o,T)` (see Sect. 1), `signat_plus_0` is the evaluation of the goal shown at the beginning of this section, and `numleaves_7` is the evaluation of the literal `numleaves(T,s(s(s(s(s(s(s(o)))))))))` having five solutions (compare Sect. 5). Since Table 1 shows the time to compute *all* answers to the given goals, the Prolog systems do not terminate due to the infinite search spaces of these goals, whereas the transformed Curry programs have a finite search space. The final benchmarks, permutation sort applied to lists containing 10, 11, and 12 decreasing Peano numbers, demonstrates the advantage of demand-driven evaluation even if the search space is finite. As discussed at various places [5,17], the functional logic version of permutation sort has the effect that permutations are explored in a demand-driven manner so that not all permutations are actually generated. Thus, our transformation maps a "generate-and-test" algorithm into a more efficient "test-of-generate-as-demanded" algorithm with a lower complexity, as apparent from the benchmarks.

---

[9] The benchmarks were executed on a Linux machine running Ubuntu 22.04 with an Intel Core i7-1165G7 (2.80 GHz) processor with eight cores. The time is the total run time of executing a binary generated with the Prolog/Curry systems.

[10] The operator "$!" is not considered by the Curry compiler KiCS2 but implemented by a specific predefined operation.

The fail-sensitive functional transformation produces functional logic programs providing the same or more general answers as the original logic programs. In the worst case, the same number of evaluation steps are performed (apart from a few additional steps to reduce occurrences of "$!" introduced due to non-total functions). In the best case, the transformation reduces infinite search spaces to finite ones. Thus, one does not get any disadvantage of this transformation but in some cases considerable advantages.

We want to remark that this improvement is not only of theoretical interest. As apparent from our examples, logic programs might have infinite search spaces when search is nested, i.e., if two predicates defined on infinite structures, like Peano numbers, lists, or trees, are sequentially called with unknown arguments, as in `plus3` or `numleaves`. When such programs are transformed into functional logic programs, a demand-driven evaluation, like needed narrowing, results in a demand-driven exploration of the search space. This is exploited in [16] to implement a domain-specific language for deep XML matching and transformation as a library in Curry. Without the demand-driven evaluation strategy, many of the library functions would not terminate. Actually, the library has similar features as the logic-based language Xcerpt [11] which uses a specialized unification procedure to ensure finite matching and unification of XML terms.

8 Related Work

The view that algorithms can be seen as declarative logic descriptions combined with appropriate control rules was introduced before decades [27]. Since the standard control rule of Prolog, left-to-right depth-first search implemented via backtracking, has the risk to loop infinitely instead of delivering an answer, there is also long history to modify the standard control rule in order to improve the operational behavior [10,33]. These proposals usually consider the operational level so that a declarative justification (soundness and completeness) is missing. This is in contrast to our work which is motivated to keep the soundness and completeness of logic programming and provide a better operational model.

There are also approaches to add functions and functional notations to logic programs and map them in various ways to logic programs, in particular, to Prolog systems with coroutining in order to have a similar effect as in a demand-driven computation model. However, one has to be careful since there are various ways and pitfalls to implement laziness in this way. For instance, [32] implements lazy evaluation in Prolog by representing closures as terms and use `when` declarations to delay insufficiently instantiated function calls. This might lead to floundering so that completeness is lost when predicates are transformed into functions. Moreover, delaying recursively defined predicates could result in infinite search spaces where a complete narrowing strategy has a finite search space [15]. A notation for functions in Prolog programs is also proposed in [12]. This syntactic extension is mapped into Prolog predicates where coroutining is used to implement lazy evaluation. Although this syntactic transformation might yield the same values and search space as functional logic languages, there are no formal results justifying this transformation.

Other approaches use Prolog as a target language to implement functional logic languages based on demand-driven narrowing strategies [3,26,30,31]. Since these approaches implement narrowing strategies, the soundness and completeness results of narrowing holds also for the generated logic programs. When implementing needed narrowing in this way [3], the resulting programs are optimal, i.e., the set of computed solutions is minimal and successful derivations have the shortest possible length [2].

In this paper we take an opposite approach since we use (pure) Prolog as the source language. This was already proposed in [19] but there it is not ensured that the set of answers is kept by the transformation, as discussed in Sect. 3. By respecting the failure behavior of the source programs, we fixed this problem and showed how to obtain an equal or better operational behavior without changing the answer behavior. The extension of this transformation to Prolog's built-in arithmetic and conditional goals is shown in [19] and also implemented in our current approach. Methods to infer result argument positions so that the transformed program is optimal (if possible) are also discussed in [19]. Since these methods can be used identically in our transformation, we omitted a detailed presentation of them in this paper.

9 Conclusions

We presented a method to transform logic programs into functional logic programs so that the transformed functional logic program always computes the same or more general answers than the original program. For a well-defined class of programs (predicates completely defined by inductively sequential rules), the set of answers and the number of evaluation steps leading to an answer are minimal [1,2]. Thus, the transformation is useful to transfer strong results from functional logic programming into logic programming.

To ensure the correctness of the transformation, we use an approximation of the failure behavior of transformed programs, which is improved by considering types provided for logic programs. As a result, the transformation yields programs that require the same number of evaluation steps in the worst case, but often require less steps and reduces infinite search spaces to finite ones. Thus, the main lesson learned from this work is that the introduction of functions and the usage of these dependencies in functional logic programs has many advantages but no disadvantages in general. We implemented our method in a tool and showed the practical advantages by evaluating a set of small but typical examples.

For future work it might be interesting to improve the presented fail-sensitive functional transformation by taking strictness information into account to avoid the introduction of strict application operations, as discussed in Sect. 7. However, the motivation of this work is not to provide a general applicable tool to transform Prolog programs, since this is difficult due to the use of non-declarative features in larger Prolog programs. Our intention is to show that the execution mechanism of functional logic programs is always equal or better than the resolution principle for pure logic programs thanks to exploiting functional dependencies. Although this could be simulated in Prolog systems with corouting, additional problems might occur due to floundering and incompleteness of executions.

References

1. Antoy, S.: Optimal non-deterministic functional logic computations. In: Hanus, M., Heering, J., Meinke, K. (eds.) ALP/HOA -1997. LNCS, vol. 1298, pp. 16–30. Springer, Heidelberg (1997). https://doi.org/10.1007/BFb0027000
2. Antoy, S., Echahed, R., Hanus, M.: A needed narrowing strategy. J. ACM **47**(4), 776–822 (2000). https://doi.org/10.1145/347476.347484
3. Antoy, S., Hanus, M.: Compiling multi-paradigm declarative programs into Prolog. In: Kirchner, H., Ringeissen, C. (eds.) FroCoS 2000. LNCS (LNAI), vol. 1794, pp. 171–185. Springer, Heidelberg (2000). https://doi.org/10.1007/10720084_12
4. Antoy, S., Hanus, M.: Set functions for functional logic programming. In: Proceedings of the 11th ACM SIGPLAN International Conference on Principles and Practice of Declarative Programming (PPDP 2009), pp. 73–82. ACM Press (2009). https://doi.org/10.1145/1599410.1599420
5. Antoy, S., Hanus, M.: Functional logic programming. Commun. ACM **53**(4), 74–85 (2010). https://doi.org/10.1145/1721654.1721675
6. Baader, F., Nipkow, T.: Term Rewriting and All That. Cambridge University Press (1998)
7. Barbosa, J., Florido, M., Santos Costa, V.: Data type inference for logic programming. In: De Angelis, E., Vanhoof, W. (eds.) LOPSTR 2021. LNCS, vol. 13290, pp. 16–37. Springer, Cham (2021). https://doi.org/10.1007/978-3-030-98869-2_2
8. Barbuti, R., Bellia, M., Levi, G., Martelli, M.: On the integration of logic programming and functional programming. In: Proceedings IEEE International Symposium on Logic Programming, Atlantic City, pp. 160–166 (1984)
9. Braßel, B., Hanus, M., Peemöller, B., Reck, F.: KiCS2: a new compiler from curry to Haskell. In: Kuchen, H. (ed.) WFLP 2011. LNCS, vol. 6816, pp. 1–18. Springer, Heidelberg (2011). https://doi.org/10.1007/978-3-642-22531-4_1
10. Bruynooghe, M., De Schreye, D., Krekels, B.: Compiling control. J. Log. Program. **6**, 135–162 (1989). https://doi.org/10.1016/0743-1066(89)90033-2
11. Bry, F., Schaffert, S.: Towards a declarative query and transformation language for XML and semistructured data: simulation unification. In: Stuckey, P.J. (ed.) ICLP 2002. LNCS, vol. 2401, pp. 255–270. Springer, Heidelberg (2002). https://doi.org/10.1007/3-540-45619-8_18
12. Casas, A., Cabeza, D., Hermenegildo, M.V.: A syntactic approach to combining functional notation, lazy evaluation, and higher-order in LP systems. In: Hagiya, M., Wadler, P. (eds.) FLOPS 2006. LNCS, vol. 3945, pp. 146–162. Springer, Heidelberg (2006). https://doi.org/10.1007/11737414_11
13. Giesl, J., Raffelsieper, M., Schneider-Kamp, P., Swiderski, S., Thiemann, R.: Automatic termination proofs for Haskell by term rewriting. ACM Trans. Programm. Lang. Syst. **33**(2) (2011). Article 7. https://doi.org/10.1145/1890028.1890030
14. González-Moreno, J., Hortalá-González, M., López-Fraguas, F., Rodríguez-Artalejo, M.: An approach to declarative programming based on a rewriting logic. J. Log. Program. **40**, 47–87 (1999). https://doi.org/10.1016/S0743-1066(98)10029-8

15. Hanus, M.: Analysis of residuating logic programs. J. Log. Program. **24**(3), 161–199 (1995). https://doi.org/10.1016/0743-1066(94)00105-F
16. Hanus, M.: Declarative processing of semistructured web data. In: Technical Communications of the 27th International Conference on Logic Programming, vol. 11, pp. 198–208. Leibniz International Proceedings in Informatics (LIPIcs) (2011). https://doi.org/10.4230/LIPIcs.ICLP.2011.198
17. Hanus, M.: Functional logic programming: from theory to Curry. In: Voronkov, A., Weidenbach, C. (eds.) Programming Logics. LNCS, vol. 7797, pp. 123–168. Springer, Heidelberg (2013). https://doi.org/10.1007/978-3-642-37651-1_6
18. Hanus, M.: Verifying fail-free declarative programs. In: Proceedings of the 20th International Symposium on Principles and Practice of Declarative Programming (PPDP 2018), pp. 12:1–12:13. ACM Press (2018). https://doi.org/10.1145/3236950.3236957
19. Hanus, M.: From logic to functional logic programs. Theory Pract. Logic Program. **22**(4), 538–554 (2022). https://doi.org/10.1017/S1471068422000187
20. Hanus, M.: Inferring non-failure conditions for declarative programs. In: Gibbons, J., Miller, D. (eds.) FLOPS 2024. LNCS, vol. 14659, pp. 167–187. Springer, Singapore (2024). https://doi.org/10.1007/978-981-97-2300-3_10
21. Hanus, M., Skrlac, F.: A modular and generic analysis server system for functional logic programs. In: Proceedings of the ACM SIGPLAN 2014 Workshop on Partial Evaluation and Program Manipulation (PEPM 2014), pp. 181–188. ACM Press (2014). https://doi.org/10.1145/2543728.2543744
22. Hanus, M. (ed.): Curry: an integrated functional logic language (vers. 0.9.0) (2016). http://www.curry-lang.org
23. Hermenegildo, M., et al.: An overview of Ciao and its design philosophy. Theory Pract. Logic Program. **12**(1–2), 219–252 (2012). https://doi.org/10.1017/S1471068411000457
24. Huet, G., Lévy, J.J.: Computations in orthogonal rewriting systems. In: Lassez, J.L., Plotkin, G. (eds.) Computational Logic: Essays in Honor of Alan Robinson, pp. 395–443. MIT Press (1991)
25. Hughes, J.: Why functional programming matters. In: Turner, D. (ed.) Research Topics in Functional Programming, pp. 17–42. Addison Wesley (1990)
26. Jiménez-Martin, J., Marino-Carballo, J., Moreno-Navarro, J.: Efficient compilation of lazy narrowing into Prolog. In: Lau, K.K., Clement, T.P. (eds.) LOPSTR 1992, pp. 253–270. Springer, London (1992). https://doi.org/10.1007/978-1-4471-3560-9_18
27. Kowalski, R.: Algorithm = logic + control. Commun. ACM **22**(7), 424–436 (1979). https://doi.org/10.1145/359131.359136
28. Lee, C., Jones, N., Ben-Amram, A.: The size-change principle for program termination. In: ACM Symposium on Principles of Programming Languages (POPL 2001), pp. 81–92 (2001). https://doi.org/10.1145/373243.360210
29. Lloyd, J.: Foundations of Logic Programming. 2nd edn. Springer, Heidelberg (1987). https://doi.org/10.1007/978-3-642-83189-8
30. Loogen, R., Fraguas, F.L., Artalejo, M.R.: A demand driven computation strategy for lazy narrowing. In: Bruynooghe, M., Penjam, J. (eds.) PLILP 1993. LNCS, vol. 714, pp. 184–200. Springer, Heidelberg (1993). https://doi.org/10.1007/3-540-57186-8_79
31. López Fraguas, F.J., Sánchez Hernández, J.: TOY: a multiparadigm declarative system. In: Narendran, P., Rusinowitch, M. (eds.) RTA 1999. LNCS, vol. 1631, pp. 244–247. Springer, Heidelberg (1999). https://doi.org/10.1007/3-540-48685-2_19

32. Naish, L.: Adding equations to NU-Prolog. In: Maluszyński, J., Wirsing, M. (eds.) PLILP 1991. LNCS, vol. 528, pp. 15–26. Springer, Heidelberg (1991). https://doi.org/10.1007/3-540-54444-5_84
33. Narain, S.: A technique for doing lazy evaluation in logic. J. Logic Programm. **3**, 259–276 (1986). https://doi.org/10.1016/0743-1066(86)90016-6
34. Partain, W.: The nofib benchmark suite of Haskell programs. In: Launchbury, J., Sansom, P. (eds.) Proceedings of the 1992 Glasgow Workshop on Functional Programming, pp. 195–202. Springer, London (1992). https://doi.org/10.1007/978-1-4471-3215-8_17
35. Peyton Jones, S. (ed.): Haskell 98 Language and Libraries—The Revised Report. Cambridge University Press (2003)
36. Reddy, U.: Transformation of logic programs into functional programs. In: Proceedings IEEE International Symposium on Logic Programming, Atlantic City, pp. 187–196 (1984)
37. Reddy, U.: Narrowing as the operational semantics of functional languages. In: Proceedings IEEE International Symposium on Logic Programming, Boston, pp. 138–151 (1985)
38. Sato, T., Tamaki, H.: Enumeration of success patterns in logic programs. Theoret. Comput. Sci. **34**, 227–240 (1984). https://doi.org/10.1016/0304-3975(84)90119-1
39. Schrijvers, T., Santos Costa, V., Wielemaker, J., Demoen, B.: Towards Typed Prolog. In: Garcia de la Banda, M., Pontelli, E. (eds.) ICLP 2008. LNCS, vol. 5366, pp. 693–697. Springer, Heidelberg (2008). https://doi.org/10.1007/978-3-540-89982-2_59
40. Slagle, J.: Automated theorem-proving for theories with simplifiers, commutativity, and associativity. J. ACM **21**(4), 622–642 (1974). https://doi.org/10.1145/321850.321859
41. Sterling, L., Shapiro, E.: The Art of Prolog, 2nd edn. MIT Press, Cambridge, Massachusetts (1994)
42. Van Roy, P., Haridi, S.: Ideas for the future of Prolog inspired by Oz. CoRR abs/2302.00558 (2023). https://doi.org/10.48550/ARXIV.2302.00558
43. Wadler, P.: How to declare an imperative. ACM Comput. Surv. **29**(3), 240–263 (1997). https://doi.org/10.1145/262009.262011

Decision Procedures

Deciding Knowledge Problems Modulo Classes of Permutative Theories

Serdar Erbatur[1], Andrew M. Marshall[2](✉), Paliath Narendran[3], and Christophe Ringeissen[4]

[1] University of Texas at Dallas, Richardson, TX, USA
serdar.erbatur@utdallas.edu
[2] University of Mary Washington, Fredericksburg, VA, USA
marshall@umw.edu
[3] University at Albany-SUNY, Albany, NY, USA
pnarendran@albany.edu
[4] Université de Lorraine, CNRS, Inria, LORIA, 54000 Nancy, France
christophe.ringeissen@loria.fr

Abstract. In the logic based approach to security protocol verification, algorithms for verifying an intruder's knowledge are critical. In this context, the capabilities of an intruder are specified by an equational theory, possibly expressed by a term rewrite system. Previous results have developed algorithms for a number of knowledge problems in many different equational and rewrite theories, such as subterm-convergent. Permutative theories such Associative-Commutative (AC) are of great interest with several procedures having been developed for AC. This leads to the question of decidability of the knowledge problems of deduction and static equivalence in permutative theories in general. It was recently shown that deduction is decidable in permutative theories. However, the decidability of static equivalence (and the related frame distinguishability problem) was still open. In this paper we show that static equivalence is undecidable in permutative theories. In addition, we show that static equivalence remains undecidable in the more restrictive case of leaf permutative theories. On the positive side, static equivalence becomes decidable for a further restricted form of permutative theories we define here.

Keywords: Permutative Equational Theories · Static Equivalence · Deduction

1 Introduction

Logic-based analysis and verification of security protocols has been a fruitful area of research, particularly for the development of formal verification tools and procedures for checking various security properties of protocols, see for example [1,8,11,12,15]. In this context, the capabilities of an intruder are specified by an equational theory, possibly expressed by a term rewrite system, and the procedure seeks to verify the knowledge a potential attacker could obtain on the protocol. Two important models of intruder knowledge are, *deduction* and *static equivalence* (or frame distinguishability) [1]. One of the most common classes of equational theories is *subterm convergent term rewrite systems*, i.e., term rewrite systems where the right-hand side of the rules are ground or strict subterms of the left-hand side. For example, see the procedures developed in [1,12]. Non-orientable equalities are also useful and further results have considered specific equational theories that often arise in protocol verification. In particular, the Associative-Commutative (AC) and Commutative (C) equational theories are of great interest and have been investigated on their own, with algorithms begin developed for both the deduction and static equivalence problems, see for example [1,5,17]. Notice that the $A = \{f(x, f(y, z)) = f(f(x, y), z)\}$ (Associativity), $C = \{f(x, y) = f(y, x)\}$ (Commutativity) and AC (Associativity-Commutativity) theories are examples of permutative theories. These are theories for which the left and right side of the equality have the same number of symbols and variables (see the next section for a complete definition). Thus, the decidability results developed for permutative cases, such as AC, lead to the question of whether the knowledge problems of deduction, frame distinguishability, and static equivalence are decidable in all permutative theories. It has already been shown that deduction is decidable in permutative theories [17]. In this paper we show that for permutative theories frame distinguishability and static equivalence are undecidable. It would then be natural to consider a more restricted form of permutative theories such as leaf permutative. Notice that this is still sufficient for the AC theory because we can reformulate the associativity axiom as $f(f(x, y), z) = f(f(z, y), x)$ in order to obtain, along with commutativity, a completely leaf permuting presentation. However, we show that even when restricted to leaf permuting theories, frame distinguishability and static equivalence are still undecidable.

Another interesting reason to investigate the decidability of these knowledge problems beyond their criticality in the formal verification of protocols, is their close connection to the problems of unification and matching. Unification and matching are well established problems that arise in many applications such as logic programming and automated theorem proving. The decidability of these problems is well studied for many types of equational theories and rewrite systems. For example, it was shown in [25] that unification is undecidable in permutative theories. Later it was shown in [22] that unification is also undecidable in variable permutative theories, a subclass of permutative theories similar to the leaf permutative theories. Thus, there is a general question about the co-decidability of unification and static equivalence, and likewise matching and deduction. While we don't answer that question here we get closer by showing, like unification, static equivalence is undecidable in permutative and leaf permutative theories.

Finally, on the positive decidability side, we are able to identify a general subclass of the permutative theories for which all three knowledge problems become decidable.

Paper Outline. The paper proceeds as follows: Section 2 introduces some background material and the knowledge problems under consideration in this paper. Section 3 recalls a decidability result for the deduction problem in permutative theories, and discusses the undecidability of static equivalence in permutative theories. Section 4 presents a new undecidability proof for static equivalence in leaf-permutative theories. This proof also subsumes the proof for permutative theories. Section 5 introduces a further restriction to the class of permutative theories in order to obtain decidability. Finally, Sect. 6 concludes the paper with a discussion, future research, and open problems.

2 Preliminaries

We use the standard notation of equational unification [7] and term rewriting systems [6]. Given a first-order signature Σ and a (countable) set of variables V, the Σ-terms over variables V are built in the usual way by taking into account the arity of each function symbol in Σ. Each Σ-term is well-formed: if it is rooted by a n-ary function symbol in Σ, then it has necessarily n direct subterms. For any term t, $|t|$ denotes the number of symbols occurring in t. The set of Σ-terms over variables V is denoted by $T(\Sigma, V)$. The set of variables from V occurring in a term $t \in T(\Sigma, V)$ is denoted by $Var(t)$. A term t is *ground* if $Var(t) = \emptyset$. For any position p in a term t (including the root position ε), $t(p)$ is the symbol at position p, $t|_p$ is the subterm of t at position p, and $t[u]_p$ is the term t in which $t|_p$ is replaced by u. A substitution is an endomorphism of $T(\Sigma, V)$ with only finitely many variables not mapped to themselves. A substitution is denoted by $\sigma = \{x_1 \mapsto t_1, \ldots, x_m \mapsto t_m\}$, where the domain of σ is $Dom(\sigma) = \{x_1, \ldots, x_m\}$ and the range of σ is $Ran(\sigma) = \{t_1, \ldots, t_m\}$. Application of a substitution σ to t is written $t\sigma$. A Σ-equation is a pair of Σ-terms denoted by $s =^? t$ or simply $s = t$ when it is clear from the context that we do not refer to an axiom.

2.1 Equational Theories

Given a set E of Σ-axioms (i.e., pairs of terms in $T(\Sigma, V)$, denoted by $l = r$), the *equational theory* $=_E$ is the congruence closure of E under the law of substitutivity (by a slight abuse of terminology, E is often called an equational theory). Equivalently, $=_E$ can be defined as the reflexive transitive closure $\leftrightarrow_E^*$ of an equational step $\leftrightarrow_E$ defined as follows: $s \leftrightarrow_E t$ if there exist a position p of s, $l = r$ (or $r = l$) in E, and substitution σ such that $s|_p = l\sigma$ and $t = s[r\sigma]_p$.

We also need to define the (sub)classes of permutative theories we consider in this paper. This is important not only for properly defining the results proven here but also because there are some previous definitions of leaf permutative theories which don't match the definition given here. For example, the definition of leaf permutative given in this paper differs from the one given in [9] (See Definition 3 below). In [9] (Definition 3) the permutation is restricted to just the variables and is only applicable to linear terms. Thus, the definition in [9] would perhaps be better named as linear variable-permuting, while the one given here is just a permutation of the leaf-nodes of the term, see Definition 2.

Definition 1 (Permutative Theory). *An equational theory E is permutative if for each axiom $l = r$ in E, l and r contain the same symbols with the same number of occurrences.*

One can easily check that $A = \{f(x, f(y,z)) = f(f(x,y), z)\}$ (Associativity), $C = \{f(x,y) = f(y,x)\}$ (Commutativity) and $AC = \{f(x, f(y,z)) = f(f(x,y),z), f(x,y) = f(y,x)\}$ (Associativity-Commutativity) are permutative.

Two important subclasses of permutative theories are given by considering the cases where the permutations only occur on leafs or on variables.

Definition 2 (Leaf Permutative Theory). *An equational theory E is Leaf permutative if for each axiom $l = r$ in E, r is a leaf permutation of l, i.e., $r = l\sigma$, where σ is a permutation of the leaf nodes of l.*

For example, C is leaf permutative, but A is not.

Definition 3 (Variable-Permuting Theory). *An axiom $l = r$ is said to be variable-permuting [22] if all the following conditions are satisfied:*

1. *the set of occurrences of l is identical to the set of occurrences of r,*
2. *for any non-variable occurrence p of l, $l(p) = r(p)$,*
3. *for any $x \in Var(l) \cup Var(r)$, the number of occurrences of x in l is identical to the number of occurrences of x in r.*

Definition 4 (Shallow Theory). *An axiom $l = r$ is shallow if variables can only occur at a position at depth at most 1 in both l and r. An equational theory is shallow if all its axioms are shallow.*

For example, C is shallow and permutative, but A is not. Note that a shallow theory is not necessarily permutative. For example, $\{x + 0 = x\}$ is shallow but not permutative.

2.2 Rewrite Relations

Given a signature Σ, an oriented Σ-axiom is called a rewrite rule of the form $l \to r$ such that $l, r \in T(\Sigma, V)$, l is not a variable and $Var(r) \subseteq Var(l)$. A finite set of rewrite rules is called a *term rewriting system* (TRS, for short). Let R be any TRS. For any Σ-terms s and t, s R-rewrites to t, denoted by $s \to_R t$, if there exist a position p of s, $l \to r \in R$, and substitution σ such that $s_{|p} = l\sigma$ and $t = s[r\sigma]_p$. The term s is said to be *R-reducible*, $s_{|p}$ is called a *redex*, and in the particular case where $s_{|p} = l\sigma$, s R-rewrites to t, denoted by $s \to_R t$. A TRS R is *terminating* if there are no infinite rewriting sequences with respect to $\to_R$. A term is an *innermost redex* if none of its proper subterms is a redex. The symmetric relation $\leftarrow_R \cup \to_R$ is denoted by $\longleftrightarrow_R$. The rewrite relation $\to_R$ is confluent if $\longleftrightarrow_R^*$ is included in $\to_R^* \circ \leftarrow_R^*$. The rewrite relation $\to_R$ is *convergent* if $\to_R$ is both terminating and convergent. When $\to_R$ is convergent, we have that for any terms t, t', $t \longleftrightarrow_R^* t'$ iff $t\!\downarrow_R = t'\!\downarrow_R$, where $t\!\downarrow_R$ (resp., $t'\!\downarrow_R$) denotes the unique normal form of t (resp., t') w.r.t $\to_R$.

Definition 5 (Unification). *Given a signature Σ, an equational theory E, and a set of Σ-equations, $\Gamma = \{s_1 =^? t_1, \ldots, s_n =^? t_n\}$. The E-unification decision problem asks if there exists a substitution σ such that $s_i\sigma =_E t_i\sigma$ for all $1 \leq i \leq n$. If E is presented as a rewrite relation R, then we ask if there exists a substitution σ such that $s_i\sigma \leftrightarrow^*_R t_i\sigma$ for all $1 \leq i \leq n$.*

2.3 Knowledge Problems

The applied pi calculus and frames are used to model attacker knowledge [2]. In this model, the set of messages or terms which the attacker knows, and which could have been obtained from observing one or more protocol sessions, are the set of terms in $Ran(\sigma)$ of the frame $\phi = \nu\tilde{n}.\sigma$, where σ is a substitution ranging over ground terms. We also need to model cryptographic concepts such as nonces, keys, and publicly known values. We do this by using names, which are essentially free constants. Here also, we need to track the names which the attacker knows, such as public values, and the names which the attacker does not know a priori, such as freshly generated nonces. $\tilde{n}$ consists of a finite set of restricted names; these names represent freshly generated names which remain secret from the attacker. The set of names occurring in a term t is denoted by $fn(t)$. For any frame $\phi = \nu\tilde{n}.\sigma$, let $fn(\phi)$ be the set of names $fn(\sigma)\setminus\tilde{n}$ where $fn(\sigma) = \bigcup_{t \in Ran(\sigma)} fn(t)$; and for any term t, let $t\phi$ denote—by a slight abuse of notation—the term $t\sigma$. For any term t, we say that t satisfies the name restriction of ϕ if $fn(t) \cap \tilde{n} = \emptyset$.

Definition 6 (Deduction). *Let $\phi = \nu\tilde{n}.\sigma$ be a frame, and t a ground term. We say that t is deduced from ϕ modulo E, denoted by $\phi \vdash_E t$, if there exists a term ζ such that $\zeta\sigma =_E t$ and $fn(\zeta) \cap \tilde{n} = \emptyset$. The term ζ is called a recipe of t in ϕ modulo E.*

Notice that deduction is modeling the ability of an adversary to deduce some secret, that is supposed to remain hidden, from a protocol. This is a critical measure of security for key establishment protocols. Another measure of security attempts to see if an adversary could tell two different runs of the same protocol apart. This security requirement is not modeled by deduction but is important for voting protocols. In such protocols you don't want an observer to be able to tell a vote for one candidate from another based on watching the protocol for multiple runs. This form of security is modeled by *statically equivalent* modulo E.

Definition 7 (Static Equivalence). *Two terms s and t are equal in a frame $\phi = \nu\tilde{n}.\sigma$ modulo an equational theory E, denoted $(s =_E t)\phi$, if $s\sigma =_E t\sigma$, and $\tilde{n} \cap (fn(s) \cup fn(t)) = \emptyset$. The set of all equalities $s = t$ such that $(s =_E t)\phi$ is denoted by $Eq(\phi)$. Given a set of equalities Eq, the fact that $(s =_E t)\phi$ for all $s = t \in Eq$ is denoted by $\phi \models Eq$. Two frames $\phi = \nu\tilde{n}.\sigma$ and $\psi = \nu\tilde{n}.\tau$ are statically equivalent modulo E, denoted as $\phi \approx_E \psi$, if $Dom(\sigma) = Dom(\tau)$, $\phi \models Eq(\psi)$ and $\psi \models Eq(\phi)$.*

There is a closely related problem, that asks if there exists a pair of recipe terms that distinguish two different frames.

Definition 8 (Frame Distinguishability). *Given frames $\phi = \nu \tilde{n}.\sigma$ and $\psi = \nu \tilde{n}.\tau$, we say that ϕ is distinguishable from ψ in theory E, denoted $\phi \not\approx_E \psi$, if there exist two terms, t and s (with $\tilde{n} \cap (fn(s) \cup fn(t)) = \emptyset$), such that $t\sigma =_E s\sigma$ and $t\tau \neq_E s\tau$.*

Notice that in this paper we are considering the decision problem form of each of the knowledge problems listed above. Thus, if an algorithm for frame distinguishability returns true for an pair of frames, we know that the frames are not static equivalent. Likewise, if and algorithm for static equivalence returns true for a pair of frames, we know that there is no pair of recipes that distinguish the frames, i.e., frame distinguishability is false. Now, if were to considered the form of the frame distinguishability problem that requires the algorithm to produce the actual witness pair of recipes then the proof given in Theorem 3 would still show the undecidability of this form of the problem. However, the relation to the static equivalence problem would be a little less clear since, since if a static equivalence algorithm returns false, we know at least of pair of witness recipes exists but to find them would possible require searching the sets $Eq(\phi)$ and $Eq(\psi)$.

Finally we can note that the above knowledge problems, deduction, static equivalence, and frame distinguishable, are known to be decidable for subterm convergent rewrite systems [1].

2.4 Turing Machines and Linear Bounded Automata

We need to recall some standard background material on Turing Machines (TMs) and Linear Bounded Automata (LBA) for the later undecidability results; see [19] for a more complete background. We will represent a TM, M, as a 7-tuple, $M = (Q, \Sigma, \Gamma, \delta, q_0, q_a, q_r)$, where Q is a finite set of states, Σ is the input alphabet, Γ is the tape alphabet. $\Sigma \subseteq \Gamma$ and $\sqcup \in \Gamma$, where $\sqcup$ represents the blank symbol. $\delta : (Q \setminus \{q_a, q_r\} \times \Gamma) \longrightarrow (Q \times \Gamma \times \{L, R\})$ is the transition function, where R represents a right move and L a left move. $q_0 \in Q$ is the unique initial state, $q_a \in Q$ is the unique accept state, and $q_r \in Q$ is the unique reject state. LBA are Turing Machines with a tape that is bounded by the size of the input string (plus two tape end-caps) [18,20]. That is, we introduce two new tape symbols, $\{<, >\} \in \Gamma$. We restrict δ so that it cannot move left of $<$ or right of $>$, $\delta : (Q \setminus \{q_a, q_r\} \times <) = (Q \times < \times \{R\})$ and $\delta : (Q \setminus \{q_a, q_r\} \times >) = (Q \times > \times \{L\})$. The initial configuration of a LBA on input word $w \in \Sigma^*$ is in state q_0, the read-write head over the left end-cap $<$ and the tape containing $< \cdot w \cdot >$. See Example 1 in Sect. 4 for a simple example.

3 Decidability of Knowledge Problems in Permutative Theories

In this section we consider the decidability of both knowledge problems in classes of permutative theories. In particular we consider permutative theories and leaf permutative theories.

3.1 Decidability of Deduction in Permutative Theories

It has already been shown in [10] that deduction is decidable in permutative theories.

Theorem 1 ([10]). *Deduction is decidable for any permutative theory.*

This is due to the fact that you can put a bound on the number of terms you need consider since permutative theories are non-size reducing. However, when considering static equivalence, the problem becomes more difficult. We briefly consider permutative theories in the next section.

3.2 Undecidability of Frame Distinguishability in Permutative Theories

While deduction is decidable it happens that frame distinguishability, and thus static equivalence, are undecidable in permutative theories.

Theorem 2. *Frame Distinguishability is undecidable in general for permutative theories.*

However, this result is subsumed by the result in the next section that shows that frame distinguishability is also undecidable in the more restrictive leaf permutative theories. Thus, we present the proof details for that more restrictive case in the following section. However, one could prove the permutative case directly by starting with the method developed in [25] for proving that the unification problem is undecidable in general for permutative theories. The key to the proof is to show how for an arbitrary deterministic Turing Machine (TM for short), M, a permutative and confluent TRS, R_M, could be created that simulates M. By starting with a suitable undecidable TM problem, one could use the construction from [25] to reduce to the frame distinguishability problem.

4 Static Equivalence in Leaf Permuting Theories

In this section we prove that the frame distinguishability problem, and thus static equivalence, is undecidable for leaf permuting theories. The results proceeds as follows, we first develop an undecidable result for Linear Bounded Automata (LBA) which we will use in the reductions. Next, we show how to create a leaf permuting TRS from a LBA. Finally, we use this TRS to show that the frame distinguishability problem is undecidable.

Lemma 1. *Given an arbitrary LBA, M, with input alphabet Σ, it's undecidable if there exists a string, $w \in \Sigma^*$, such that M accepts w.*

Proof. Easy reduction from the LBA empty language problem proved undecidable in [3]. There it is shown that it is undecidable if $L(M) = \emptyset$ for an arbitrary deterministic LBA M.

Next, we need to show how to obtain a leaf permuting TRS from a LBA.

Lemma 2. *Given a deterministic LBA, M, one can construct a leaf permuting TRS R such that if M accepts a string w then there exists a term t which encodes the initial configuration of M on input w and a term s that encodes the final accepting configuration of M on w such that $t \downarrow_R = s$.*

Proof. Here we modify the encoding from [25] to obtain a conversion from LBA to leaf permuting TRS. Let $M = (Q, \Sigma, \Gamma, q_0, q_a, q_r, \delta)$. Assume $\Sigma = \{a_1, a_2, \ldots, a_n\}$ and $\Gamma = \{<, >, \sqcup\} \cup \Sigma$, where $<, >$ are the left and right end caps respectively, and $\sqcup$ is the blank symbol. We now need to construct the terms that will represent the tape of the LBA. We introduce three new non-constant function symbols, f, g, h and three new constants, a, b, and P. We use each as follows:

- h has arity $|Q|+1$ and is used to represent the state of the LBA. To represent state q_i, a constant b is placed at the ith position with the remaining positions containing a constants. For example, if $|Q| = 2$ then q_0 is represented as $h(b, a, a)$. The final configuration, with constant b in the final position, is used to represent a non-state or "dummy state", the use of which is described below. We denote this dummy state as q_d.
    - **We use the notation $h(q_i)$ to abbreviate the encoding of the state q_i using h.**
- g has arity $|\Gamma|$ and is used to encode the alphabet characters. We place a constant b at position i in g with a constants placed at all other positions to encode a_i. Positions $n+1, \ldots n+3$ are used to encode $\{<, >, \sqcup\}$ in the same way.
    - **We use the notation $g(a_i)$ to abbreviate the encoding of the character a_i using g.**
- f is used to form terms which consist of an encoding of a state, an encoding of a single character, and an f-rooted subterm or the constant P. The subterm is used to encode the rest of the string. The constant P is used to stop the encoding. For example, the dummy state and the character a_i can be encoded as the term $f(h(q_d), g(a_i), P)$.

We now form terms representing the configurations of the LBA and its tape as follows:

- For any string $w \in \Gamma^+$ we represent w as a layered f-term, one layer per alphabet character. The last layer is ended using the P constant. The dummy state is used by default for each of the state positions in the f-terms.

- We use the notation $f(\overline{w})$ to abbreviate the encoding of the string w using f-terms.
- For example, let $<w_1 q_i a_i w_2>$ be a configuration of the LBA. We represent this by an f rooted term as $f(\overline{<w_1}, f(h(q_i), g(a_i), f(\overline{w_2>})))$.

We now need to construct the leaf permuting TRS. Let's consider the moves of the transition function, δ, and construct from them a TRS R:

- For each right move, $\delta(q_i, a_i) = (q_j, a_j, R)$, we create a rule for each possible character $a_k \in \Gamma \setminus \{>\}$ of the form:

$$f(h(q_i), g(a_i), f(h(q_d), g(a_k), x)) \to f(h(q_d), g(a_j), f(h(q_j), g(a_k), x))$$

- For each left move, $\delta(q_i, a_i) = (q_j, a_j, L)$, we create a rule for each possible character $a_k \in \Gamma \setminus \{<\}$ of the form:

$$f(h(q_d), g(a_k), f(h(q_i), g(a_i), x)) \to f(h(q_j), g(a_k), f(h(q_d), g(a_j), x))$$

Finally, we need to describe the initial configurations for the LBA. The LBA will start in the configuration $q_o <w>$ for input w. We encode this as an f-term, where all the states are initially $h(q_d)$ accept the first (most left h). That is, we encode using the term $t = f(h(q_0), g(<), f(\overline{w>}))$.

Notice that every rule in R is leaf permuting. In addition, the LBA accepts the string w iff $f(h(q_0), g(<), f(\overline{w>})) \to_R^* s$ such that s is a term that encodes a final accepting configuration of the LBA. This final configuration will have just a single state $h(q_a)$ with all other state positions being dummy states $h(q_i)$.

Example 1. Let's consider the following toy LBA and construct the corresponding TRS.

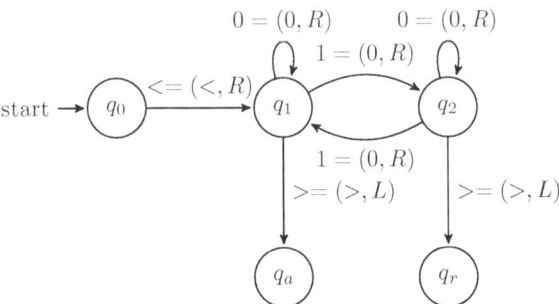

Notice for this simple example $\Gamma = \{<, >, 0, 1, \sqcup\}$ and $Q = \{q_0, q_1, q_2, q_r, q_a\}$. To encode the symbol $<$ we use $g(b, a, a, a, a)$, and 0 is $g(a, a, b, a, a)$. Now to encode the state q_0 we use $h(b, a, a, a, a, a)$, and we use $h(a, b, a, a, a, a)$ for q_1. Notice the additional final position in $h()$, this is to encode the dummy, q_d, state that is just used to track the current state of the LBA during rewriting, and we encode this state with $h(a, a, a, a, a, b)$.

Now let's construct one of the rewrite rules based on the above transition function in full without any of the shorthand developed above. We can then construct several others using the shorthand. Consider $\delta(q_0, <) = (q_1, <, R)$ in TRS rule form one of the instances of this rule would be:

$$f(h(b,a,a,a,a,a), g(b,a,a,a,a,a), f(h(a,a,a,a,a,b), g(a,a,b,a,a,a), x)) \to$$
$$f(h(a,a,a,a,a,b), g(b,a,a,a,a,a), f(h(a,b,a,a,a,a), g(a,a,b,a,a,a), x))$$

Notice that this is for the rule where 0 (encoded as $g(a,a,b,a,a)$) is on the right of $<$ on the tape. We would also need the same rule for each of the other symbols.

Let's construct some of the rewrite rules for the above LBA using the more compact notation. for the transition $\delta(q_0, <) = (q_1, <, R)$ some of the rules are:

$$f(h(q_0), g(<), f(h(q_d), g(0), x)) \to f(h(q_d), g(<), f(h(q_1), g(0), x))$$
$$f(h(q_0), g(<), f(h(q_d), g(>), x)) \to f(h(q_d), g(<), f(h(q_1), g(>), x))$$
$$f(h(q_0), g(<), f(h(q_d), g(1), x)) \to f(h(q_d), g(<), f(h(q_1), g(1), x))$$

For the transition $\delta(q_2, 1) = (q_1, 0, R)$, some of the resulting rules are:

$$f(h(q_2), g(1), f(h(q_d), g(1), x)) \to f(h(q_d), g(0), f(h(q_1), g(1), x))$$
$$f(h(q_2), g(1), f(h(q_d), g(0), x)) \to f(h(q_d), g(0), f(h(q_1), g(0), x))$$

For the transition $\delta(q_2, >) = (q_r, >, L)$, some of the resulting rules are:

$$f(h(q_d), g(0), f(h(q_2), g(>), x)) \to f(h(q_r), g(0), f(h(q_d), g(>), x))$$
$$f(h(q_d), g(1), f(h(q_2), g(>), x)) \to f(h(q_r), g(1), f(h(q_d), g(>), x))$$

Finally let's show an encoding of a string and a rewrite step. Assume we have an initial configuration of q_0 <01>. This configuration is encoded as

$$f(h(q_0), g(<), f(h(q_d), g(0), f(h(q_d), g(1), f(h(q_d), g(>), P))))$$

Notice that for any configuration encoding there is always only one position with a non-dummy state. For this term we can apply the rule

$$f(h(q_0), g(<), f(h(q_d), g(0), x)) \to f(h(q_d), g(<), f(h(q_1), g(0), x))$$

which results in the rewrite step:

$$f(h(q_0), g(<), f(h(q_d), g(0), f(h(q_d), g(1), f(h(q_d), g(>), P)))) \to_R$$
$$f(h(q_d), g(<), f(h(q_1), g(0), f(h(q_d), g(1), f(h(q_d), g(>), P))))$$

Notice that since the LBA is deterministic we obtain the following result.

Lemma 3. *Let M be a deterministic LBA. Then the variable permuting TRS, R, constructed from M is locally confluent. Furthermore, if M is both deterministic and terminating then the rewrite sequence $t \to_R^* s$, terminates when t is the term encoding of the start configuration for M on some input string w.*

We now need to describe the LBA and TRS we will use in the Reduction. First, from a LBA M_1 we can easily construct the following LBA.

Lemma 4. *Let M_1 be a LBA that before accepting and halting it replaces the tape (except the end caps) with blank symbols and stops (enters the accept or reject state) with the tape head over the left end-cap. One can construct a LBA M_2 from M_1 such that $L(M_2) = \emptyset$ and every transition of M_1 and M_2 are the same except the accepting transitions from M_1 are now rejecting in M_2.*

Notice that for each LBA M_1 and M_2 there is a corresponding leaf permuting TRS, R_1 and R_2 respectively. However, we need to combine these two TRS into a single TRS.

Lemma 5. *Given LBAs M_1 and M_2 and their leaf permutative TRSs R_1 and R_2 respectively. Let q_{0_i} be the initial state for M_i. One can construct a leaf permutative TRS $R_{1,2}$ such that for input string w and term $t_i = f(h(q_{0_i}), g(<), f(\overline{w>}))$, $t_i \to^*_{R_{1,2}} s_i$ iff $t_i \to^*_{R_i} s_i$, $i \in \{1, 2\}$.*

Proof. One just needs to ensure that $\{Q_1 \setminus \{q_a, q_r\}\} \cap \{Q_2 \setminus \{q_a, q_r\}\} = \emptyset$. Then, the rules of R_1 and R_2 are disjoint.

We can now combine all the above into the final undecidability proof.

Theorem 3. *Frame distinguishability is undecidable for leaf permuting theories.*

Proof. We can proceed by reduction using Lemma 1. Assume we are given a deterministic LBA M_1. Without loss of generality assume that before halting and entering q_a or q_r, M_1 erases its tape and halts with the head over the left end-cap. Furthermore, assume that as an initial step M_1 scans the tape from left to right end-cap and then back to the left. Finally, we can assume q_0 is the unique start state of M_1 and after leaving this state the LBA never returns to it. Since these steps could be added to any LBA without changing its language, they don't represent a restriction. We proceed as follows.

1. From M_1 construct the always reject LBA M_2 as in Lemma 4. We can assume w.l.o.g., that $\{Q_1 \setminus \{q_a, q_r\}\} \cap \{Q_2 \setminus \{q_a, q_r\}\} = \emptyset$. This can be done by simply creating a marked version of $Q_1 \setminus \{q_a, q_r\}$, s.t. a $q_{i_2} \in Q_2$ for each $q_{i_1} \in Q_1$. Let q_{0_i} be the start state for M_i.
   - For any transition from M_1, $\delta(q_{i_1}, a_i) = (q_{j_1}, a_j, x)$, if $q_{j_1} \neq q_a$ then for M_2, $\delta(q_{i_2}, a_i) = (q_{j_2}, a_j, x)$. If $q_{j_1} = q_a$ then $\delta(q_{i_2}, a_i) = (q_r, a_j, x)$.
2. From M_1 and M_2 construct R_1 and R_2 as in Lemma 2.
   - Notice if a term t_1 encodes a configuration and $t_1 \to^+_{R_1} t_2$ s.t. t_2 doesn't encode an accepting configuration, then $t'_1 \to^+_{R_2} t'_2$, where $t_i = t'_i$ except each state q_{i_1} is swapped for q_{i_2}.
3. Construct $R_{1,2}$ from R_1 and R_2 as in Lemma 5.
4. Construct two frames, such that $\tilde{n} = \emptyset$ for each frame::

    a) $\phi_1 = \nu\tilde{n}.\sigma_1 = \nu\tilde{n}.\{x \mapsto h(q_{0_1})\}$ b) $\phi_2 = \nu\tilde{n}.\sigma_2 = \nu\tilde{n}.\{x \mapsto h(q_{0_2})\}$

Assume that we have an algorithm for the frame distinguishability problem and it returns true. Then, there exists a recipe pair (t, s) such that $t\sigma_1 \leftrightarrow^*_{R_{1,2}} s\sigma_1$ and $t\sigma_2 \not\leftrightarrow^*_{R_{1,2}} s\sigma_2$. We can assume the following about these two recipe terms:

(i) It can't be that $s = t$, nor can t and s both be ground, otherwise $t\sigma_2 \leftrightarrow^*_{R_{1,2}} s\sigma_2$. Thus, $t\sigma_1 \leftrightarrow^+_{R_{1,2}} s\sigma_1$.

(ii) For at least one of the terms, t and s, we have that $t\{x \mapsto h(q_{0_1})\} \neq t$ and (or) $s\{x \mapsto h(q_{0_1})\} \neq s$. That is, we use the substitution on x. If this was not the case, then $t\sigma_2 \leftrightarrow^*_{R_{1,2}} s\sigma_2$.

(iii) At least one of, $s\sigma_1$ (and $s\sigma_2$) or $t\sigma_1$ (and $t\sigma_2$), encodes a valid, single, starting configuration, and this starting configuration is the entire term (not a subterm of a larger term). This is due to the following (we consider $t\sigma_i$ but the same applies for $s\sigma_i$):

– Notice the rules of $R_{1,2}$ can only be applied to redexes encoding valid configurations and we have that $t\sigma_1 \leftrightarrow^+_{R_{1,2}} s\sigma_1$. Since M_1 (and thus M_2) first scan the tape from left to right, the rewrite derivation on $t\sigma_1$ would stop when any malformed subterm was reached and before the final configuration. However, since the steps of M_2, except the step entering the final configuration, are the same as those for M_1, M_2 would stop at the same point in the derivation from $t\sigma_2$. That is, $t\sigma_2 \to^*_{R_{1,2}} s\sigma_2$. Thus, $t\sigma_i$ must contain a valid configuration.

– Suppose all the redex for the rewrite derivation $t\sigma_i \leftrightarrow^*_{R_{1,2}} s\sigma_i$ are contained in a proper subterm of t, t'. Then we could just consider the recipe pair (t', s). Likewise, if t encodes multiple disjoint configurations as subterms, then there must be at least one of these subterms, say t', such that $t'\sigma_1 \leftrightarrow^*_{R_{1,2}} s\sigma_1$ and $t'\sigma_2 \not\leftrightarrow^*_{R_{1,2}} s\sigma_2$, otherwise $t\sigma_2 \leftrightarrow^*_{R_{1,2}} s\sigma_2$. Thus, we could just consider the recipe pair (t', s). Thus, $t\sigma_i$ encodes a single valid configuration which is the entire term.

– $t\sigma_i$ encodes an initial configuration since, by construction, q_{0_i} is only used for the initial starting configuration of M_i and never reused. Thus, if $t\sigma_i \neq t$ then $t\sigma_i$ encodes an initial configuration. Therefore, $t\sigma_i$ encodes a single valid initial configuration.

Let $t\sigma_1 \to^*_{R_{1,2}} r \leftarrow^*_{R_{1,2}} s\sigma_1$. Without loss of generality we can assume the following for this derivation:

(a) r encodes an accepting configuration, if not then $t\sigma_2 \to^*_{R_{1,2}} r \leftarrow^*_{R_{1,2}} s\sigma_2$.

Notice that (a) allows us to simplify the rewrite derivation. Without loss of generality we can assume that $s\sigma_i = r$. Since r must be an accepting configuration and at least one side of the derivation $t\sigma_i \to^*_{R_{1,2}} r \leftarrow^*_{R_{1,2}} s\sigma_i$ must use the substitution on x, we can assume that side to be $t\sigma_i \to^*_{R_{1,2}} r$ (swap if not). Now the pair (t, r) form a recipe pair s.t. $t\sigma_1 \leftrightarrow^*_{R_{1,2}} r\sigma_1$ and $t\sigma_2 \not\leftrightarrow^*_{R_{1,2}} r\sigma_2$. Therefore, we can assume that $t\sigma_1 \to^+_{R_{1,2}} s\sigma_1$ and $t\sigma_2 \not\to^+_{R_{1,2}} s\sigma_2$. Finally, from (i)–(iii) above, $t\sigma_i$ is a well formed term encoding an initial configuration of an LBA M_i for some initial input string w. From Lemma 2, $t\sigma_1 \to^+_{R_{1,2}} s\sigma_1$ iff M_1 accepts some string w.

Therefore, a frame distinguishability algorithm would allow us to decide if for an arbitrary LBA M_1 if there exists some string w, accepted by M_1, violating Lemma 1.

If the frame distinguishability problem is undecidable then we also get undecidability of the static equivalence problem and since distinguishability is undecidable for leaf permutative theories it is also undecidable for the more general permutative theories.

Corollary 1. *Frame distinguishability is undecidable for permutative theories.*

5 Decidable Static Equivalence in Subclasses of Permutative Theories

A decision procedure for static equivalence in some permutative theories E can obtained by computing a finite set of equalities bounded by $|E|$, where $|E| = \max_{\{l \mid l=r \in E\}} |l|$. To define this particular finite set of equalities, we first construct a frame saturation dedicated to permutative theories. Then, the recipes of the (deducible) terms included in that saturation are used to build an appropriate set of bounded equalities. Given a term t, $St(t)$ is the smallest set of terms including t such that

- if $u' =_E u$ and $u \in St(t)$, then $u' \in St(t)$,
- if $u \in St(t)$ and p is a non-root position of u, then $u|_p \in St(t)$.

Notice that $St(t)$ is finite since E is permutative. For a set of terms T, $St(T) = \bigcup_{t \in T} St(t)$, and for a substitution σ, $St(\sigma) = St(Ran(\sigma))$.

Definition 9 (Frame Saturation for Permutative Theories). *Let E be a permutative theory. For any frame $\phi = \nu\tilde{n}.\sigma$, let $ESt(\sigma) = St(\sigma) \cup \bigcup_{t \in ET} St(t)$ where ET is the set of terms $s[\boldsymbol{d}]$ such that $|s| \leq |E|$, $fn(s) \cap \tilde{n} = \emptyset$, $\boldsymbol{d} \in St(\sigma)$ and $s[\boldsymbol{d}] \hookrightarrow^\epsilon_E t'$ for some term t'.*

The set of terms $sat_E(\phi)$ is the smallest set D such that:

(1) $Ran(\sigma) \subseteq D$,
(2) if $t_1, \ldots, t_n \in D$ and $f(t_1, \ldots, t_n) \in ESt(\sigma)$ then $f(t_1, \ldots, t_n) \in D$,
(3) if $t \in D$, $t' \in ESt(\sigma)$, $t =_E t'$, then $t' \in D$,

For any frame $\phi = \nu\tilde{n}.\sigma$, σ_* denotes the substitution defined as follows:

$$\sigma\{\chi_u \mapsto u \mid u \in sat_E(\phi) \backslash Ran(\sigma)\}$$

where χ_u is a fresh variable for any $u \in sat_E(\phi) \backslash Ran(\sigma)$, and ϕ_* denotes the frame $\nu\tilde{n}.\sigma_*$. Given a recipe ζ_u for each $u \in sat_E(\phi) \backslash Ran(\sigma)$, the substitution $\{\chi_u \mapsto \zeta_u \mid u \in sat_E(\phi) \backslash Ran(\sigma)\}$ is called a recipe substitution of ϕ and is denoted by ζ_ϕ.

For any set Eq of equalities between two terms satisfying the name restriction of ϕ, we denote $\phi \models Eq$ if for any $t = t' \in Eq$ we have $t\phi =_E t'\phi$. The set $Eq^B(\phi)$ is the set of equalities $t\zeta_\phi = t'\zeta_\phi$ such that t, t' are terms satisfying the name restriction of ϕ, $|t| \leq |E|$, $|t'| \leq |E|$ and $(t\zeta_\phi)\phi =_E (t'\zeta_\phi)\phi$.

Definition 10. *A permutative theory E is said to be* locally stable permutative *(LSP, for short), if the following property holds: for any frame ϕ and any terms s and t satisfying the name restriction of ϕ, if $s\phi_* =_E t\phi_*$ and $\psi \models Eq^B(\phi)$ then $(s\zeta_\phi)\psi =_E (t\zeta_\phi)\psi$.*

LSP theories bear similarities with locally stable theories as defined in [1]. In the context of this paper, we consider a slight adaptation of local stability by assuming a fixed finite set of deducible terms for all permutative theories we want to target. Then, Definition 10 captures the main property to satisfy in order to get local stability and then a decision procedure for static equivalence based on checking finitely many equalities.

Lemma 6 ([17]). *Any shallow permutative theory is LSP.*

Remark 1. Unification in shallow permutative theories is decidable and finitary since there exists a sound, complete and terminating unification procedure for shallow theories [14].

Particular variable-permuting theories provide other examples of *LSP* theories.

Definition 11 (Separate Variable-Permuting Theory). *An equational theory E is said to be* separate variable-permuting *(SVP, for short) if E is variable-permuting and for any $l = r \in E$, l and r are rooted by the same function symbol that does not occur elsewhere in E.*

Lemma 7 ([10]). *Any SVP theory is LSP.*

Remark 2. Unification in *SVP* theories is decidable and finitary since all *SVP* theories are closed by paramodulation and there exists a sound, complete and terminating unification procedure for theories closed by paramodulation [21,23].

Example 2. Consider

$$K = \{keyexch(x, pk(x'), y, pk(y')) = keyexch(x', pk(x), y', pk(y))\}.$$

K is a permutative theory which is not shallow. However, it is a *SVP* theory, and so it is a *LSP* theory.

The equational theory K, combined with a subterm convergent term rewrite system, is useful in practice to model a group messaging protocol [13]. One can notice that the *pk* symbol occurs in K as an "inner" constructor. Thus, K can be combined with a term rewrite system R sharing *pk*, provided that *pk* is a constructor of R [16]. Consequently, variable-permuting theories like K provide good examples to constructor-sharing combination. The application of shallow permutative theories to constructor-sharing combination is quite limited: in that case, the inner constructors can only occur in ground terms. For this reason, it is clearly interesting to identity *LSP* theories beyond the case of shallow permutative. We conjecture the existence of *LSP* theories that are neither shallow permutative nor variable-permuting.

A decision procedure for static equivalence in *LSP* theories is given by the following lemma:

Lemma 8. *Let E be any LSP theory. For any frames ϕ and ψ, $\phi \approx_E \psi$ iff $\psi \models Eq^B(\phi)$ and $\phi \models Eq^B(\psi)$.*

Proof. (If direction) Let $Eq(\phi)$ be the set of all equalities $s\zeta_\phi = t\zeta_\phi$ such that $(s\zeta_\phi =_E t\zeta_\phi)\phi$.

Consider any $s\zeta_\phi = t\zeta_\phi \in Eq(\phi)$. By definition of ζ_ϕ and ϕ_*, we have $(s\zeta_\phi)\phi =_E s\phi_*$ and $(t\zeta_\phi)\phi =_E t\phi_*$. Since $(s\zeta_\phi)\phi =_E (t\zeta_\phi)\phi$, we obtain $s\phi_* =_E t\phi_*$. Then, by Definition 10, we get $(s\zeta_\phi)\psi =_E (t\zeta_\phi)\psi$, which means that $\psi \models Eq(\phi)$. In a symmetric way, we show that $\phi \models Eq(\psi)$. Then, we can conclude since $\phi \approx_E \psi$ iff $\psi \models Eq(\phi)$ and $\phi \models Eq(\psi)$.

Theorem 4. *Static equivalence is decidable in any LSP theory.*

6 Conclusion

We have shown in this paper that static equivalence is undecidable in permutative theories and that it remains undecidable even if one restricts the equational theory to just leaf permutative. It is interesting to note that the problem of unification is also undecidable in these theories and has a close connection to static equivalence by definition. Thus, it would be interesting to know the relation between these two problems with respect to decidability. There are a number of results on this question already, For example it's shown in [2] that both static equivalence and deduction are undecidable in general and that there are theories for which deduction is decidable but static equivalence is undecidable. It's shown in [10] that deduction is decidable in permutative theories but we have in [22] that unification is undecidable for both permutative and variable permutative theories. Interestingly, in [4] a theory is developed for which unification is decidable but deduction is undecidable. Perhaps it would be possible to extend this result in order to show that static equivalence is also undecidable.

With respect to positive results, we have also identified a class of permutative theories for which all knowledge problems are decidable.

We plan to continue the study of the knowledge problems in equational theories given by rewrite systems modulo permutative axioms. On the one hand, it is possible to consider extensions of subterm rewrite systems, such as contracting rewrite systems [10,24]. On the other hand, static equivalence is decidable in particular permutative theories defined by shallow permutative axioms or by some specific "disjoint" variable-permuting axioms. We are working on finding additional permutative theories with decidable static equivalence. Due to the undecidability result reported here, it is clear now that we cannot consider any arbitrary set of permutative axioms.

References

1. Abadi, M., Cortier, V.: Deciding knowledge in security protocols under equational theories. Theor. Comput. Sci. **367**(1–2), 2–32 (2006)
2. Abadi, M., Fournet, C.: Mobile values, new names, and secure communication. In: Hankin, C., Schmidt, D. (eds.) Conference Record of POPL 2001: The 28th ACM SIGPLAN-SIGACT Symposium on Principles of Programming Languages, London, UK, 17–19 January 2001, pp. 104–115. ACM (2001)
3. Akçam, Z., Cornell, K.A., Hono II, D.S., Narendran, P., Pulver, A.: On problems dual to unification: the string-rewriting case (2023). https://doi.org/10.48550/arXiv.2103.00386
4. Anantharaman, S., Narendran, P., Rusinowitch, M.: Intruders with caps. In: Baader, F. (ed.) RTA 2007. LNCS, vol. 4533, pp. 20–35. Springer, Heidelberg (2007). https://doi.org/10.1007/978-3-540-73449-9_4
5. Ayala-Rincón, M., Fernández, M., Nantes-Sobrinho, D.: Intruder deduction problem for locally stable theories with normal forms and inverses. Theor. Comput. Sci. **672**, 64–100 (2017)
6. Baader, F., Nipkow, T.: Term Rewriting and All That. Cambridge University Press, New York, NY, USA (1998)
7. Baader, F., Snyder, W.: Unification theory. In: Robinson, J.A., Voronkov, A. (eds.) Handbook of Automated Reasoning, pp. 445–532. Elsevier and MIT Press (2001)
8. Baudet, M., Cortier, V., Delaune, S.: YAPA: a generic tool for computing intruder knowledge. ACM Trans. Comput. Log. **14**(1), 4 (2013)
9. Boy de la Tour, T., Echenim, M., Narendran, P.: Unification and matching modulo leaf-permutative equational presentations. In: Armando, A., Baumgartner, P., Dowek, G. (eds.) IJCAR 2008. LNCS (LNAI), vol. 5195, pp. 332–347. Springer, Heidelberg (2008). https://doi.org/10.1007/978-3-540-71070-7_30
10. Bunch, C., Satterfield, S.D., Erbatur, S., Marshall, A.M., Ringeissen, C.: Knowledge problems in protocol analysis: extending the notion of subterm convergent (2024). https://doi.org/10.48550/arXiv.2401.17226
11. Chadha, R., Cheval, V., Ciobâcă, S., Kremer, S.: Automated verification of equivalence properties of cryptographic protocols. ACM Trans. Comput. Log. **17**(4), 23:1–23:32 (2016)
12. Ciobâcă, S., Delaune, S., Kremer, S.: Computing knowledge in security protocols under convergent equational theories. J. Autom. Reasoning **48**(2), 219–262 (2012)
13. Cohn-Gordon, K., Cremers, C., Garratt, L., Millican, J., Milner, K.: On ends-to-ends encryption: asynchronous group messaging with strong security guarantees. In: Lie, D., Mannan, M., Backes, M., Wang, X. (eds.) Proceedings of the 2018 ACM SIGSAC Conference on Computer and Communications Security, CCS 2018, Toronto, ON, Canada, 15–19 October 2018, pp. 1802–1819. ACM (2018)
14. Comon, H., Haberstrau, M., Jouannaud, J.: Syntacticness, cycle-syntacticness, and shallow theories. Inf. Comput. **111**(1), 154–191 (1994)
15. Dreier, J., Duménil, C., Kremer, S., Sasse, R.: Beyond subterm-convergent equational theories in automated verification of stateful protocols. In: Maffei, M., Ryan, M. (eds.) POST 2017. LNCS, vol. 10204, pp. 117–140. Springer, Heidelberg (2017). https://doi.org/10.1007/978-3-662-54455-6_6
16. Erbatur, S., Marshall, A.M., Ringeissen, C.: Notions of knowledge in combinations of theories sharing constructors. In: de Moura, L. (ed.) CADE 2017. LNCS (LNAI), vol. 10395, pp. 60–76. Springer, Cham (2017). https://doi.org/10.1007/978-3-319-63046-5_5

17. Erbatur, S., Marshall, A.M., Ringeissen, C.: Computing knowledge in equational extensions of subterm convergent theories. Math. Struct. Comput. Sci. **30**(6), 683–709 (2020)
18. Hopcroft, J.E., Ullman, J.D.: Some results on tape-bounded Turing machines. J. ACM **16**(1), 168–177 (1969)
19. Hopcroft, J.E., Ullman, J.D.: Introduction to Automata Theory, Languages, and Computation. Addison-Wesley (1979)
20. Kuroda, S.Y.: Classes of languages and linear-bounded automata. Inf. Control **7**(2), 207–223 (1964)
21. Lynch, C., Morawska, B.: Basic syntactic mutation. In: Voronkov, A. (ed.) CADE 2002. LNCS (LNAI), vol. 2392, pp. 471–485. Springer, Heidelberg (2002). https://doi.org/10.1007/3-540-45620-1_37
22. Narendran, P., Otto, F.: Single versus simultaneous equational unification and equational unification for variable-permuting theories. J. Autom. Reason. **19**(1), 87–115 (1997)
23. Nieuwenhuis, R.: Decidability and complexity analysis by basic paramodulation. Inf. Comput. **147**(1), 1–21 (1998)
24. Satterfield, S.D., Erbatur, S., Marshall, A.M., Ringeissen, C.: Knowledge problems in security protocols: going beyond subterm convergent theories. In: Gaboardi, M., van Raamsdonk, F. (eds.) 8th International Conference on Formal Structures for Computation and Deduction, FSCD 2023, 3–6 July 2023, Rome, Italy. LIPIcs, vol. 260, pp. 30:1–30:19. Schloss Dagstuhl - Leibniz-Zentrum für Informatik (2023)
25. Schmidt-Schauß, M.: Unification in permutative equational theories is undecidable. J. Symb. Comput. **8**(4), 415–421 (1989)

Binary Implication Hypergraphs for the Representation and Simplification of Propositional Formulae

Jordina Francès de Mas

School of Computer Science, University of St Andrews, St Andrews, Scotland, UK
jfdm2@st-andrews.ac.uk

Abstract. Propositional simplification and preprocessing are of key importance in the fields of automated reasoning and theorem proving. We present a novel propositional formula representation able to capture all of its n-ary implication information in a tractable manner. From the exploration of this novel encoding in the form of a directed hypergraph, we derive a novel simplification rule which is guaranteed to be equivalence-preserving, monotonically decreasing in the size of the problem, and capable of deep inference. We are not aware of any such generic preprocessing framework capable of systematically simplifying propositional problems in arbitrary form. Interestingly, our rule effectively generalises and streamlines most of the known equivalence-preserving SAT preprocessing techniques. Additionally, since our problem transformations are domain- and application-independent, they can be used in combination with any propositional-logic-based techniques, including those currently used in automated reasoning, solving and optimisation tools.

Keywords: Knowledge representation · Propositional logic simplification · Hypergraphs · Implication graphs · Existential graphs

1 Introduction

Propositional logic (PL) simplification is the process of reducing a PL formula to a simplified expression that is easier to understand, reason about, or work with. PL simplification has many practical as well as theoretical applications. For example, simplifying PL problems is of key importance in the minimisation of logic circuits (aka logic optimisation), in logic synthesis, and even in cryptography (see, e.g., [14,28,29,31]). PL simplification is also closely related to the reduction of complex Boolean algebraic expressions and to preprocessing techniques in Boolean satisfiability (SAT). Simplifying PL formulae can also help find tighter lower and upper bounds on the complexity of functions [19], and it has an essential role in the development and performance of automated theorem proving and in automated reasoning in general [5].

Intuitively, all equivalence-preserving PL simplifications can eventually be achieved by applying the axioms of Boolean algebra, known logical equivalences, and inference rules (such as those in sequent calculus and natural deduction) in nontrivial ways; however, there are currently no known *generic* or *systematic* methods of doing so. For example, given the school-level PL problem $(((A \vee B) \rightarrow (C \leftrightarrow A)) \wedge ((A \leftrightarrow B) \rightarrow ((\neg A \vee B) \wedge C)) \wedge (((A \wedge B) \vee (\neg C)) \wedge (B \rightarrow C))) \rightarrow (C \rightarrow B)$, how can we simplify it? From the extensive set of known logical identities, inference rules, and transformations, which ones do we use and in what order? Current theorem provers and solvers cannot *automatically* detect the triviality of this statement—which can be reduced to $\top$ by means of equivalence-preserving simplification alone—unless they are interactively guided by the user to try certain tactics or resort to 'brute-force' search. As a result of these non-systematic heuristic approaches, the generated proofs are usually riddled with 'bureaucratic' or unnecessary steps which do not help with characterising or better understanding the problem in order to improve future reasoning (see Fig. 1). For example, it is known that theorem provers output the *first* proof they find, which is rarely an optimal proof by any measure of optimality [18].

```
(fun (A B C : Prop)
  (H : (A \/ B -> C <-> A) /\
       (A <-> B -> (~ A \/ B) /\ C) /\ (A /\ B \/ ~ C) /\ (B -> C))
  (H0 : C) =>
and_ind
  (fun (H1 : A \/ B -> C <-> A)
       (H2 : (A <-> B -> (~ A \/ B) /\ C) /\ (A /\ B \/ ~ C) /\ (B -> C)) =>
  and_ind
    (fun (H3 : A <-> B -> (~ A \/ B) /\ C) (H4 : (A /\ B \/ ~ C) /\ (B -> C)) =>
    and_ind
      (fun (H5 : A /\ B \/ ~ C) (H6 : B -> C) =>
      or_ind
        (fun H7 : A /\ B =>
        and_ind
          (fun (H8 : A) (H9 : B) =>
          let H10 : C := H6 H9 in
          let H11 : (A -> B) -> (B -> A) -> (~ A \/ B) /\ C :=
            fun (H11 : A -> B) (H12 : B -> A) => H3 (conj H11 H12) in
          let H12 : A -> C <-> A := fun H12 : A => H1 (or_introl H12) in
          let H13 : B -> C <-> A := fun H13 : B => H1 (or_intror H13) in
          let H14 : C <-> A := H12 H8 in
          and_ind
            (fun (H15 : C -> A) (H16 : A -> C) =>
            let H17 : A := H15 H0 in
            let H18 : C := H16 H8 in
            let H19 : C <-> A := H13 H9 in
            and_ind
              (fun (H20 : C -> A) (H21 : A -> C) =>
              let H22 : A := H20 H0 in let H23 : C := H21 H8 in H9) H19) H14)
          H7)
        (fun H7 : ~ C =>
        let H8 : False := H7 H0 in
        let H9 : (A -> B) -> (B -> A) -> (~ A \/ B) /\ C :=
          fun (H9 : A -> B) (H10 : B -> A) => H3 (conj H9 H10) in
        let H10 : A -> C <-> A := fun H10 : A => H1 (or_introl H10) in
        let H11 : B -> C <-> A := fun H11 : B => H1 (or_intror H11) in
        False_ind B H8) H5) H4) H2) H)
```

Fig. 1. Proof of $(((A \vee B) \rightarrow (C \leftrightarrow A)) \wedge ((A \leftrightarrow B) \rightarrow ((\neg A \vee B) \wedge C)) \wedge (((A \wedge B) \vee (\neg C)) \wedge (B \rightarrow C))) \rightarrow (C \rightarrow B)$ obtained by the proof assistant (aka interactive theorem prover) Coq [33] after hinting it to use the tactic `tauto`, which requires the user to know in advance that the given statement is a tautology.

On the other hand, most automated reasoning techniques—including pre-processing and simplification—require a problem to be transformed to a normal form (e.g. CNF, DNF) which is supported by a given solver/prover due to its particular inner implementation and features. For instance, the automatic theorem prover Vampire [34] needs to translate formulae and subgoals into several normal forms in order to automatically proof Peirce's Law $(((P \to Q) \to P) \to P)$ by contradiction [24, p. 5]. In fact, a lot of currently available model refinement techniques are problem-, solver- or normal-form-dependent. Besides some of these automated transformations being 'bureaucratic' or heuristic, many are non-equivalence-preserving (they preserve equisatisfiability instead), can result in bigger problems, introduce unnecessary redundancy, or lead to the loss of important structural properties of the original problem, such as symmetries, which can make it more difficult to find a proof or solution [1,11].

In this paper we offer a novel, systematic, equivalence-preserving PL preprocessing approach based on existential graphs and implication (hyper)graphs which is capable of systematically simplifying PL formulae in arbitrary form. These theoretical findings build upon our previous work on PL simplification presented in [12], where our rules use binary implication information to detect redundancies, whilst here we make use of n-ary implication knowledge. We are not aware of any such generic preprocessing framework and, in fact, our novel approach unifies and extends most known equivalence-preserving PL preprocessing techniques, including our previous generalisations of these, introduced in [12]. Additionally, since our problem reformulations are domain- and application-independent, they can potentially benefit any PL-based techniques, including those currently used in automated reasoning, solving and optimisation tools.

The remainder of this paper is structured as follows. Section 2 introduces basic concepts and notation on existential graphs and implication graphs that will be used in the rest of the paper. In Sect. 3, we define our notion of flatness and introduce a novel representation for PL formulae able to capture all of its implication information in a tractable manner, which we name binary implication hypergraphs (BIHs). From the study of BIHs' connectivity properties, we derive a simplification rule called Strong-Hyper-TWSR, which we present in Sect. 4, and suggest a simplification procedure based on it. Section 5 ends the paper with some concluding remarks.

2 Background

We assume the reader is familiar with basic notions of *propositional logic* (PL) and *graph theory*. In addition to the classical PL symbolic linear representation, we will also make use of Peirce's **existential graphs** (EGs) [8,22,27,30], which are a non-symbolic and non-linear alternative system equivalent to sentential languages[1] [25,26,35] which represent logical formulae in a two-dimensional fashion. In particular, negation is represented by oval lines (aka *cuts*), and all elements contained in the same area (delimited by negation lines) are in implicit

---

[1] Extensions of EGs allow for the representation of first-order logic and higher-order logic formulas, but these are beyond the scope of this paper (for more details, see e.g. [8,26]).

conjunction. We will refer to an area being enclosed by an even (resp. odd) number of cuts as *evenly* (resp. *oddly*) *nested*. The blank sheet of assertion denotes True ($\top$), and it is assumed to be evenly enclosed with nesting level 0. Figure 2 offers a few introductory examples of EGs which illustrate how the combination of these simple primitives suffices to express any propositional logic formula. The most straightforward way to convert an arbitrary formula to EGs is to translate it to its equivalent form restricted to the negation ($\neg$) and conjunction ($\wedge$) operators. Then each $\neg$ corresponds to a cut line and all $\wedge$ operators can be erased to obtain its EG representation. This transformation can clearly be done in linear time, and results in a formula of the same length as the original one (except for XOR constraints or equivalence clauses).

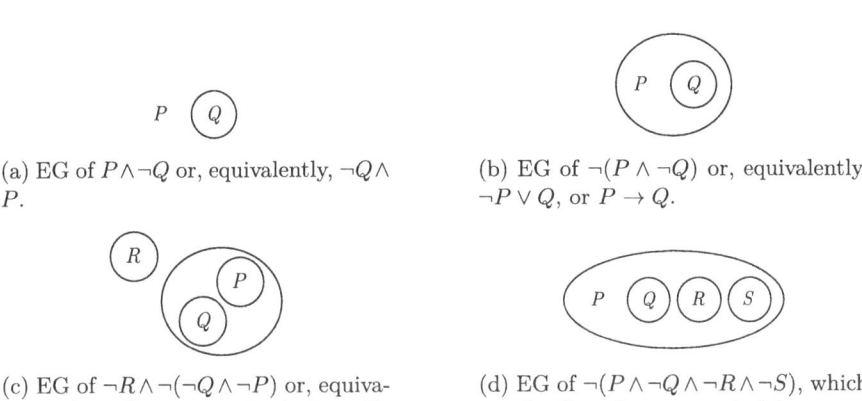

(a) EG of $P \wedge \neg Q$ or, equivalently, $\neg Q \wedge P$.

(b) EG of $\neg(P \wedge \neg Q)$ or, equivalently, $\neg P \vee Q$, or $P \to Q$.

(c) EG of $\neg R \wedge \neg(\neg Q \wedge \neg P)$ or, equivalently, $(P \vee Q) \wedge \neg R$, or $\neg R \wedge (\neg P \to Q)$, or $(\neg Q \to P) \wedge (R \to \bot)$.

(d) EG of $\neg(P \wedge \neg Q \wedge \neg R \wedge \neg S)$, which can be linearly represented by many equivalent formulations, such as $P \to (Q \vee R \vee S)$, or $(P \wedge \neg Q) \to (R \vee S)$, or $(P \wedge \neg Q \wedge \neg R) \to S$, or $\neg P \vee Q \vee R \vee S$.

Fig. 2. Existential graphs representing different propositional logic formulae, adapted from [12].

Peirce gave special attention to the syntactical device of two nested cuts (aka *scroll*), that is, subgraphs of the form ⦅A⦆B⦆, where A and B can be arbitrary EGs themselves. Since a cut negates the enclosed subgraph, this diagram can be read as 'it is not true that A holds and B does not hold', which is equivalent to 'A implies B'. This last interpretation equips us with a logical device to interpret any EG—and so any propositional logic formula—in terms of implication, which is particularly useful for the study of reasoning and deduction. For example, even when reading the very simple propositional formula $\neg P \vee Q \vee R \vee S$, all of its semantic interpretations in terms of implication are not immediately apparent, but they become obvious and salient when expressed as an EG and read or interpreted with the scroll implication pattern in mind (revisit Fig. 2d).

Even if this diagrammatic notation is less commonly used, it offers a simple, elegant, and easy-to-learn alternative to symbolic logic which exploits geometric intuition and makes it easier to follow complex proofs and inferences [8,12,30]. Moreover, EGs come with a sound and complete deductive system which has many advantages over the traditional Gentzen's rules of *natural deduction* and *sequent calculus* and, thus, over all rewriting methods based on these. Most significantly, EGs' inference rules are symmetric, and no bookkeeping of previous steps is required in the process of inferring a graph from another. Another advantage is that the visual representation allows for the recognition of nested patterns that would otherwise be obscured in symbolic linear form. Furthermore, as illustrated in Fig. 2, EGs' graphical non-ordered nature provides a canonical representation of a logical formula that, in linear notation, can be written in many equivalent forms. Thus, we will use EGs throughout this paper, but if a linear symbolic notation is preferred, one only needs to translate any arbitrary PL formula or any EG to its equivalent linear form restricted to the negation ($\neg$) and conjunction ($\wedge$) operators, and think of every $\neg$ together with its implicit or explicit parenthesis as a cut line.

Whatever notation we use, it is easy to see that every binary clause ($x \vee y$) admits the following two interpretations: ($\neg x \to y$) and its logically equivalent contrapositive ($\neg y \to x$). Thus, binary clauses provide the necessary information to build what is known as the **binary implication graph** (BIG) [2] of a formula, which is a skew-symmetric[2] directed graph where edges represent implication, and nodes are all the variables occurring in binary clauses and their negations (see Fig. 3). Note that BIGs need to be skew-symmetric in order to capture all binary implications together with their contrapositives to link both the positive and negative instances of each variable.

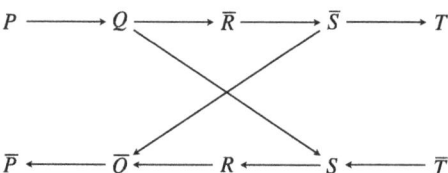

Fig. 3. BIG of $\varphi = (\overline{P} \vee Q) \wedge (Q \to \overline{R}) \wedge (S \to R) \wedge (S \vee T) \wedge (S \vee \overline{Q})$.

3 Implication Hypergraphs

As we saw in Fig. 2, any clause or subgraph of size n can be interpreted as an implication, which can be linearly expressed in many equivalent forms. In order to capture this n-ary implication information, we need to extend BIGs, which are only capable of storing binary implications, that is, a literal implying another

---

[2] A directed graph is *skew-symmetric* if it is isomorphic to its transpose graph under an involution function with no fixed points, that is, where a vertex cannot be mapped to itself.

literal. A straightforward extension is to allow the nodes of a BIG to be clauses of arbitrary size instead of just unit clauses. We refer to such implication graph as n-ary implication graph (n-IG).

For example, let $\varphi = (\overline{P} \vee \overline{Q} \vee Y) \wedge (\overline{Y} \vee X) \wedge (\overline{X} \vee R \vee S) \wedge (\overline{P} \vee \overline{Q} \vee R \vee S)$. Note that the first clause can be interpreted as $(P \wedge Q) \to Y$, the second as $Y \to X$, the third as $X \to (R \vee S)$, and the fourth as $(P \wedge Q) \to (R \vee S)$, amongst many other alternative readings. These particular interpretations give rise to the following (skew-symmetric) n-IG:

$$P \wedge Q \longrightarrow Y \longrightarrow X \longrightarrow R \vee S \qquad \overline{P \wedge Q} \longleftarrow \overline{Y} \longleftarrow \overline{X} \longleftarrow \overline{R \vee S}$$

However, the clause $\overline{X} \vee R \vee S$, for example, can also be interpreted as $\overline{R} \to (\overline{X} \vee S)$, $\overline{S} \to (\overline{X} \vee R)$, $(X \wedge \overline{R}) \to S$, $(X \wedge \overline{S}) \to R$ or $(\overline{R} \wedge \overline{S}) \to \overline{X}$. This means that the n-IG above is not unique. Two of the many other possible n-IG(φ) are:

$$\overline{Q} \vee Y \longleftarrow P \longrightarrow \overline{Q} \vee R \vee S \qquad \overline{P} \vee Y \longleftarrow Q \longrightarrow \overline{P} \vee R \vee S$$
$$\overline{R} \wedge \overline{S} \longrightarrow \overline{X} \longrightarrow \overline{Y} \qquad \overline{R} \wedge X \longrightarrow S$$
$$\qquad\qquad\qquad\qquad\qquad Y \longrightarrow X$$

where we have omitted their skew-symmetric halves. In fact, for each clause of size n, there are

$$\sum_{r=1}^{n-1} \frac{n!}{r!(n-r)!} = 2^n - 2 \tag{1}$$

equivalent implication readings, and so for a formula φ with m clauses of different sizes, there are

$$\prod_{i=1}^{m} 2^{c_i} - 2 \tag{2}$$

possible n-IG representations, where c_i is the size of each clause. This means that there are exponentially many n-IGs corresponding to any given formula, and so it is unfeasible to explore all of them in order to find potential reductions. Thus, any method based on the exploration of n-ary implication graphs as defined above is clearly intractable. Nevertheless, it is still worthwhile to attempt to go beyond binary implications. To make this possible, let us introduce a novel characterisation and representation of all n-ary implications of a given formula.

Definition 1 (Implication Hypergraph). *An* implication hypergraph *of a propositional formula is a directed hypergraph (V, E), where:*

- *V is the set of nodes,*
- *$E \subseteq (V \times V)$ is the set of hyperedges,*
- *each element of V is a propositional (sub)formula, and*
- *each hyperedge $(D, C) \in E$ corresponds to a subclause of the formula, where the vertex subset D (aka domain) corresponds to the negation of its negative components, and C (aka codomain) corresponds to its positive components.*

Even if this novel representation reduces the number of potential n-ary implication graphs of a given formula significantly, it is still not unique, given that nodes can be arbitrary propositional clauses. The next example illustrates this.

Example 1. Let $\varphi = (V \vee W \vee (U \wedge S)) \wedge (A \vee M \vee \overline{U} \vee \overline{S} \vee \overline{P}) \wedge (V \rightarrow A) \wedge (W \rightarrow M)$. As opposed to n-IGs, our novel implication hypergraph representation does not require a skew-symmetric counterpart—since each hyperedge encodes all implication readings of a clause—which simplifies the graph. However, since nodes can be arbitrary clauses, there is still a hypergraph for each possible partition of the formula. Figure 4 shows two possible implication hypergraphs of φ.

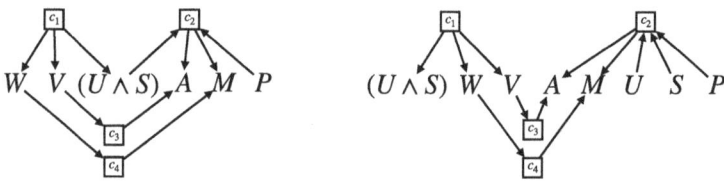

Fig. 4. Two possible implication hypergraphs representing φ, where square nodes represent hyperedges (i.e. clauses). Note that both hypergraphs contain a 2-ary clause node, namely $(U \wedge S)$.

In order to ensure uniqueness, we need the nodes of the hypergraph to be restricted to single variables. This corresponds to formulae whose (sub)clauses contain only literals or negated literals. We refer to such formulae as being in *flat* form, defined formally below.

Definition 2 (Flatness). *We say that a propositional (sub)formula is flat if it corresponds to an EG or a subset of EGs (subgraph) where each cut (oval) can only contain literals (or their negation), or cuts containing only literals (or their negation). That is, a flat level-n cut can contain cuts of level $n+1$, but can only have cuts of level $n+2$ if these contain a single literal, and any further nesting is not allowed.*

Example 2. Let $\varphi = \neg(A \wedge \overline{B} \wedge \neg(D \wedge \overline{C}) \wedge \neg(P \wedge \overline{Q} \wedge \neg(R \wedge \overline{S}) \wedge \neg(X \wedge Y \wedge \overline{Z})))$. φ is not a flat formula, given that it contains non-singleton elements at nesting level 2 (see Fig. 5). However, it does contain many flat subformulae. In particular, we can partition φ into two maximal flat subformulae: $(A \wedge \overline{B} \wedge \neg(D \wedge \overline{C}))$ at level 1 and $(P \wedge \overline{Q} \wedge \neg(R \wedge \overline{S}) \wedge \neg(X \wedge Y \wedge \overline{Z}))$ at level 2.

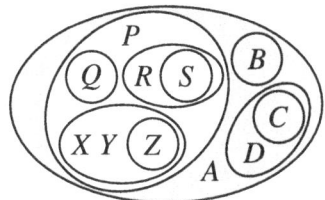

Fig. 5. EG of $\varphi = \neg(A \wedge \overline{B} \wedge \neg(D \wedge \overline{C}) \wedge \neg(P \wedge \overline{Q} \wedge \neg(R \wedge \overline{S}) \wedge \neg(X \wedge Y \wedge \overline{Z})))$, where the gray area constitutes a flat subformula at level 1, and the level-2 biggest cut in white also contains a flat subformula.

Remark 1. Note that this definition encompasses any formula or *nested* subformula in DNF or CNF. In fact, we can see arbitrarily nested formulae as a succession of DNF and CNF subformulae, which can be equally processed at a local level.

Thus, we restrict our novel implication graph characterisation to *flat* (sub)formulae so that the nodes of the hypergraph are guaranteed to be a single variable. This results in a representation able to capture n-ary implications in a unique manner, and so tractability is regained.

Definition 3 (Binary Implication Hypergraph (BIH)). *The* binary implication hypergraph *of a* flat *propositional (sub)formula is a directed hypergraph (V, E), where:*

- *V is the set of nodes,*
- *$E \subseteq (V \times V)$ is the set of hyperedges,*
- *each element of V is a propositional variable, and*
- *each hyperedge $(D, C) \in E$ corresponds to a subclause of the formula, where the vertex subset D (aka domain) corresponds to the negation of its negative literals, and C (aka codomain) corresponds to its positive literals.*

More intuitively, a BIH has as many nodes as variables has the flat (sub)formula and as many hyperedges as clauses. Each hyperedge has *order* equal to the number of literals in the clause, i.e. $|D| + |C|$, and points to the variable nodes of its positive literals, and away from the variable nodes corresponding to its negative literals. That is, D contains the variables which occur negatively in a clause and C is the subset of variables with positive occurrences in the clause. For ease of representation, we will depict a hyperedge as a square node from where its multiple directed edges begin and/or end (see Fig. 6).

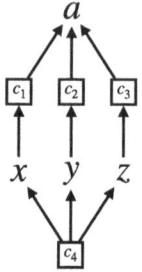

Fig. 6. Example of a simple BIH corresponding to the propositional formula $(x \to a) \wedge (y \to a) \wedge (z \to a) \wedge (x \vee y \vee z)$.

Note that any hyperedge with a simple loop, that is, pointing to and from the same variable node, is a tautology, and so we require that $D \cap C = \emptyset$ in order to keep the BIH tautology-free. Notice that if we restrict the BIH to binary hyperedges only, then we have an alternative (non-skew-symmetric) representation of the BIG (see Fig. 7).

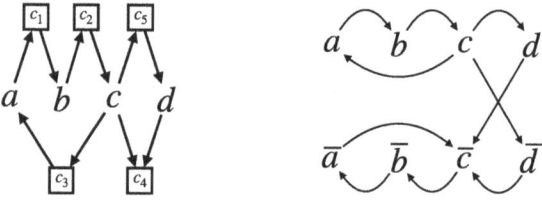

Fig. 7. BIH of the binary formula $\varphi = (\neg a \vee b) \wedge (\neg b \vee c) \wedge (\neg c \vee a) \wedge (\neg c \vee \neg d) \wedge (\neg c \vee d)$ (left), having 4 variable nodes and 5 directed hyperedges (which correspond to the 5 square clausal nodes and the 10 arrows). The BIG of φ has 8 literal nodes and 10 edges (right).

3.1 Implication Hyperpaths

In order to study connectivity over BIHs, we first need to characterise its *reachability* conditions and define what constitutes a *(hyper)path* in it. Several definitions of *hyperpaths* over *directed hypergraphs* exist in the literature (see, e.g., [3,32]), but they do not account for permissible path polarity changes present in BIHs—where edge direction is used to store sign (positive or negative) information of each literal occurrence—and do not allow both the source and target of a path to be a set of nodes. To the best of our knowledge, only hyperpaths with a single source node and a single target node have ever been studied, which are only capable of analysing propositional Horn formulae [13]. In order to use the existing definitions and techniques, we would need to define a skew-symmetric version of our BIH, which would not only double the nodes (a positive and a negative for each variable), but also explode in terms of the number of hyperedges that would then be required to encode all possible implication readings, as in n-IGs. Thus, we need a new definition to identify the hyperpaths of a BIH. To do so, we will first define the concepts of *adjacent hyperedges*, *adjacent nodes*, *source set*, *target set* and *flat implication hyperpaths* of a BIH.

Definition 4 (Adjacent hyperedges). *Let* $\mathrm{BIH}(\varphi) = (V, E)$. *Let* $e_i = (D_i, C_i)$ *be the* i^{th} *hyperedge in* E, *with* $D_i, C_i \subseteq V$. *We say that two hyperedges* $e_i, e_j \in E$ *are* adjacent *if* $D_i \cap C_j \neq \emptyset$ *or* $C_i \cap D_j \neq \emptyset$.

Definition 5 (Adjacent nodes, source set and target set). *Let* $\mathrm{BIH}(\varphi) = (V, E)$. *Let* $\varepsilon = (D, C)$ *be a hyperedge in* E, *with* $D, C \subseteq V$. *Let* $S = (\mathcal{D}, \mathcal{C})$ *be a subset of nodes of* ε *with* $\mathcal{D} \subseteq D$ *and* $\mathcal{C} \subseteq C$. *We say that the subset of nodes* S *is* adjacent *to the subset of nodes* $T = (\mathcal{D}', \mathcal{C}')$, *with* $\mathcal{D}' = D \setminus \mathcal{D}$ *and* $\mathcal{C}' = C \setminus \mathcal{C}$. *We call* S *a* source set *of* ε *and* T *the corresponding* target set.

Definition 6 (Flat implication hyperpath). *Let* $\mathrm{BIH}(\varphi) = (V, E)$. *Let* $e_i = (D_i, C_i)$ *be the* i^{th} *hyperedge in* E, *with* $D_i, C_i \subseteq V$. *Let* $S_i = (\mathcal{D}_i, \mathcal{C}_i)$ *be a (source) subset of nodes of* e_i *with* $\mathcal{D}_i \subseteq D_i$ *and* $\mathcal{C}_i \subseteq C_i$. *A* flat implication hyperpath *from a (source) subset of nodes* S_0 *of* $e_0 \in E$ *to a (target) subset of nodes* T *of* $e_n \in E$ *is an alternating sequence of nodes and adjacent binary hyperedges* $P = [S_0, e_0, v_1, e_1, \ldots, v_n, e_n, T]$ *such that:*

- $|(\mathcal{D}_0 \cup \mathcal{C}_0) \setminus (\mathcal{D}_0 \cup \mathcal{C}_0)| = 1$,
- $v_1 = (D_0 \cup C_0) \setminus (\mathcal{D}_0 \cup \mathcal{C}_0)$,
- $|D_i \cup C_i| = 2$ for all $i \in \{1, \ldots, n-1\}$,
- $v_i = (D_i \cup C_i) \setminus \{v_{i-1}\}$ for all $i \in \{2, \ldots, n\}$, and
- either $v_i \in D_{i-1}$ and $v_i \in C_i$, or $v_i \in C_{i-1}$ and $v_i \in D_i$, and
- $T = (D_n \setminus \{v_n\}, C_n \setminus \{v_n\})$.

More intuitively, two hyperedges are connected by a flat implication hyperpath if one can be reached from the other by following binary hyperedges having the same direction along the shared variable node (see Fig. 8).

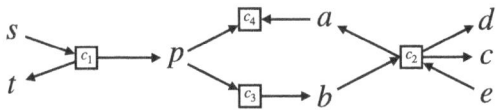

Fig. 8. BIH of $\varphi = (\overline{s} \vee t \vee p) \wedge (a \vee \overline{b} \vee c \vee d \vee \overline{e}) \wedge (p \rightarrow b) \wedge (\overline{p} \vee \overline{a})$, where we can see two flat implication hyperpaths between c_1 and c_2, in particular, between the source set $(\{s\}, \{t\})$ and the target set $(\{e\}, \{a, c, d\})$, and between $(\{s\}, \{t\})$ and $(\{b, e\}, \{c, d\})$. Note that there are also flat implication paths between $(\{s\}, \{t\})$ and $(\{a\}, \{\})$, between $(\{s\}, \{t\})$ and $(\{\}, \{b\})$, between $(\{p\}, \{\})$ and $(\{b, e\}, \{c, d\})$, and between $(\{p\}, \{\})$ and $(\{e\}, \{a, c, d\})$.

However, this notion of hyperpath is still not enough to capture arbitrary n-ary propositional implications, since it can only detect implication paths of the form $S \implies x_1 \implies \cdots \implies x_n \implies T$, where x_i are single literals, S is a conjunction of literals, and T is a disjunction of literals. To be able to capture more complex implication chains, we introduce the following concept.

Definition 7 (Implication hyperpath). Let $\text{BIH}(\varphi) = (V, E)$. Let $e_i = (D_i, C_i)$ be the ith hyperedge in E, with $D_i, C_i \subseteq V$. Let $S_i = (\mathcal{D}_i, \mathcal{C}_i)$ be a subset of nodes of e_i with $\mathcal{D}_i \subseteq D_i$ and $\mathcal{C}_i \subseteq C_i$. Let T_i be the rest of the nodes of e_i, that is, $T_i = (\mathcal{D}'_i, \mathcal{C}'_i) = (D_i \setminus \mathcal{D}_i, C_i \setminus \mathcal{C}_i)$.

- We say that a node $v_n \in S_n$ is reachable from a node $v_{n-1} \in S_{n-1}$, with $|S_{n-1}| = 1$ if for every node $v_j \in T_{n-1}$ there is a flat implication hyperpath from $e_{n-1} \setminus \{v_j\}$ to $e_n \setminus \{v_n\}$.
- We say that a node $v_n \in S_n$ is reachable from a subset of nodes S_0 if v_n is reachable from every node in T_0.
- We say that there is an implication hyperpath from a subset of nodes S_0 to a subset of nodes T_n if every node in S_n is reachable from every node in T_0.
- More generally, we can say that there is an implication hyperpath from S_0 to $\bigcup_{k=1}^{|T_0|} \mathcal{T}_k$, where $\mathcal{T}_k$ is a subset of nodes reachable from $v_k \in T_0$. We equivalently call $\bigcup_{k=1}^{|T_0|} \mathcal{T}_k$ a set of BIH-descendants of S_0.

More intuitively, an implication hyperpath can be thought of as a flow that can only travel along same-direction node-crossing edges, and can only travel through a hyperedge (square) node e_n whose source nodes (in S_n) have all been reached by flows coming from each of their ancestors already 'touched' by the flow (see Fig. 9).

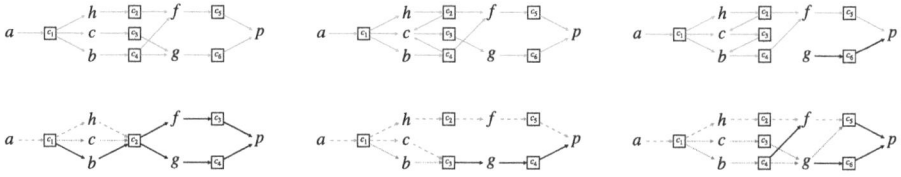

Fig. 9. Examples of implication hyperpaths from a to p highlighted in light blue (top three BIHs), and BIHs without an implication hyperpath from a to p (bottom three BIHs), where green dashed arrows indicate the search for a path, which is aborted as soon as the red dotted edges are found. (Color figure online)

4 BIH-Based Simplification

In this section we present a simplification rule which uses the implication knowledge stored in BIHs and can be nestedly applied to PL formulae in arbitrary form, not only in flat form. However, note that BIHs—like BIGs—are only defined for flat (sub)formulae. Thus, we build the BIH of each nesting level with the information provided in flat form and use this information to guide our reductions.

Definition 8 (Strong Hyper-Tuple Wipe and Sublip Rule (Strong-Hyper-TWSR)). Let x be a literal in nesting level c. If there are m literals $\{y_1, \ldots, y_m\}$ at levels $c \leq d_1 \leq \cdots \leq d_m$ such that $\bigcup_{i=1}^{m}\{\overline{y_i}\}$ is a set of BIH-descendants of x, then the d_m^{th} cut, which contains y_m, can be deleted. If an empty cut is generated, the area containing the empty negation oval shall be deleted.

Note that if we have an occurrence of l in level L, by the double negation insertion rule we can interpret it as having an occurrence of $\bar{l}$ in level $L+1$. Thus, if we are looking for a literal l but we encounter its negation, we know that this is equivalent to having an occurrence of l at one level deeper.

Example 3. Let $\varphi = (A \vee B \vee C) \wedge (A \to D) \wedge (B \to E) \wedge (C \to F) \wedge (D \vee E \vee x \vee (\overline{F} \wedge y \wedge \neg(z)))$, where x, y and z stand for arbitrary (possibly empty) subformulae. If we apply Strong-Hyper-TWSR to φ, we obtain the much simpler equivalent formula $\varphi' = $ Strong-Hyper-TWSR$(\varphi) = (A \vee B \vee C) \wedge (A \to D) \wedge (B \to E) \wedge (C \to F) \wedge (D \vee E \vee x)$. This is because, from the BIH of the subformula at level 0, we can see that $\overline{D} \to (E \vee F)$, and so the cut containing $\overline{F}$ can be deleted (see Fig. 10). More precisely, the BIH at level 0 contains a hyperpath from $(\{\}, \{D\})$ to $(\{\}, \{E\}) \cup (\{\}, \{F\})$, that is, $(\{\}, \{E\}) \cup (\{\}, \{F\})$ is a set of BIH-descendants of $(\{\}, \{D\})$. The cut containing $\overline{F}$ can hence be deleted by Strong-Hyper-TWSR.

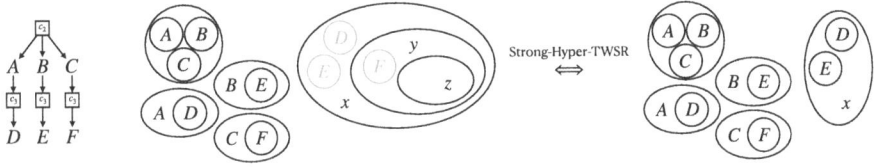

Fig. 10. Application of Strong-Hyper-TWSR to φ. The EG representation of φ is given in the middle diagram, where x, y and z represent arbitrary (possibly empty) subformulae. The BIH on the left-hand side corresponds to the four level-0 clauses. The EG of φ' is displayed on the right-hand side.

Theorem 1. Strong-Hyper-TWSR *is reversible, and so equivalence-preserving.*

Proof. In order to prove this result, we will make use of Peirce's sound and complete deductive system for EGs, which consists of six inference rules: insertion, erasure, iteration, deiteration, double cut introduction and double cut elimination. For their definitions and more details, we refer the reader to [8,22,27,30].

$\Longrightarrow$) By the definition of Strong-Hyper-TWSR, we know that $\bigcup_{i=1}^{m} \overline{y_i}$ is a set of BIH-descendants of x at level d_m or at a level d such that $c \leq d \leq d_m$. Thus, the clause $C = (x \vee y_1 \vee \cdots \vee y_m)$ can be inferred at level d_m. Then, by Peirce's deiteration rule, we can deiterate all literals in C, since they exist in levels $c \leq d_1 \leq \cdots \leq d_m$, which leaves an empty cut at level d_m, meaning that the surrounding cut (i.e., the d_m^{th} cut which contains y_m) can be deleted.

$\Longleftarrow$) Any cut deleted by Strong-Hyper-TWSR's application can be recovered as follows. If the deleted cut was in an oddly nested area, i.e., $d_m = 2k+1, k \in \mathbb{N}$, then the d_m^{th} cut with all of its contents can be introduced by simply applying Peirce's insertion rule. If the deleted cut was in an evenly nested area, i.e., $d_m = 2k, k \in \mathbb{N}$, then we can derive the clause $C = (x \vee y_1 \vee \cdots \vee y_m)$ at level $d_m - 1$, from where we can deiterate all the literals present at lower levels d_i with $c \leq d_i < d_m$, and since d_m is an oddly nested area, we can recover the rest of the original contents by applying Peirce's insertion rule.

Theorem 2. Strong-Hyper-TWSR *is monotonically non-increasing in the number of variables, in the number of clauses and in the number of literals of a propositional formula.*

Proof. Trivial, since either Strong-Hyper-TWSR does not apply so the formula stays the same, or it applies and an inner cut is deleted together with all of its contents, and so the number of literals and (sub)clauses is at least reduced by one.

Proposition 1. *Repeatedly applying* Hyper-Strong-TWSR *over an arbitrary pro-positional formula with an acyclic BIH until fixpoint is a confluent procedure.*

Proof. By Theorem 1, we know that this rule is equivalence-preserving and so locally confluent when no cycles are present in the BIH. By Theorem 2, we know that repeatedly applying Hyper-Strong-TWSR is a terminating procedure. Thus, by Newman's lemma, Hyper-Strong-TWSR is confluent.

Interestingly, Strong-Hyper-TWSR strictly generalises and streamlines most of the known equivalence-preserving SAT preprocessing techniques, including *unit propagation* [9], *failed-literal probing* [6,20], *subsumption, self-subsuming resolution* [10], *hidden literal elimination* [17], *hidden subsumption elimination* [16], *hidden tautology elimination* [16], *hyper-binary resolution* [4], *asymmetric literal elimination* [7], *asymmetric subsumption elimination* [16], *asymmetric tautology elimination* [16], *distillation* [15] and *vivification* [23], and their combinations, with the added advantages of never increasing the size of the problem, not having to store decisions (i.e. assigned literals) nor backtrack their propagation (e.g. many successive rounds of modus ponens or unit propagation), and it can be applied in a nested manner. No existing propositional preprocessing rules we are aware of can automatically achieve the reductions that Strong-Hyper-TWSR effectively detects without flattening the formula first, adding redundant literals or clauses to the formula, or arbitrarily assigning literals and backtracking.

4.1 The Strong-Hyper-TWSR Simplification Procedure

The systematic simplification procedure stemming from the Strong-Hyper-TWSR rule is as follows: parse the given PL formula and store it as a tree of BIHs, one for each flat subformula. Then repeatedly apply Strong-Hyper-TWSR until fixpoint, starting from the outermost literals, which have a greater reduction power. After reaching a fixpoint using the BIH at level 0, we move to level 1 and apply Strong-Hyper-TWSR using the union of the BIH at level 1 and the BIH at level 0. More generally, at each level c, Strong-Hyper-TWSR uses the rolling union of all prior BIHs together with the current-level BIH. The procedure terminates after either each literal of the formula has been checked or as soon as the procedure derives ⊤ or ⊥.

Let us illustrate the potential of our novel approach by revisiting Peirce's Law, which Vampire was able to automatically prove by contradiction in 7 steps, 3 of which were transformations to a normal form. Note that other state-of-the-art tools such as the interactive theorem prover Coq and the MiniZincIDE [21]— which is an IDE to run high-level, solver-independent constraint models—are not able to *automatically* prove or simplify this tautology. In contrast, our novel systematic approach can automatically reduce Peirce's Law to ⊤ in as little as 1 application of Strong-Hyper-TWSR, while avoiding any form of flattening nor requiring bookkeeping of rule applications (since all of our transformations are equivalence-preserving), and therefore requiring less space too. Figure 11 displays the corresponding transformations both in terms of BIHs and in terms of EGs, for comparison. Starting from the outermost singleton, $\overline{P}$, since it trivially implies $\overline{P}$, Strong-Hyper-TWSR first deletes any deeper cuts containing P. This results in an empty cut at level 1, and so the whole formula can be deleted, resulting in a blank sheet of assertion, which signifies ⊤.

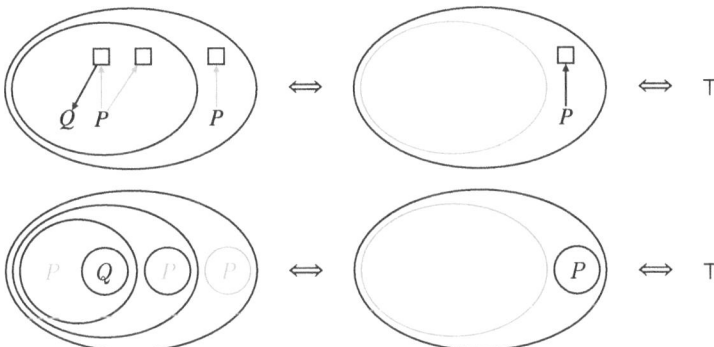

Fig. 11. Proof of Peirce's Law, which states that $((P \to Q) \to P) \to P$, obtained from an application of Strong-Hyper-TWSR, which prioritises the propagation of the outermost unprocessed literals. The top graphs represent the transformations in terms of nested BIHs, and the bottom ones in terms of EGs. The subgraphs involved in each reduction step are highlighted in orange. Recall that the automatic theorem prover Vampire required 7 steps with 3 heuristic flattening transformations to prove this simple statement. (Color figure online)

Applying our Strong-Hyper-TWSR-based procedure, we can also automatically prove that the PL problem $(((A \lor B) \to (C \leftrightarrow A)) \land ((A \leftrightarrow B) \to ((\neg A \lor B) \land C)) \land (((A \land B) \lor (\neg C)) \land (B \to C))) \to (C \to B)$ introduced in Sect. 1 is in fact a tautology. The corresponding proof requires only 2 applications of the Strong-Hyper-TWSR rule, one for the propagation of C and one for the propagation of $\neg B$, which are the two outermost literals. After propagating these two literals (in any order), an empty cut is generated at level 1 and the whole formula reduces to ⊤.

4.2 Discussion

Even though EGs are useful to visualise and facilitate the understanding of our redundancy detection methods, we do not need to store nor work with both the EG and the BIH representations of a formula. Storing a BIH for each flat area and computing a rolling union of BIHs suffices for the application of Strong-Hyper-TWSR, as we have seen above. This explicit-BIH approach is also pedagogically useful to depict and better understand our novel procedure, its power and theoretical interest, but we will need a more efficient data structure in order to efficiently implement it. Thus, in practice, we use a (less human-readable) version of BIHs which allows for node-sharing across levels, has a single node for each problem variable, a hyperedge (square) node for each level-0 cut, and the rest of the levels are indicated as edge attributes (see Fig. 12). This data structure is able to capture all information stored in EGs and BIHs in a very succinct form, which is more conducive to an efficient implementation. After we finish the implementation of Strong-Hyper-TWSR over this structure, we plan to offer it as a domain-, application- and technology-independent technique to be used on its own or integrated with existing preprocessing, inprocessing and automated reasoning pipelines, which will in turn enable the corresponding comparative study.

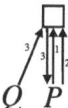

Fig. 12. Representation of Peirce's Law, which states that $((P \to Q) \to P) \to P$, as a unified BIH data structure, where the numerical edge attributes indicate the nesting level of each literal.

5 Conclusion

In this paper we introduce binary implication hypergraphs (BIHs), which are a novel PL formula representation able to capture all of its n-ary implication information in a tractable manner. From the study of BIHs' connectivity, we derive a novel simplification rule, called Strong-Hyper-TWSR, which is guaranteed to be equivalence-preserving and monotonically decreasing in the size of the problem, and can be applied over PL formulae in arbitrary form. We then suggest an automatic simplification procedure based on the systematic application of Strong-Hyper-TWSR which prioritises processing the outermost unprocessed literals and uses the union of each level's BIH together with its ancestor-level BIHs to find deeply nested redundancies.

Our approach turns out to strictly generalise and streamline the combination of most of the known equivalence-preserving PL preprocessing techniques, with the added advantages of never increasing the size of the problem, not having to store decisions nor backtrack their propagation, and it can be applied in a nested manner. We are not aware of any such generic preprocessing framework capable of automatically detecting deeply nested redundancies without flattening the formula first, adding redundant literals or clauses to the formula, or arbitrarily assigning literals and backtracking. Additionally, since our problem transformations are domain-, application- and technology-independent, they can potentially benefit any propositional-logic-based techniques, including those currently used in automated reasoning, solving and optimisation tools.

Besides the theoretical interest of such a novel and powerful technique, we think that Strong-Hyper-TWSR and its systematic application can be of great practical interest too. Thus, we are currently working on its implementation using the less-human-readable but computationally more efficient BIH counterpart discussed at the end of the previous section. The next obvious steps are the integration of our procedure with state-of-the-art tools and an experimental comparative study, which currently stand as important future work.

References

1. Anders, M.: SAT preprocessors and symmetry. In: Meel, K.S., Strichman, O. (eds.) 25th International Conference on Theory and Applications of Satisfiability Testing, SAT 2022, Haifa, Israel, 2–5 August 2022. LIPIcs, vol. 236, pp. 1:1–1:20. Schloss Dagstuhl - Leibniz-Zentrum für Informatik (2022). https://doi.org/10.4230/LIPIcs.SAT.2022.1
2. Aspvall, B., Plass, M.F., Tarjan, R.E.: A linear-time algorithm for testing the truth of certain quantified boolean formulas. Inf. Process. Lett. **8**(3), 121–123 (1979). https://doi.org/10.1016/0020-0190(79)90002-4
3. Ausiello, G., Laura, L.: Directed hypergraphs: introduction and fundamental algorithms - a survey. Theor. Comput. Sci. **658**, 293–306 (2017). https://doi.org/10.1016/j.tcs.2016.03.016
4. Bacchus, F.: Enhancing Davis Putnam with extended binary clause reasoning. In: Dechter, R., Kearns, M.J., Sutton, R.S. (eds.) Proceedings of the Eighteenth National Conference on Artificial Intelligence and Fourteenth Conference on Innovative Applications of Artificial Intelligence, Edmonton, Alberta, Canada, July 28 - August 1, 2002, pp. 613–619. AAAI Press / The MIT Press (2002). http://www.aaai.org/Library/AAAI/2002/aaai02-092.php
5. Bachmair, L.: Paramodulation, superposition, and simplification. In: Gottlob, G., Leitsch, A., Mundici, D. (eds.) KGC 1997. LNCS, vol. 1289, pp. 1–3. Springer, Heidelberg (1997). https://doi.org/10.1007/3-540-63385-5_28
6. Berre, D.L.: Exploiting the real power of unit propagation lookahead. Electron. Notes Discret. Math. **9**, 59–80 (2001). https://doi.org/10.1016/S1571-0653(04)00314-2
7. Biere, A., Järvisalo, M., Kiesl, B.: Preprocessing in SAT solving. In: Biere, A., Heule, M., van Maaren, H., Walsh, T. (eds.) Handbook of Satisfiability - Second Edition, Frontiers in Artificial Intelligence and Applications, vol. 336, pp. 391–435. IOS Press (2021). https://doi.org/10.3233/FAIA200992
8. Dau, F.: Mathematical logic with diagrams – based on the existential graphs of Peirce. Habilitation thesis. TU Darmstadt, Germany (2008). http://www.dr-dau.net/Papers/habil.pdf

9. Davis, M., Putnam, H.: A computing procedure for quantification theory. J. ACM (JACM) **7**(3), 201–215 (1960)
10. Eén, N., Biere, A.: Effective preprocessing in SAT through variable and clause elimination. In: Bacchus, F., Walsh, T. (eds.) SAT 2005. LNCS, vol. 3569, pp. 61–75. Springer, Heidelberg (2005). https://doi.org/10.1007/11499107_5
11. Eén, N., Mishchenko, A., Sörensson, N.: Applying logic synthesis for speeding up SAT. In: Marques-Silva, J., Sakallah, K.A. (eds.) SAT 2007. LNCS, vol. 4501, pp. 272–286. Springer, Heidelberg (2007). https://doi.org/10.1007/978-3-540-72788-0_26
12. Francès de Mas, J., Bowles, J.: A novel EGs-based framework for systematic propositional-formula simplification. In: Glück, R., Kafle, B. (eds.) Logic-Based Program Synthesis and Transformation, pp. 169–187. Springer, Cham (2023). https://link.springer.com/chapter/10.1007/978-3-031-45784-5_11
13. Gallo, G., Longo, G., Pallottino, S., Nguyen, S.: Directed hypergraphs and applications. Discrete Appl. Math. **42**(2), 177–201 (1993). https://www.sciencedirect.com/science/article/pii/0166218X9390045P
14. Haaswijk, W., Soeken, M., Mishchenko, A., Micheli, G.D.: SAT-based exact synthesis: encodings, topology families, and parallelism. IEEE Trans. Comput. Aided Des. Integr. Circuits Syst. **39**(4), 871–884 (2020). https://doi.org/10.1109/TCAD.2019.2897703
15. Han, H., Somenzi, F.: Alembic: an efficient algorithm for CNF preprocessing. In: Proceedings of the 44th Design Automation Conference, DAC 2007, San Diego, CA, USA, 4–8 June 2007, pp. 582–587. IEEE (2007). https://doi.org/10.1145/1278480.1278628
16. Heule, M., Järvisalo, M., Biere, A.: Clause elimination procedures for CNF formulas. In: Fermüller, C.G., Voronkov, A. (eds.) LPAR 2010. LNCS, vol. 6397, pp. 357–371. Springer, Heidelberg (2010). https://doi.org/10.1007/978-3-642-16242-8_26
17. Heule, M.J.H., Järvisalo, M., Biere, A.: Efficient CNF simplification based on binary implication graphs. In: Sakallah, K.A., Simon, L. (eds.) SAT 2011. LNCS, vol. 6695, pp. 201–215. Springer, Heidelberg (2011). https://doi.org/10.1007/978-3-642-21581-0_17
18. Kinyon, M.: Proof simplification and automated theorem proving. Phil. Trans. R. Soc. A **377**(2140), 20180034 (2019)
19. Kulikov, A.S.: Improving circuit size upper bounds using SAT-solvers. In: Madsen, J., Coskun, A.K. (eds.) 2018 Design, Automation & Test in Europe Conference & Exhibition, DATE 2018, Dresden, Germany, 19–23 March 2018, pp. 305–308. IEEE (2018). https://doi.org/10.23919/DATE.2018.8342026
20. Lynce, I., Silva, J.P.M.: Probing-based preprocessing techniques for propositional satisfiability. In: 15th IEEE International Conference on Tools with Artificial Intelligence (ICTAI 2003), Sacramento, California, USA, 3–5 November 2003, p. 105. IEEE Computer Society (2003). https://doi.org/10.1109/TAI.2003.1250177
21. MiniZinc: a high-level constraint modelling language that allows you to easily express and solve discrete optimisation problems. https://www.minizinc.org/. Accessed 18 May 2024
22. Peirce, C.: Existential graphs: manuscript 514, with commentary by JF Sowa. Self-published by JF Sowa [online] (1909). http://www.jfsowa.com/peirce/ms514.htm
23. Piette, C., Hamadi, Y., Sais, L.: Vivifying propositional clausal formulae. In: Ghallab, M., et al. (eds.) ECAI 2008 - 18th European Conference on Artificial Intelligence, Patras, Greece, 21–25 July 2008, Proceedings. Frontiers in Artificial Intelli-

gence and Applications, vol. 178, pp. 525–529. IOS Press (2008). https://doi.org/10.3233/978-1-58603-891-5-525
24. Reger, G.: Better proof output for Vampire. In: Kovács, L., Voronkov, A. (eds.) Vampire@IJCAR 2016. Proceedings of the 3rd Vampire Workshop, Coimbra, Portugal, 2 July 2016. EPiC Series in Computing, vol. 44, pp. 46–60. EasyChair (2016). https://doi.org/10.29007/5dmz
25. Roberts, D.D.: The existential graphs of Charles S. Peirce. Ph.D. thesis, University of Illinois at Urbana-Champaign (1963)
26. Roberts, D.D.: The Existential Graphs of Charles S. Peirce. Mouton (1973)
27. Shin, S.: Reconstituting beta graphs into an efficacious system. J. Log. Lang. Inf. **8**(3), 273–295 (1999). https://doi.org/10.1023/A:1008303204427
28. Soeken, M., et al.: Practical exact synthesis. In: Madsen, J., Coskun, A.K. (eds.) 2018 Design, Automation & Test in Europe Conference & Exhibition, DATE 2018, Dresden, Germany, 19–23 March 2018, pp. 309–314. IEEE (2018). https://doi.org/10.23919/DATE.2018.8342027
29. Soos, M., Nohl, K., Castelluccia, C.: Extending SAT solvers to cryptographic problems. In: Kullmann, O. (ed.) SAT 2009. LNCS, vol. 5584, pp. 244–257. Springer, Heidelberg (2009). https://doi.org/10.1007/978-3-642-02777-2_24
30. Sowa, J.F.: Peirce's tutorial on existential graphs. Semiotica **2011**(186), 347–394 (2011)
31. Stoffelen, K.: Optimizing S-Box implementations for several criteria using SAT solvers. In: Peyrin, T. (ed.) FSE 2016. LNCS, vol. 9783, pp. 140–160. Springer, Heidelberg (2016). https://doi.org/10.1007/978-3-662-52993-5_8
32. Thakur, M., Tripathi, R.: Linear connectivity problems in directed hypergraphs. Theor. Comput. Sci. **410**(27-29), 2592–2618 (2009). https://doi.org/10.1016/j.tcs.2009.02.038
33. The Coq Development Team: The Coq reference manual — release 8.18.0 (2023). https://coq.inria.fr/doc/
34. Vampire: an automatic theorem prover for first-order classical logic. https://vprover.github.io/. Accessed 18 May 2024
35. Zeman, J.J.: The graphical logic of C. S. Peirce. Ph.D. thesis, The University of Chicago (1964)

Combined Abstract Congruence Closure for Theories with Associativity or Commutativity

Christophe Ringeissen[ID] and Laurent Vigneron[✉][ID]

Université de Lorraine, CNRS, Inria, Loria, 54000 Nancy, France
{Christophe.Ringeissen,Laurent.Vigneron}@loria.fr

Abstract. Formal verification techniques, such as deductive verification, require a logic-based tool support to discharge proof obligations, in other words, satisfiability procedures. We design satisfiability procedures thanks to congruence closure methods applied to unions of axiomatized theories, targeting equational axioms such as Associativity or Commutativity. In the proposed approach, any function symbol can be uninterpreted, associative only, commutative only, but also associative and commutative. To tackle the union of these theories, we introduce a combined congruence closure procedure that can be viewed as a particular Nelson-Oppen combination method using particular congruence closure procedures for the individual theories. In this context, we consider terminating congruence closure procedures, but also non-terminating ones. Hence, we have terminating ones for Commutativity and Associativity-Commutativity, while the one for Associativity is non-terminating. We show how all the congruence closure procedures, including the combined one, can be presented in a uniform and abstract way.

Keywords: Satisfiability Procedure · Congruence Closure · Commutativity · Associativity · Union of Theories

1 Introduction

Formal verification techniques, such as deductive verification, require a logic-based tool support to discharge proof obligations, in other words, decision procedures to check whether a logic formula is satisfiable or not. Satisfiability Modulo Theories [5] (SMT, for short) is nowadays a very active research field, where many efficient SMT solvers are developed concurrently following a common specification format, the SMT-Lib. In this context, several theories are very useful in practice, such as the theory of equality, also called the empty theory or the theory of uninterpreted function symbols. This particular theory admits an efficient satisfiability procedure based on the construction of a congruence closure [14,15].

For sake of expressiveness, it is highly desirable to cope with function symbols that are interpreted in some given theory. For instance, this given theory could be a large fragment of arithmetic, such as Linear Rational Arithmetic.

But this theory could be also specified by some equational axioms. For example, Associativity-Commutativity is particularly interesting to reason modulo (multi)sets, Associativity allows us to reason modulo lists, and Commutativity is useful to reason modulo a symmetric relation.

Compared to Associativity-Commutativity and Commutativity where unification is finitary, Associativity is a challenging theory since associative unification is known to be infinitary. Despite the lack of an associative unification algorithm, it is possible to get useful procedures to reason modulo Associativity, as shown in [12] for the variant-based unification and satisfiability problems. Analogously, we believe that Associativity is a theory deserving to be investigated in the context of a combined congruence closure procedure.

We design rule-based congruence closure procedures modulo union of axiomatized theories, targeting equational axioms such as Associativity or Commutativity. In the proposed approach, any function symbol can be uninterpreted, associative only, flat permutative only (e.g. commutative), but also associative and commutative. To tackle the union of these theories, we introduce a combined congruence closure procedure that can be viewed as a particular Nelson-Oppen combination method [13] using particular congruence closure procedures for the individual theories. The combined congruence procedure is based on the ping-ponging of entailed equalities between (shared) constants. Actually, the congruence closure procedures used for the individual theories allow us to deduce these equalities. In this context, we consider terminating congruence closure procedures, but also non-terminating ones. Hence, we have terminating congruence closure procedures for flat permutation and Associativity-Commutativity, while the one for Associativity is non-terminating. We show how all the congruence closure procedures, including the combined one, can be presented in a uniform and abstract way along the lines of [3, 8–11].

Related Work. Congruence closure modulo Associativity-Commutativity has been successfully investigated in [3, 8]. It has been revisited more recently, showing how the method can be extended to take into account additional orientable axioms, for instance to handle the theory of Abelian Groups [9]. The case of flat permutative axioms, such as Commutativity, has been considered in [11]. The theory of Groups and all of its subtheories including Associativity is considered in [10], where the related congruence closure procedure is not necessarily terminating, contrarily to the one known for Associativity-Commutativity. In these papers, some particular unions of theories are studied, for instance to handle several symbols following the same equational axioms.

In our paper, we clearly focus on the combination of congruence closure procedures to cope with arbitrary unions of theories (sharing only constants). This combination of congruence closure procedures can be seen as a particular case of combination of deduction-complete satisfiability procedures, already investigated in [20]. In addition to Associativity-Commutativity, we believe that it is interesting to consider Associativity alone and flat permutation alone. On one hand, Associativity provides a significant case study of a non-terminating congruence closure procedure. On the other hand, flat permutation (such as Commutativity) leads to a simple extension of the congruence closure procedure known for the theory of equality as done in [11].

Paper Outline. The paper is organized as follows. Section 2 introduces the main notions and notations related to terms, term rewriting and satisfiability problems. In Sect. 3, we describe our combination method using two kinds of processes: the orchestrator whose role is to prepare and handle a combination of theories; a theory process whose role is to complete the set of rewrite rules for a specific theory. The supported theories are detailed in Sect. 4. The completeness of the method is discussed in Sect. 5. The method is not necessarily terminating, as shown in Sect. 6 in the case of Associativity. In Sect. 7, we detail a running example involving an associative-commutative operator and a commutative one. Eventually, Sect. 8 concludes discussing some possible future developments.

2 Preliminaries

We assume the reader familiar with the notions of terms and term rewriting [1].

We consider n theories $\mathcal{E}_1, \ldots, \mathcal{E}_n$ such that each theory $\mathcal{E}_i$ is defined by a set of equalities over a signature Σ_i.[1] The theories $\mathcal{E}_1, \ldots, \mathcal{E}_n$ are assumed to be pairwise signature-disjoint or to share only constants, meaning that $\Sigma_i \cap \Sigma_j$ is a set of constants which can be empty, for any $i, j \in [1, n], i \neq j$. The union of theories $\mathcal{E}_1 \cup \cdots \cup \mathcal{E}_n$ is denoted by $\mathcal{E}$ and the union of signatures $\Sigma_1 \cup \cdots \cup \Sigma_n$ is denoted by Σ. We assume a set of ground equalities Γ and a set of ground disequalities Δ, where both Γ and Δ are expressed over the signature Σ. Given any theory $\mathcal{E}'$, $=_{\mathcal{E}'}$ denotes the congruence closure of the equalities in $\mathcal{E}'$ under the law of substitutivity.

Terms. We denote $T(\Sigma)$ the set of *ground* (without variables) terms built over the signature Σ. A *constant* is a function symbol of Σ with arity 0. The set of positions in a term t is written $O(t)$. The subterm of a term t at position p is written $t|_p$. A subterm of t is *strict* if its position is not at the top of t. The term obtained from t by replacing $t|_p$ by a term s is written $t[s]_p$.

The process described in this paper relies on a flattening of terms. For theory $\mathcal{E}_i$ including an operator, say $+$, such that $(x + y) + z \approx x + (y + z)$ occurs in $\mathcal{E}_i$, this flattening will be performed using $+$ as a variadic operator, e.g. $a + (b + c)$ is flattened into $+(a, b, c)$.

Equalities and Rewrite Rules. The initial set of ground equalities Γ will be purified via flattening thanks to the introduction of new constants as in [7] (K denotes the set of used new constants taken from an infinite countable set U disjoint from Σ), generating pure flat rewrite rules for each theory $\mathcal{E}_i$ (denoted by the set R_i); and further deductions between those rules may generate flat equalities in this theory (denoted by the set E_i).

The rewrite rules in R_i can have two shapes: *D-rules* denoted by $f(c_1, \ldots, c_n) \to c$, where $f \in \Sigma_i$ and $c_1, \ldots, c_n, c \in K$; *E-rules* denoted by $f(c_1, \ldots, c_m) \to$

---

[1] The empty theory is defined by a signature, but with an empty set of equalities.

$f(d_1,\ldots,d_n)$, where $f \in \Sigma_i$ is a variadic operator and c_1, ..., c_m, d_1, ..., $d_n \in K$. The equalities in E_i have the same two shapes: *D-equalities* denoted by $f(c_1,\ldots,c_n) \approx c$; *E-equalities* denoted by $f(c_1,\ldots,c_m) \approx f(d_1,\ldots,d_n)$. An equality $c_1 \approx c_2$ between two constants of K is called a *C-equality*. E-rules and E-equalities will be generated only for variadic operators by the Superposition inference rule, because of the use of extended rewrite rules (see Sect. 3.2).

Considering a theory $\mathcal{E}_i$, if two terms t_1 and t_2 are $\mathcal{E}_i$-equal, we write $t_1 \leftrightarrow^*_{\mathcal{E}_i} t_2$. By $(R_i, \mathcal{E}_i)$ we denote the rewriting system defined by $\{u' \to v \mid u \to v \in R_i \text{ and } u' \leftrightarrow^*_{\mathcal{E}_i} u\}$. And by $(R_i^e, \mathcal{E}_i)$ we denote the rewriting system extending $(R_i, \mathcal{E}_i)$ with all possible extended rewrite rules from R_i.

Ordering. For any rewrite rule $t \to s$, t has to be greater than s ($t \succ s$); the definition of an ordering may be difficult for deduction systems modulo equational theories; but in our case the ordering is very simple as we only have to consider D-rules and E-rules: for D-rules, it suffices to assume $\forall f \in \Sigma, \forall c \in K, f \succ c$; for E-rules, we have to compare lists of constants: if of the same length, this can be done with a lexicographic or a multiset extension of an arbitrary ordering comparing two constants of K (the choice is done for each theory), and if of different length, the longest is the greatest. For example, for an associative theory the lexicographic extension will be used, and for an associative-commutative theory the multiset extension will be used.

The *lexicographic extension* of an ordering $\succ$ is defined by: $(t_1,\ldots,t_n) \succ^{lex} (s_1,\ldots,s_n)$ if $\exists i \in [1,n], \forall j < i, t_j =_{\mathcal{E}} s_j$ and $t_i \succ s_i$.

The *multiset extension* of an ordering $\succ$ is defined by: $T = \{t_1,\ldots,t_n\} \succ^{mult} \{s_1,\ldots,s_n\} = S$ if, given I the multiset $\mathcal{E}$-intersection of T and S (where the equality is checked w.r.t. $=_{\mathcal{E}}$), $\forall s_i \in S \setminus I, \exists t_j \in T \setminus I, t_j \succ s_i$.

Satisfiability Problems. In the context of this paper, a satisfiability problem is given by a finite set of ground literals, where a literal is either an equality or a disequality. Given a theory T, a satisfiability problem φ expressed over the signature of T is said to be *T-satisfiable* if $T \cup \varphi$ is consistent, i.e. $T \cup \varphi$ admits a model. We say that a model is *trivial* when its domain is of cardinality 1. A consistent theory is *trivial* when all its models are trivial. Notice that any equational theory is consistent since it admits a trivial model. The equational theories considered in this paper are non-trivial.

3 Combined Satisfiability Procedure

We describe in this section a procedure that aims at (semi-)deciding the satisfiability of any set of ground equalities Γ together with any set of ground disequalities Δ, modulo a combination of signature-disjoint equational theories $\mathcal{E}_i$ (or sharing only constants). This procedure, called CombCC, is based on congruence closure and involves two kinds of processes: an orchestrator decomposing the problem to separate the different theories, and theory processes that complete rewrite rules, one process for each theory.

3.1 The Orchestrator

The role of the orchestrator is to purify and flatten the problem to be solved, to send each theory process the rewrite rules it has to handle, and to detect if any contradiction w.r.t. Δ is generated by one of the theory processes. For this purpose, several sets are handled in addition to Γ and Δ: the set of new constants K, and for each equational theory $\mathcal{E}_i$ a set of rewrite rules R_i and a set of equalities E_i. We denote RE the set of (R_i, E_i) for all theories $\mathcal{E}_i$. In the following inference rules, we will only indicate the involved sets.

The first task of the orchestrator is to transform the disequalities for "hiding" the theories involved. This is done with the following inference rule that replaces an arbitrary disequality by a disequality between two new constants together with the equalities associating each of these constants to the corresponding term:

Splitting: $\dfrac{K, \Delta \cup \{t_1 \not\approx t_2\}, \Gamma}{K \cup \{c_1, c_2\}, \Delta \cup \{c_1 \not\approx c_2\}, \Gamma \cup \{t_1 \approx c_1, t_2 \approx c_2\}}$

if $t_1, t_2 \notin K$, $c_1, c_2 \in U \setminus K$.

Once all disequalities have been decomposed, the second task of the orchestrator is to purify the equalities of Γ, by generating rewrite rules that are purely in one theory. For this purpose, it applies the following inference rules:

Flattening: $\dfrac{K, \Gamma[t], R_i}{K \cup \{c\}, \Gamma[c], R_i \cup \{t \to c\}}$

if $t \to c$ is a D-rule, $c \in U \setminus K$, t occurs in some equality in Γ that is not a D-equality, and t is Σ_i-rooted.

Orientation: $\dfrac{K \cup \{c\}, \Gamma \cup \{t \approx c\}, R_i}{K \cup \{c\}, \Gamma, R_i \cup \{t \to c\}}$ if $t \approx c$ is a D-equality and t is Σ_i-rooted.

When all equations have been transformed ($\Gamma = \emptyset$), the orchestrator runs one process per equational theory $\mathcal{E}_i$, providing it two sets of information: the set of new constants K and the set of D-rules R_i defined over Σ_i and K.

Its final task is to manage equalities between new constants, when generated by a theory process in some set E_i; there are two possibilities: if the equality contradicts a disequality of Δ then the system stops, otherwise one of the constants has to be replaced by the other one in all sets.

Contradiction: $\dfrac{K \cup \{c, d\}, \Delta \cup \{c \not\approx d\}, RE \cup \{(R_i, E_i \cup \{c \approx d\})\}}{\bot}$

Compression: $\dfrac{K \cup \{c, d\}, \Delta, RE \cup \{(R_i, E_i \cup \{c \approx d\})\}}{K \cup \{d\}, \Delta\langle c \mapsto d\rangle, RE\langle c \mapsto d\rangle \cup \{(R_i\langle c \mapsto d\rangle, E_i\langle c \mapsto d\rangle)\}}$

if $c \approx d \notin \Delta$ and $c \succ d$; the notation $\langle c \mapsto d \rangle$ denotes the homomorphic extension of the mapping σ defined as $\sigma(c) = d$ and $\sigma(x) = x$ for $x \neq c$, and $S\langle c \mapsto d\rangle$ denotes the set obtained by applying the mapping $\langle c \mapsto d\rangle$ to each term in set S.

The strategy of the orchestrator can therefore be described by: **Split**$^*\cdot$(**Flat**$^*\cdot$ **Ori**)$^* \cdot$ (**Cont**|**Comp**)*.

3.2 A Theory Process

A process run for an equational theory $\mathcal{E}_i$ will use inference rules to complete its term rewriting system R_i. Some inference rules are used for transforming the rewrite rules (Composition), for deducing new equalities added to a set E_i (Collapse, Superposition), or for handling those new equalities (Simplification, Orientation, Deletion). The set of new constants K is never modified, so not indicated, but it is useful in the process for checking if a constant is a new one or belongs to the theory.

For some theories, the inference system has to consider extended rewrite rules as we do not explicitly use the axioms of a theory: an extension is built w.r.t. a context defined from the theory axioms; a context is a triplet: a term *Cont*, a non variable position p of a strict subterm in *Cont*, and a set of $\mathcal{E}_i$-unification constraints UC. Let us denote $Cont_{\mathcal{E}_i}$ the set of contexts for the theory $\mathcal{E}_i$; given a D-rule or a E-rule $u \to v$, its extended version by a context $(Cont, p, UC) \in Cont_{\mathcal{E}_i}$ is written $Cont[u]_p\sigma \to Cont[v]_p\sigma$, where σ is the ground substitution in a minimal complete set of $\mathcal{E}_i$-unifiers of $Cont|_p =^?_{\mathcal{E}_i} u$ and UC. These notions of extended rewrite rules and contexts were already mentioned in [16,17], and the construction of contexts for generating extensions has been explained in [21]. The principle is to start by identifying non variable positions in the left-hand and right-hand sides of axioms of the theory, giving an initial set of contexts. Then those contexts are combined by unification from one context into the marked position of another context, and so on; each combination adds unification constraints in the new context. A notion of redundant context has been defined for keeping only useful contexts: a redundant context will create extended rewrite rules that will generate by deduction either redundant equalities/rules, or equalities/rules that can be generated using already existing (simpler) contexts.

In this paper, as we want to handle only flat rewrite rules, we will consider only theories for which extended rewrite rules have a flat form. For example, if an operator f is associative, from the axiom of this theory $f(f(x,y),z) \approx f(x,f(y,z))$, we can build three contexts:

$$(f(f(y_1,y_2),x_1),1,\emptyset),\ (f(x_2,f(y_1,y_2)),2,\emptyset) \text{ and } (f(x_3,f(y_1,y_2),x_4),2,\emptyset)$$

In each one, the left-hand side of the rewrite rule to be extended will have to unify with $f(y_1,y_2)$. So, a rewrite rule $f(a,b) \to c$ has three extensions, written in flat form: $f(a,b,x_1) \to f(c,x_1)$, $f(x_2,a,b) \to f(x_2,c)$ and $f(x_3,a,b,x_4) \to f(x_3,c,x_4)$.

The inference rules used by a theory process are the following.

Simplification: $\dfrac{R_i, E_i[t]}{R_i, E_i[s]}$ where t occurs in some equality of E_i, and $t \to_{(R_i^e, \mathcal{E}_i)} s$.

Orientation: $\dfrac{R_i, E_i \cup \{t \approx s\}}{R_i \cup \{t \to s\}, E_i}$ if $t \succ s$ and $t \to s$ is a D-rule or a E-rule.

Deletion: $\dfrac{R_i, E_i \cup \{t \approx s\}}{R_i, E_i}$ if $t \leftrightarrow^*_{\mathcal{E}_i} s$.

Composition: $\dfrac{R_i \cup \{t \to s, u \to v\}, E_i}{R_i \cup \{t \to s', u \to v\}, E_i}$ if $s \to_{(\{u \to v\}^e, \mathcal{E}_i)} s'$.

Collapse: $\dfrac{R_i \cup \{t \to s, u \to v\}, E_i}{R_i \cup \{u \to v\}, E_i \cup \{t' \approx s\}}$ if $t \to_{(\{u \to v\}^e, \mathcal{E}_i)} t'$, and if $t \leftrightarrow^*_{\mathcal{E}_i} u$ then $s \succ v$.

Superposition: $\dfrac{R_i \cup \{t_1 \to s_1, t_2 \to s_2\}, E_i}{R_i \cup \{t_1 \to s_1, t_2 \to s_2\}, E_i \cup \{Cont_1[s_1]_{p_1}\sigma \approx Cont_2[s_2]_{p_2}\sigma\}}$

if $(Cont_1, p_1, UC_1), (Cont_2, p_2, UC_2) \in Cont_{\mathcal{E}_i}$, and the substitution σ is the ground substitution in a minimal complete set of $\mathcal{E}_i$-unifiers of $Cont_1[t_1]_{p_1} =^?_{\mathcal{E}_i} Cont_2[t_2]_{p_2}$, $Cont_1|_{p_1} =^?_{\mathcal{E}_i} t_1$, $Cont_2|_{p_2} =^?_{\mathcal{E}_i} t_2$, UC_1 and UC_2; the resulting equality will be written in flat form. Note that according to the theory $\mathcal{E}_i$, the contexts may be selected to guarantee a useful ground new equality (see Sect. 4.2).

A strategy for combining all those inference rules is: $(\mathbf{Com}^* \cdot (\mathbf{Col}|\mathbf{Sup}) \cdot \mathbf{Sim}^* \cdot (\mathbf{Del}|\mathbf{Ori}))^*$

So this process handles a pair (R_i, E_i): it starts with $(R_i, \emptyset)$ and, if terminating, it ends with $(R_i^\infty, \emptyset)$ where there is no more possible inference rule involving rules of R_i^∞. If an equality between two constants of K is generated, it will be handled by the orchestrator.

For applying inference rules, this theory process has to use a $\mathcal{E}_i$-matching algorithm for applying rewriting steps w.r.t. $(R_i^e, \mathcal{E}_i)$. It also needs a $\mathcal{E}_i$-unification algorithm, but which will have to solve only simple unification problems, of the shape $Cont_1[t_1]_{p_1} =^?_{\mathcal{E}_i} Cont_2[t_2]_{p_2}$, where t_1 and t_2 are ground; if there is a solution, it will be the unique most general unifier since the variables occurring in $Cont_i[\cdot]_{p_i}$ will be instantiated by subterms of the ground term t_{3-i}.

4 Supported Theories

We detail now some of the input theories $\mathcal{E}_i$ that allow us to get a complete combined satisfiability procedure. We consider three kinds of theories (in addition to the empty theory of course), and discuss also about additional ones.

4.1 Flat Permutative Theories

A flat permutative theory is represented by a set of axioms

$$\{f(x_1, \ldots, x_k) \approx f(x_{\sigma(1)}, \ldots, x_{\sigma(k)}) \mid \sigma \text{ is a permutation of } \{1, \ldots, k\}\}$$

where x_i are variables that do not need to be all distinct.

With such a theory, there is no extension of rewrite rules to be considered, so the Superposition inference rule cannot apply. For the ordering, we can compare the arguments of two f-terms with a lexicographic extension of the ordering between constants of K, and always use the biggest term of the equivalence class w.r.t. this theory.

Commutative theories are a particular case of flat permutative theories where all x_i are distinct variables and any permutation is possible; for the ordering we can use a multiset extension of the ordering between constants of K.

4.2 Associative Theories

An associative theory is represented by the axiom

$$f(f(x_1, x_2), x_3) \approx f(x_1, f(x_2, x_3))$$

This axiom generates three possible extensions of rewrite rules whose left-hand side is a f-term, with the contexts $(f(f(y_1, y_2), x_1), 1, \emptyset)$, $(f(x_2, f(y_1, y_2)), 2, \emptyset)$ and $(f(x_3, f(y_1, y_2), x_4), 2, \emptyset)$. Those three contexts can be used for applying term rewriting steps w.r.t. $(R_i^e, \mathcal{E}_i)$. But for the Superposition inference rule between two rules $t_1 \to s_1$ and $t_2 \to s_2$, we only need to consider their extensions $f(t_1, x_1) \to f(s_1, x_1)$ and $f(x_2, t_2) \to f(x_2, s_2)$ because this is the only combination of contexts for which the unification problem $(f(t_1, x_1) =^?_{\mathcal{E}_i} f(x_2, t_2))$ can generate a ground most general unifier, and any other combination of contexts would generate a redundant equation. For the ordering, the arguments of an associative operator are compared with a lexicographic extension of the ordering between constants of K.

4.3 Associative-Commutative Theories

An associative-commutative theory is represented by axioms

$$f(x_1, x_2) \approx f(x_2, x_1) \text{ and } f(f(x_1, x_2), x_3) \approx f(x_1, f(x_2, x_3))$$

It generates only one possible extension of rewrite rules whose left-hand side is a f-term, with the context $(f(f(y_1, y_2), x_1), 1, \emptyset)$, used for applying term rewriting steps w.r.t. $(R_i^e, \mathcal{E}_i)$, and the Superposition inference rule. For the ordering, the arguments of an associative-commutative operator are compared with a multiset extension of the ordering between constants of K.

4.4 Additional Theories

The theory process defined in the previous section is very general and can apply to many theories. We have given above three examples of theories that are often necessary to theorem provers or SMT solvers.

It is possible to define some extensions to those three theories by considering additional axioms that will be represented by a set of *flat collapsing* rewrite rules

$R_{\mathcal{E}_i}$ which is confluent and terminating, and so that those axioms do not introduce new contexts w.r.t. $Cont_{\mathcal{E}_i}$, and extended rules from $R_{\mathcal{E}_i}$ are redundant: if $t \to^*_{(R^e_{\mathcal{E}_i}, \mathcal{E}_i)} s$ and s is in normal form w.r.t. $(R^e_{\mathcal{E}_i}, \mathcal{E}_i)$, then $t \to^*_{(R_{\mathcal{E}_i}, \mathcal{E}_i)} s$. The right-hand side of a rewrite rule of $R_{\mathcal{E}_i}$ is either a subterm of the left-hand side, or a constant. So their orientation from left to right does not need a specific ordering.

The building of $R_{\mathcal{E}_i}$ from the added axioms uses the Flattening and Orientation inference rules of the orchestrator; so, non-variable subterms in $R_{\mathcal{E}_i}$ are new constants of K and they may be concerned by the Compression inference rule. The use of $R_{\mathcal{E}_i}$ in the theory process consists in normalizing w.r.t. $(R_{\mathcal{E}_i}, \mathcal{E}_i)$ every equation or rule deduced.

The conditions defining those extensions are very restrictive. The reason is that therefore they imply only minor modifications in our procedure. Without such conditions, we would have either to handle additional contexts together with extended matching and unification algorithms, or to add many rewrite rules for each deduced rule as done in [9] for considering idempotency or nilpotency in addition to associativity-commutativity. But even with those conditions, some interesting examples can be considered. For theories generating contexts (such as those including associativity of an operator f), extensions can include axioms such as $f(x,1) \approx x$ or $f(x,0) \approx 0$. For theories that do not generate contexts (such as the commutativity of an operator f), extensions can include axioms like $f(x,x) \approx x$ or $f(x,x) \approx 0$.

5 Completeness of CombCC

The CombCC procedure is refutationally complete, provided that deductions are fairly applied. Moreover, if the CombCC procedure terminates without finding a contradiction with disequalities of Δ, it generates a terminating confluent term rewriting system for the equational theory $\mathcal{E} \cup \Gamma$.

Theorem 1. *Let $\mathcal{E}$ be any disjoint union of empty, flat permutative, associative, and associative-commutative theories over the combined signature Σ which may include uninterpreted function symbols and constants. Consider Γ is any set of ground Σ-equalities and Δ is any set of ground Σ-disequalities. Given the input $\Gamma \cup \Delta$, the CombCC procedure halts on $\bot$ iff $\Gamma \cup \Delta$ is $\mathcal{E}$-unsatisfiable. If the CombCC procedure halts on an output distinct from $\bot$, then $\Gamma \cup \Delta$ is $\mathcal{E}$-satisfiable, and the output provides a rewriting system R such that (1) R is terminating and confluent modulo $\mathcal{E}$ on $T(\Sigma \cup K)$, and (2) any two ground terms in $T(\Sigma)$ are $\mathcal{E} \cup R$-equal iff they are $\mathcal{E} \cup \Gamma$-equal. Moreover, the CombCC procedure is necessarily terminating if $\mathcal{E}$ does not involve associative theories.*

To prove the completeness of CombCC, we can rely on a Nelson-Oppen combination method [13] based on the ping-ponging of entailed equalities between (shared) constants. This combination method is applicable without loss of completeness because one can rely on a union of convex and stably infinite theories.

The same proof idea as the one initiated in [2] can be reused, by considering two cases. In the simple case of the trivial model, a set of literals is satisfiable if and only if it is a set of equalities. In the general case excluding the trivial model, the underlying theories are stably infinite.

Lemma 1 ([2]). *Let $\mathcal{E}_i$ be any non-trivial equational Σ_i-theory, Γ any finite set of ground Σ_i-equalities, Δ any finite set of ground Σ_i-disequalities, and a, b two distinct fresh constants not occurring in $\mathcal{E}_i$, Γ and Δ. Then, we have:*

(i) $\mathcal{E}_i \cup \{a \not\approx b\}$ is convex and stably infinite,
(ii) $\Gamma \cup \Delta$ is $(\mathcal{E}_i \cup \{a \not\approx b\})$-unsatisfiable iff $\Gamma \cup \Delta \cup \{a \not\approx b\}$ is $\mathcal{E}_i$-unsatisfiable.

Proof. The statement (ii) is straightforward, so let us focus on (i): $\mathcal{E}_i \cup \{a \not\approx b\}$ is convex since it is a Horn theory and any Horn theory is known to be convex [19]. Moreover, $\mathcal{E}_i \cup \{a \not\approx b\}$ is stably infinite since any convex theory with no trivial models is known to be stably infinite [4].

The convexity induces a particular way to decide the satisfiability of equalities plus a conjunction of disequalities: it allows us to consider each disequality separately. This is a direct rewording of the definition of convexity.

Lemma 2. *Let T be any convex theory. For any T-satisfiable finite set Γ of ground equalities, and any finite set Δ of ground disequalities, both expressed over the signature of T, we have that $\Gamma \cup \Delta$ is T-unsatisfiable iff there exists some disequality $s \not\approx t$ in Δ such that $\Gamma \cup \{s \not\approx t\}$ is T-unsatisfiable.*

In CombCC, the input satisfiability problem is transformed via Splitting and Flattening into an equisatisfiable problem including only flat literals, meaning that all the disequalities in Δ are of the form $c \not\approx d$ where c and d are constants. Thus, we are looking for inference systems with the property of being deduction-complete [20], in order to derive each equality $c \approx d$ such that $\Gamma \to c \approx d$ is valid in the underlying theory. This is exactly the purpose of a congruence closure procedure when it applies to an input set of flat equalities Γ. It generates all the equalities between constants that are logically entailed by Γ.

Compared to a classical application of the Nelson-Oppen combination method, we have to accommodate congruence closure procedures that are not necessarily terminating since $\mathcal{E}$ may include associative theories for which non-terminating derivations are shown in Sect. 6. Let us shortly explain why CombCC is refutationally complete. According to the completeness of the Nelson-Oppen method, the satisfiability problem in any disjoint union of stably infinite theories is reducible to the satisfiability problems in the component theories, provided that all possible arrangements are guessed. Consequently, given any disjoint union of stably infinite theories, using refutationally complete procedures for the satisfiability problems in the component theories allows us to get a refutationally complete procedure for the satisfiability problem in the union. In our context, stably infinite theories are also convex and so the guessing of all possible arrangements can be replaced by a ping-ponging of entailed equalities between constants. Then, we use the property that all the entailed equalities between constants are eventually generated since our congruence closure procedures are deduction-complete.

Remark 1 (About theories sharing constants). Our combination procedure permits theories to share constants, as exemplified in Sect. 4.4 by the addition of axioms. Let us briefly explain how those shared constants are handled. Assume that a constant a is shared by two theories $\mathcal{E}_1$ and $\mathcal{E}_2$. The orchestrator will replace this constant by a new one by Flattening, generating a D-rule $a \to c$ ($c \in K$). Then it will include this D-rule in the set of rewrite rules of each theory (R_1 and R_2). The theory process of $\mathcal{E}_i$ may have to use this D-rule for transforming rules of R_i and equations in E_i, eliminating all other occurrences of constant a. And if later on an equation $c \approx d$ is generated, it will be propagated in all other theory processes by the Compression inference rule of the orchestrator. Thus, considering theories with shared constants preserves the way our CombCC procedure works and its refutational completeness.

6 On Non-termination for Associative Theories

Our procedure may not terminate (if no contradiction exists) with associative theories. For example, we consider the combination of two theories, one with an associative operator f, and the empty theory with constants a, b, c and d, and we are given the equalities

$$\Gamma = \{f(a,b) \approx c, f(d,a) \approx c, f(a,c) \approx f(c,a)\}$$

For any precedence ordering between the four constants, an infinite number of rewrite rules is generated as illustrated below by the following case study, where traces are rebuilt from the output of our implementation.[2]

If $d \prec c$, the E-rules $f(c, d^n, c) \to f(d^n, c, c)$ are generated as follows:

```
> Equality: f(a,b) = c
  D-rule f(a,b) → c generated (Orientation)
> Equality: f(d,a) = c
  D-rule f(d,a) → c generated (Orientation)
> Equality: f(a,c) = f(c,a)
  E-rule f(a,c) → f(c,a) generated (Orientation)
> Equality f(c,b) = f(d,c) generated
  (Superposition between f(d,a) → c and f(a,b) → c)
> Equality: f(c,b) = f(d,c)
  E-rule f(c,b) → f(d,c) generated (Orientation)
> Equality f(c,c) = f(d,c,a) generated
  (Superposition between f(d,a) → c and f(a,c) → f(c,a))
> Equality: f(c,c) = f(d,c,a)
  E-rule f(d,c,a) → f(c,c) generated (Orientation)
> Equality f(c,c,b) = f(d,c,c) generated
  (Superposition between f(d,c,a) → f(c,c) and f(a,b) → c)
```

---

[2] For clarity, constants a, b, c, d have not been replaced by new constants in the traces.

> Equality: $f(c,c,b) = f(d,c,c)$
  Subterm $f(c,c,b)$ simplified into $f(c,d,c)$ (with $f(c,b) \to f(d,c)$)
  E-rule $f(c,d,c) \to f(d,c,c)$ generated (Orientation)
> Equality $f(d,c,c,b) = f(c,d,d,c)$ generated
  (Superposition between $f(c,d,c) \to f(d,c,c)$ and $f(c,b) \to f(d,c)$)
> Equality: $f(d,c,c,b) = f(c,d,d,c)$
  Subterm $f(d,c,c,b)$ simplified into $f(d,c,d,c)$ (with $f(c,b) \to f(d,c)$)
  Subterm $f(d,c,d,c)$ simplified into $f(d,d,c,c)$ (with $f(c,d,c) \to f(d,c,c)$)
  E-rule $f(c,d,d,c) \to f(d,d,c,c)$ generated (Orientation)
...
> Equality $f(d^n,c,c,b) = f(c,d^{n+1},c)$ generated
  (Superposition between $f(c,d^n,c) \to f(d^n,c,c)$ and $f(c,b) \to f(d,c)$)
  %% extended term $f(c,d^n,c,b)$ generates $f(d^n,c,c,b) = f(c,d^n,d,c)$
> Equality $f(d^n,c,c,b) = f(c,d^{n+1},c)$
  Subterm $f(d^n,c,c,b)$ simplified into $f(d^n,c,d,c)$ (with $f(c,b) \to f(d,c)$)
  Subterm $f(d^n,c,d,c)$ simplified into $f(d^{n+1},c,c)$ (with $f(c,d,c) \to f(d,c,c)$)
  E-rule $f(c,d^{n+1},c) \to f(d^{n+1},c,c)$ generated (Orientation)

Else ($d \succ c$), if $c \prec b$, the E-rules $f(c,b^n,c) \to f(c,c,b^n)$ are generated as follows:

> Equality: $f(a,b) = c$
  D-rule $f(a,b) \to c$ generated (Orientation)
> Equality: $f(d,a) = c$
  D-rule $f(d,a) \to c$ generated (Orientation)
> Equality: $f(a,c) = f(c,a)$
  E-rule $f(a,c) \to f(c,a)$ generated (Orientation)
> Equality $f(c,b) = f(d,c)$ generated
  (Superposition between $f(d,a) \to c$ and $f(a,b) \to c$)
> Equality: $f(c,b) = f(d,c)$
  E-rule $f(d,c) \to f(c,b)$ generated (Orientation)
> Equality $f(c,c) = f(d,c,a)$ generated
  (Superposition between $f(d,a) \to c$ and $f(a,c) \to f(c,a)$)
> Equality: $f(c,c) = f(d,c,a)$
  Subterm $f(d,c,a)$ simplified into $f(c,b,a)$ (with $f(d,c) \to f(c,b)$)
  E-rule $f(c,b,a) \to f(c,c)$ generated (Orientation)
> Equality $f(c,c,b) = f(c,b,c)$ generated
  (Superposition between $f(c,b,a) \to f(c,c)$ and $f(a,b) \to c$)
> Equality: $f(c,c,b) = f(c,b,c)$
  E-rule $f(c,b,c) \to f(c,c,b)$ generated (Orientation)
> Equality $f(c,b,b,c) = f(d,c,c,b)$ generated
  (Superposition between $f(d,c) \to f(c,b)$ and $f(c,b,c) \to f(c,c,b)$)
> Equality: $f(c,b,b,c) = f(d,c,c,b)$
  Subterm $f(d,c,c,b)$ simplified into $f(c,b,c,b)$ (with $f(d,c) \to f(c,b)$)
  Subterm $f(c,b,c,b)$ simplified into $f(c,c,b,b)$ (with $f(c,b,c) \to f(c,c,b)$)
  E-rule $f(c,b,b,c) \to f(c,c,b,b)$ generated (Orientation)
...
> Equality $f(c,b^{n+1},c) = f(d,c,c,b^n)$ generated
  (Superposition between $f(d,c) \to f(c,b)$ and $f(c,b^n,c) \to f(c,c,b^n)$)
  %% extended term $f(d,c,b^n,c)$ generates $f(c,b,b^n,c) = f(d,c,c,b^n)$
> Equality: $f(c,b^{n+1},c) = f(d,c,c,b^n)$

Subterm $f(d,c,c,b^n)$ simplified into $f(c,b,c,b^n)$ (with $f(d,c) \to f(c,b)$)
Subterm $f(c,b,c,b^n)$ simplified into $f(c,c,b^{n+1})$ (with $f(c,b,c) \to f(c,c,b)$)
E-rule $f(c,b^{n+1},c) \to f(c,c,b^{n+1})$ generated (Orientation)

Else ($d \succ c \succ b$), the E-rules $f(c,b^n,c,b) \to f(c,b,b^n,c)$ are generated as follows:

> Equality: $f(a,b) = c$
D-rule $f(a,b) \to c$ generated (Orientation)
> Equality: $f(d,a) = c$
D-rule $f(d,a) \to c$ generated (Orientation)
> Equality: $f(a,c) = f(c,a)$
E-rule $f(a,c) \to f(c,a)$ generated (Orientation)
> Equality $f(c,b) = f(d,c)$ generated
(Superposition between $f(d,a) \to c$ and $f(a,b) \to c$)
> Equality: $f(c,b) = f(d,c)$
E-rule $f(d,c) \to f(c,b)$ generated (Orientation)
> Equality $f(c,c) = f(d,c,a)$ generated
(Superposition between $f(d,a) \to c$ and $f(a,c) \to f(c,a)$)
> Equality: $f(c,c) = f(d,c,a)$
Subterm $f(d,c,a)$ simplified into $f(c,b,a)$ (with $f(d,c) \to f(c,b)$)
E-rule $f(c,b,a) \to f(c,c)$ generated (Orientation)
> Equality $f(c,c,b) = f(c,b,c)$ generated
(Superposition between $f(c,b,a) \to f(c,c)$ and $f(a,b) \to c$)
> Equality: $f(c,c,b) = f(c,b,c)$
E-rule $f(c,c,b) \to f(c,b,c)$ generated (Orientation)
> Equality $f(c,b,c,b) = f(d,c,b,c)$ generated
(Superposition between $f(d,c) \to f(c,b)$ and $f(c,c,b) \to f(c,b,c)$)
> Equality: $f(c,b,c,b) = f(d,c,b,c)$
Subterm $f(d,c,b,c)$ simplified into $f(c,b,b,c)$ (with $f(d,c) \to f(c,b)$)
E-rule $f(c,b,c,b) \to f(c,b,b,c)$ generated (Orientation)
> Equality $f(c,b,b,c,b) = f(d,c,b,b,c)$ generated
(Superposition between $f(d,c) \to f(c,b)$ and $f(c,b,c,b) \to f(c,b,b,c)$)
> Equality: $f(c,b,b,c,b) = f(d,c,b,b,c)$
Subterm $f(d,c,b,b,c)$ simplified into $f(c,b,b,b,c)$ (with $f(d,c) \to f(c,b)$)
E-rule $f(c,b,b,c,b) \to f(c,b,b,b,c)$ generated (Orientation)
...
> Equality $f(c,b^{n+1},c,b) = f(d,c,b^{n+1},c)$ generated
(Superposition between $f(d,c) \to f(c,b)$ and $f(c,b^n,c,b) \to f(c,b,b^n,c)$)
%% extended term $f(d,c,b^n,c,b)$ generates $f(c,b,b^n,c,b) = f(d,c,b,b^n,c)$
> Equality: $f(c,b^{n+1},c,b) = f(d,c,b^{n+1},c)$
Subterm $f(d,c,b^{n+1},c)$ simplified into $f(c,b,b^{n+1},c)$ (with $f(d,c) \to f(c,b)$)
E-rule $f(c,b^{n+1},c,b) \to f(c,b,b^{n+1},c)$ generated (Orientation)

There are also shorter examples, for instance with two associative operators f and g and the equalities

$$\Gamma = \{f(a,b) \approx c, f(a,c) \approx f(c,a), g(b,a) \approx c, g(a,c) \approx g(c,a)\}$$

Either the theory process of f, or the one of g, will generate an infinite number of rewrite rules, depending on the chosen ordering between constants a and c deciding of the orientation of the second and fourth equalities.

For associativity, it is well-known that unification is infinitary. But with those examples we illustrate that it is possible to generate infinite derivations, even if we consider only ground equalities and unification problems with finitely many most general unifiers.

7 Running Example

We have implemented the combination procedure described in this paper for several theories: the empty theory, commutative theories, associative theories, associative-commutative theories. It is written in C (6000 lines of code).

We detail below the trace obtained in 6 ms for an example combining three theories: one where the operator cp (a short-cut for *compatible*) is commutative, one where the operator and is associative-commutative, and the empty theory with all the constants and the $owns$ operator. The initial set of equalities Γ is:

$$\begin{cases} owns(Ali, and(cp(boat, engine), cp(engine, captain), cp(captain, boat))) \approx true \\ and(cp(boat, captain), cp(boat, engine)) \approx and(boat, cp(captain, engine)) \\ and(cp(captain, engine), cp(captain, engine)) \approx cp(captain, engine) \\ and(cp(engine, captain), boat) \approx ready\_boat \\ owns(Ali, ready\_boat) \approx false \end{cases}$$

and there is one disequality: $\Delta = \{true \not\approx false\}$.

A contradiction is derived by the following deductions;[3] new constants (of K) are written $\_i$, where i is an integer.

```
> Disequality: true ≠ false
  C-disequality true ≠ false generated (Decomposition)
> Equality: owns(Ali, rdy) = false
  D-rule owns(Ali, rdy) → false generated (Orientation)
> Equality: and(cp(eng, cap), boat) = rdy
  D-rule cp(cap, eng) → _1 generated (Extension)
  D-rule and(boat, _1) → rdy generated (Orientation)
> Equality: and(cp(cap, eng), cp(cap, eng)) = cp(cap, eng)
  Subterm cp(cap, eng) simplified into _1 (with cp(cap, eng) → _1)
  Subterm cp(cap, eng) simplified into _1 (with cp(cap, eng) → _1)
  Subterm cp(cap, eng) simplified into _1 (with cp(cap, eng) → _1)
  D-rule and(_1, _1) → _1 generated (Orientation)
> Equality: and(cp(boat, cap), cp(boat, eng)) = and(boat, cp(cap, eng))
  D-rule cp(boat, cap) → _2 generated (Extension)
  D-rule cp(boat, eng) → _3 generated (Extension)
  Subterm cp(cap, eng) simplified into _1 (with cp(cap, eng) → _1)
  Subterm and(boat, _1) simplified into rdy (with and(boat, _1) → rdy)
  D-rule and(_3, _2) → rdy generated (Orientation)
> Equality: owns(Ali, and(cp(boat, eng), cp(eng, cap), cp(cap, boat))) = true
  Subterm cp(boat, eng) simplified into _3 (with cp(boat, eng) → _3)
  Subterm cp(cap, eng) simplified into _1 (with cp(cap, eng) → _1)
```

---
[3] For clarity, in the trace the initial constants have been shortened: "*eng*" stands for "*engine*", "*cap*" stands for "*captain*", "*rdy*" stands for "*ready_boat*".

```
  Subterm cp(boat, cap) simplified into _2 (with cp(boat, cap) → _2)
  Subterm and(_3, _2, _1) simplified into and(_1, rdy) (with and(_3, _2) → rdy)
  D-rule and(_1, rdy) → _4 generated (Extension)
  D-rule owns(Ali, _4) → true generated (Orientation)
> Equality and(boat, _1) = and(_1, rdy) generated
  (Superposition between and(boat, _1) → rdy and and(_1, _1) → _1)
> Equality: and(boat, _1) = and(_1, rdy)
  Subterm and(boat, _1) simplified into rdy (with and(boat, _1) → rdy)
  Subterm and(_1, rdy) simplified into _4 (with and(_1, rdy) → _4)
  C-rule _4 → rdy generated (Orientation)
> Compression with _4 → rdy
  D-rule and(_1, rdy) → _4 replaced by and(_1, rdy) → rdy
  (Composition with _4 → rdy)
  D-rule owns(Ali, _4) → true replaced by owns(Ali, rdy) → true
  (Collapse with _4 → rdy)
  D-rule owns(Ali, rdy) → false replaced by false → true
  (Collapse with owns(Ali, rdy) → true)
> Compression with false → true
  Disequality: true ≠ false simplified into true ≠ true (with false → true)
*** A contradiction has been found! ***
```

8 Conclusion

We have defined CombCC as an orchestrator that does not need to handle specific algorithms related to theories $\mathcal{E}_i$. It could be more efficient using some inference rules of theory processes, like Simplification and Deletion. But we did this on purpose for the clarity of the paper and to show that the orchestrator can be defined independently from the theories. We are considering several extensions of our procedure, to apply it to any theory having a deduction system preserving the groundness of generated rules/equalities. This applies to flat permutative theories, an extension of commutative theories. It also applies to extensions of associative and/or commutative theories with axioms that can be used as flat collapsing rewrite rules.

Our new implementation allows us to experiment extensions of Associativity and/or Commutativity with orientable equational axioms specifying for instance an idempotent or nilpotent operator, or an operator with a neutral or absorbent element. This implementation is very helpful to identify counter-examples and non-terminating examples.

In the future, we plan to investigate congruence closure procedures for the unions of theories possibly sharing constructor symbols with associative and/or commutative equational properties. It is clearly challenging to try to go beyond the simple case of shared constants.

Currently, we consider theories where classical equational completion techniques are applicable. An interesting future work would be to study other completion techniques, such as unfailing completion, in the context of congruence closure. In order to integrate various completion techniques within a uniform framework, we could envision to reuse the notion of normalizing mapping initially introduced for the (combined) word problem [6], or study the combination of extended canonizers [18] to go beyond the classical rewrite-based normalization. Applying this notion to congruence closure remains to be studied. This would pave the way of extending our combined abstract framework for congruence closure to new equational theories.

References

1. Baader, F., Nipkow, T.: Term Rewriting and All That. Cambridge University Press, New York (1998)
2. Baader, F., Tinelli, C.: A new approach for combining decision procedures for the word problem, and its connection to the Nelson-Oppen combination method. In: McCune, W. (ed.) CADE 1997. LNCS, vol. 1249, pp. 19–33. Springer, Heidelberg (1997). https://doi.org/10.1007/3-540-63104-6_3
3. Bachmair, L., Tiwari, A., Vigneron, L.: Abstract congruence closure. J. Autom. Reason. **31**(2), 129–168 (2003)
4. Barrett, C.W., Dill, D.L., Stump, A.: A generalization of Shostak's method for combining decision procedures. In: Armando, A. (ed.) FroCoS 2002. LNCS (LNAI), vol. 2309, pp. 132–146. Springer, Heidelberg (2002). https://doi.org/10.1007/3-540-45988-X_11
5. Barrett, C.W., Sebastiani, R., Seshia, S.A., Tinelli, C.: Satisfiability modulo theories. In: Biere, A., Heule, M., van Maaren, H., Walsh, T. (eds.) Handbook of Satisfiability, 2nd Edn., Frontiers in Artificial Intelligence and Applications, vol. 336, pp. 1267–1329. IOS Press (2021)
6. Erbatur, S., Marshall, A.M., Ringeissen, C.: Combined hierarchical matching: the regular case. In: Felty, A.P. (ed.) 7th International Conference on Formal Structures for Computation and Deduction, FSCD, Haifa, Israel, August 2022. LIPIcs, vol. 228, pp. 6:1–6:22. Schloss Dagstuhl - Leibniz-Zentrum für Informatik (2022)
7. Kapur, D.: Shostak's congruence closure as completion. In: Comon, H (ed.) RTA 1997. LNCS, vol. 1232, pp. 23–37. Springer, Heidelberg (1997). https://doi.org/10.1007/3-540-62950-5_59
8. Kapur, D.: A modular associative commutative (AC) congruence closure algorithm. In: Kobayashi, N. (ed.) 6th International Conference on Formal Structures for Computation and Deduction, FSCD, Buenos Aires, Argentina, July 2021 (Virtual Conference). LIPIcs, vol. 195, pp. 15:1–15:21. Schloss Dagstuhl - Leibniz-Zentrum für Informatik (2021)
9. Kapur, D.: Modularity and combination of associative commutative congruence closure algorithms enriched with semantic properties. Log. Methods Comput. Sci. **19**(1) (2023)
10. Kim, D.: Congruence closure modulo groups. CoRR **abs/2310.05014** (2023). https://doi.org/10.48550/arXiv.2310.05014
11. Kim, D., Lynch, C.: Congruence closure modulo permutation equations. In: Kutsia, T. (ed.) 9th International Symposium on Symbolic Computation in Software Science, SCSS, Hagenberg, Austria, September 2021, Proceedings. EPTCS, vol. 342, pp. 86–98 (2021)
12. Meseguer, J.: Variants and satisfiability in the infinitary unification wonderland. J. Log. Algebr. Methods Program. **134**, 100877 (2023)

13. Nelson, G., Oppen, D.C.: Simplification by cooperating decision procedures. ACM Trans. Program. Lang. Syst. **1**(2), 245–257 (1979)
14. Nelson, G., Oppen, D.C.: Fast decision procedures based on congruence closure. J. ACM **27**(2), 356–364 (1980)
15. Nieuwenhuis, R., Oliveras, A.: Fast congruence closure and extensions. Inf. Comput. **205**(4), 557–580 (2007)
16. Peterson, G.E., Stickel, M.E.: Complete sets of reductions for some equational theories. J. ACM **28**(2), 233–264 (1981)
17. Plotkin, G.: Building-in equational theories. Mach. Intell. **7**, 73–90 (1972)
18. Ranise, S., Ringeissen, C., Tran, D.-K.: Nelson-Oppen, Shostak and the extended Canonizer: a family picture with a newborn. In: Liu, Z., Araki, K. (eds.) ICTAC 2004. LNCS, vol. 3407, pp. 372–386. Springer, Heidelberg (2005). https://doi.org/10.1007/978-3-540-31862-0_27
19. Tinelli, C.: Cooperation of background reasoners in theory reasoning by residue sharing. J. Autom. Reason. **30**(1), 1–31 (2003)
20. Tran, D., Ringeissen, C., Ranise, S., Kirchner, H.: Combination of convex theories: modularity, deduction completeness, and explanation. J. Symb. Comput. **45**(2), 261–286 (2010)
21. Vigneron, L.: Positive deduction modulo regular theories. In: Kleine Büning, H. (ed.) CSL 1995. LNCS, vol. 1092, pp. 468–485. Springer, Heidelberg (1996). https://doi.org/10.1007/3-540-61377-3_54

A Certifying Algorithm for Linear (and Integer) Feasibility in Horn Constraint Systems

Piotr Wojciechowski[ORCID] and K. Subramani[✉][ORCID]

LDCSEE, West Virginia University, Morgantown, WV, USA
{pwojciec,k.subramani}@mail.wvu.edu

Abstract. In this paper, we discuss a certifying version of the Lifting Algorithm for Horn constraint systems. Recall that a Horn constraint system (HCS) is a specialized polyhedral system that finds application in a number of domains such as program verification (abstract interpretation) and operations research. HCSs are closely related to Leontief substitution systems. In previous work, it was established that the problem of checking if a Horn polytope is non-empty is polynomial time solvable. However, that algorithm is not certifying in that if the input HCS is infeasible, it does not provide a certificate which attests to the infeasibility of the HCS. Consequently, the output provided by an implementation of the algorithm is not *trustworthy*. Trustworthiness is an integral aspect of AI systems. The current paper rectifies this issue by modifying the lifting algorithm to make it certifying. In particular, both "yes"-instances and "no"-instances of input HCSs will be certified through appropriate Farkas' variables. However, the increased trustworthiness of the algorithm comes at a cost; the new algorithm is less efficient than its non-certifying counterpart.

Keywords: Horn Constraints · Certifying Algorithm · Farkas' Lemma · Integer Feasibility · Linear Feasibility

1 Introduction

In this paper, we design a certifying variant of the Lifting Algorithm for checking linear (and integer) feasibility in Horn Constraint Systems (HCSs) [4]. Note that an HCS is a system of constraints in which each coefficient belongs to the set $\{0, 1, -1\}$, and each constraint has at most one positive coefficient. Throughout this paper, we use n to refer to the number of variables in the input HCS and m to refer to the number of constraints. The certificates returned by this algorithm come in two forms depending on if the input HCS is feasible or infeasible. If the input HCS is feasible, then the algorithm returns a feasible solution $\mathbf{x}$ and a certificate that $\mathbf{x}$ is the solution for which the optimization function $\sum_{i=1}^{n} x_i$ is minimized. If the input HCS is infeasible, then the algorithm returns a certificate of infeasibility.

An HCS can be represented in matrix form as a system $\mathbf{A} \cdot \mathbf{x} \geq \mathbf{b}$, $\mathbf{x} \geq \mathbf{0}$. In the matrix $\mathbf{A}$, each entry is 0, 1, or -1 and each row contains at most one 1. This is closely related to Leontief substitution systems [7]. A Leontief substitution system is a system of constraints of the form $\mathbf{A} \cdot \mathbf{x} = \mathbf{b}$, $\mathbf{x} \geq \mathbf{0}$ where each column of $\mathbf{A}$ has at most one positive entry. A matrix with this property is called Leontief. Note that the transpose of the matrix corresponding to an HCS is Leontief. In particular, it is gainfree Leontief [14].

Certificates, like those given in this paper, are used to verify the correctness of the algorithm. Note that a proof of correctness for a theoretical algorithm is insufficient for this purpose. It may be the case that a valid theoretical algorithm is implemented incorrectly. This exact situation happened in version 2.0 of the LEDA software [20]. Note that the process of verifying the output using the certificate should be straightforward, since a complicated verification process may introduce its own errors.

2 Preliminaries

In this section, we formally describe terms that will be used in this paper. Our principal focus is on a specialized linear program called a Horn program or Horn Constraint System (HCS).

Definition 1. *A linear constraint system* $\mathbf{A} \cdot \mathbf{x} \geq \mathbf{b}$, $\mathbf{x} \geq \mathbf{0}$ *is said to be a* **Horn Constraint System (HCS)** *if:*

1. *The entries in* $\mathbf{A}$ *belong to the set* $\{0, 1, -1\}$.
2. *Each row of* $\mathbf{A}$ *contains at most one positive entry.*

If a Horn constraint has only one non-zero coefficient, then it is called an **absolute constraint**. If this coefficient is positive, then this constraint is called a **positive absolute constraint**. If a Horn constraint has no variables with positive coefficient, then it is called an **all-negative** constraint.

Example 1. System (1) is an HCS with 2 constraints over 3 variables.

$$x_1 - x_2 - x_3 \geq -1 \qquad x_2 - x_3 \geq 3 \qquad (1)$$

Recall that the transpose of a matrix defining an HCS is Leontief. As a result, the dual of an optimization problem over an HCS without non-negativity constraints is an optimization problem over a Leontief substitution system [15].

Let $\mathbf{H}$ be an HCS with m constraints over n variables. For the algorithms and proofs in this paper, we use the following notation from [26]:

1. $N(\mathbf{H})$ is the set of all-negative constraints in $\mathbf{H}$.
2. $P(\mathbf{H})$ is the set of positive absolute constraints in $\mathbf{H}$.
3. For a constraint $l_k \in \mathbf{H}$, P_k is the set of variables with positive coefficient and N_k is the set of variables with negative coefficient. Note that $|P_k| \leq 1$.

The **Linear Refutability Problem (LRP)** is concerned with providing proof that a given HCS does not have any linear (rational) solutions. Such a proof is known as a refutation. Refutations consist of a sequence of inferences that results in a contradiction of the form $0 \geq b$ where $b > 0$. In case of LRP, we use a single inference rule (Rule 2).

$$\text{ADD} : \frac{\sum_{i=1}^{n} a_i \cdot x_i \geq b_1 \quad \sum_{i=1}^{n} a'_i \cdot x_i \geq b_2}{\sum_{i=1}^{n} (a_i + a'_i) \cdot x_i \geq b_1 + b_2} \quad (2)$$

We refer to Rule (2) as the **ADD rule**. It is easy to see that Rule (2) is sound since any assignment that satisfies the hypotheses also satisfies the consequent. Additionally, the ADD rule is complete [10].

Example 2. Consider the constraints $x_1 - x_2 - x_3 \geq -1$ and $x_2 - x_3 \geq 4$. Applying the ADD rule to these constraints results in the constraint $x_1 - 2 \cdot x_3 \geq 3$.

We distinguish between constraints present in the original HCS (input constraints) and those obtained through the application of an inference rule (derived constraints). Depending on how constraints are rederived, we focus on three different types of refutations.

Definition 2. *A refutation is **dag-like**, if any constraint, input or derived, can be used multiple times without needing to be rederived.*

Definition 3. *A refutation is **tree-like**, if any input constraint can be used multiple times and each derived constraint can be used only once.*

Note that in a tree-like refutation a derived constraint can be reused, if it is rederived. This increases the length (number of inference steps) of the refutation.

Definition 4. *A refutation is **read-once**, if any constraint, input or derived, can be used only once.*

The infeasibility of linear systems is commonly established using Farkas' lemma [10,18].

Lemma 1. *Let $\mathbf{A}$ denote an $m \times n$ matrix and let $\mathbf{b}$ denote an m-vector. Then, either*

$$\mathbf{I} : \quad \exists \mathbf{x} \in \mathbb{R}_+^n \quad \mathbf{A} \cdot \mathbf{x} \geq \mathbf{b}$$

or *(mutually exclusively)*

$$\mathbf{II} : \quad \exists \mathbf{y} \in \mathbb{R}_+^m \quad \mathbf{y} \cdot \mathbf{A} \leq \mathbf{0}, \quad \mathbf{y} \cdot \mathbf{b} > 0.$$

As a result of Farkas' lemma, a proof of linear infeasibility of an HCS $\mathbf{H} : \mathbf{A} \cdot \mathbf{x} \geq \mathbf{b}$, $\mathbf{x} \geq \mathbf{0}$ can simply be a non-negative vector $\mathbf{y}$, such that $\mathbf{y} \cdot \mathbf{A} \leq \mathbf{0}$, $\mathbf{y} \cdot \mathbf{b} > 0$. This vector $\mathbf{y}$ is called the Farkas witness of the infeasibility of $\mathbf{H}$. The elements of $\mathbf{y}$ are called Farkas variables (dual variables).

Note that a Farkas vector **y** corresponds to a summation of the constraints in **H**. Such a refutation can be broken up into individual summations, each between a pair of constraints. These summations correspond to applications of the ADD inference rule. The algorithm described in this paper, always returns a Farkas vector. If the input HCS is feasible, then the algorithm returns a feasible solution $\mathbf{x}^*$ and a Farkas vector $\mathbf{y}^* \geq \mathbf{0}$ such that $\mathbf{y}^* \cdot \mathbf{A} \leq \mathbf{1}$ and $\mathbf{y}^* \cdot \mathbf{b} = \sum_{i=1}^{n} x_i^*$. In this case, the principle of weak duality establishes that the solution $\mathbf{x}^*$ minimizes the objective function $\sum_{i=1}^{n} x_i^*$. If the input HCS is infeasible, then the algorithm returns a Farkas vector $\mathbf{y}' \geq \mathbf{0}$ such that $\mathbf{y}' \cdot \mathbf{A} \leq \mathbf{0}$ and $\mathbf{y}' \cdot \mathbf{b} > 0$.

The principal contribution of this paper is a certifying version of the Lifting Algorithm [4].

3 Motivation

In this section, we provide motivation for our work. In a Difference Constraint System (DCS), every constraint is a difference constraint, that is a constraint of the form $x_i - x_j \geq b_{ij}$. It is easy to see that every difference constraint is also a Horn constraint. Note that a DCS has a linear solution if and only if it has an integer solution. This follows from the structure of the underlying constraint matrix. In a DCS $\mathbf{D} : \mathbf{A} \cdot \mathbf{x} \geq \mathbf{b}$, the matrix $\mathbf{A}$ is Totally Unimodular (TU), that is a matrix in which every square submatrix has a determinant of 1, -1, or 0 [21]. Thus, if **b** is integral, then every extreme point of the polyhedron $\mathbf{A} \cdot \mathbf{x} \geq \mathbf{b}$ is integral.

In this paper, we focus on HCS. HCSs are a generalization of DCSs. Note that the underlying constraint matrix of an HCS is not TU. However, it is known that if $\mathbf{A} \cdot \mathbf{x} \geq \mathbf{b}$ is feasible, then it has a minimal element, which is integral [27]. Horn constraint systems have been used as domains in abstract interpretation [2,6]. They are also used in Satisfiability Modulo Theories (SMT) solvers. These SMT solvers, in turn, are increasingly used in the field of program verification [8,11]. SMT solvers are also part of procedures for bounded model checking, infinite state systems, and test-case generation [9]. Horn systems also find applications in declarative programming [13,17]. The applications of Horn constraints to program verification has been discussed extensively in [3,16]. In all these applications, certificates are useful since they provide either proof that an HCS is infeasible, or proof that the value returned by an algorithm is optimal.

As mentioned previously, a certificate is more useful if it is easy to verify. If the verification of a certificate involves a complex procedure, then the verification process is also susceptible to errors. If there is a mistake in the process of verifying a certificate, then neither the certificate, nor the original answer can be trusted. This makes the certificate useless. Thus, certificates should be intuitive and easily checkable. However, this is not always the case [19]. As a result, it is important to find "natural" certificates. The problem of identifying such certificates for well known problems that can be solved in polynomial time is one of the primary concerns of the field of certifying algorithms. For constraint systems, the structure of a certificate depends heavily on the associated proof system. In fact, it is possible for a problem to have short, easily verifiable certificates in one proof system, whereas even asking whether a certificate exists is **NP-complete** in another proof system [12]. Even when short certificates are easily deducible, the problem of finding certificates of "optimal length" may be difficult [1].

4 Related Work

In this section, we examine work on related problems and give a detailed description of the Lifting Algorithm [4].

Certificates and certifying algorithms have been examined for various constraint systems. For DCSs, the problem of finding certificates of infeasibility is equivalent to the problem of finding negative cycles in weighted, directed graphs [5]. Every DCS can be converted into a weighted directed graph. This graph has a negative cycle, if and only if the original DCS is infeasible. Thus, this negative cycle can be used as a certificate that the corresponding DCS is infeasible. Note that for a DCS with m constraints over n variables, such a certificate can be found in $O(m \cdot n)$ time. Additional work has focused on finding the shortest such certificates for DCSs. The first polynomial time algorithm for this problem was proposed in [22] and runs in $O(n^3 \cdot \log K)$ time, where K is the length of the negative cycle. This time was later improved to $O(m \cdot n \cdot K)$ [23].

Previous work has examined the length of shortest tree-like and dag-like refutations of HCSs [29]. In that paper, it was shown that there exist HCSs for which the shortest tree-like refutation using only the add rule has length exponential in the size of the HCS. The approximation complexity of finding the shortest read-once refutations of HCS has also been studied [28]. This problem is **NPO PB-complete**.

Certifying algorithms have also been developed for a class of constraints known as Unit Two Variable Per Inequality (UTVPI) constraints. A UTVPI constraint is a constraint of the form $a_i \cdot x_i + a_j \cdot x_j \leq b_{ij}$ where $a_i, a_j \in \{0, 1, -1\}$. It is easy to see that UTVPI Constraint Systems (UCS) generalize DCSs. Certifying algorithms have been developed for both the linear feasibility problem for UCSs [24], and the integer feasibility problem for UCSs [25].

4.1 The Lifting Algorithm

Let **H** be an HCS with m constraints over n variables. For each variable x_i, the lifting algorithm maintains a value b_i such that $x_i \geq b_i$ in any solution to **H**. The Lifting Algorithm in [4] is based on the following properties of Horn constraint systems:

1. Let $l_k \in \mathbf{H}$ be a Horn constraint of the form: $x_i - \sum_{x_j \in N_k} x_j \geq b_k$. If $x_j \geq b_j$ for each $x_j \in N_k$, then $x_i \geq b_k + \sum_{x_j \in N_k} b_j$.
2. Let $l_k \in \mathbf{H}$ be a Horn constraint of the form: $-\sum_{x_i \in N_k} x_i \geq b_k$. If $x_i \geq b_i$ for each $x_i \in N_k$, and $\sum_{x_i \in N_k} b_i > -b_k$, then **H** is infeasible.

These properties are handled by the following proof system:

$$\frac{x_i - \sum_{x_j \in N_k} x_j \geq b_k \qquad x_j \geq b_j,\, j \in N_k}{x_i \geq b_k + \sum_{x_j \in N_k} b_j} \qquad (3)$$

$$\frac{-\sum_{x_i \in N_k} x_i \geq b_k \qquad x_i \geq b_i,\, i \in N_k \;:\; \sum_{x_i \in N_k} b_i > -b_k}{\bot} \qquad (4)$$

In the proof system, Rule (3) is used to update the lower bound on x_i. If this increases the lower bound on x_i, then this is known as a **lift**. A lift is **greedy** if, out of the set of currently possible lifts, it results in the greatest increase in a lower bound. On an input HCS **H** with m constraints over n variables, the Lifting Algorithm performs n^2 greedy lifts. If any lifts are still possible, then **H** is declared infeasible. Finding the greedy lift takes $O(m)$ time. Thus the Lifting Algorithm in [4] runs in time $O(m \cdot n^2)$.

The Lifting Algorithm implements a computationally restricted version of the proof system consisting of Rules (3) and (4). This proof system is both sound and complete for HCSs [4]. Algorithms 1 and 2 detail the Lifting Algorithm [4]. Note that the Lifting Algorithm returns a feasible solution if the input HCS is feasible, but does not return any proof of infeasibility if the HCS is infeasible. Thus, the Lifting Algorithm is *not* certifying.

Algorithm 1. The Lift Procedure for One Variable

Input: HCS **H** : $\mathbf{A} \cdot \mathbf{x} \geq \mathbf{b}$, $\mathbf{x} \geq \mathbf{0}$, vector **o**, index i, and value c.

1: **procedure** LIFT(**H**, **o**, i, c)
2:     $o_i := o_i + c$.
3:     **for** (each constraint $l_k \in \mathbf{H}$) **do**
4:        **if** ($x_i \in P_k$) **then**
5:           $b_k := b_k - c$.
6:        **if** ($x_i \in N_k$) **then**
7:           $b_k := b_k + c$.

Algorithm 2. The Lifting Algorithm

Input: HCS $\mathbf{H}$: $\mathbf{A} \cdot \mathbf{x} \geq \mathbf{b}$, $\mathbf{x} \geq \mathbf{0}$.
Output: Feasible solution $\mathbf{x}$ if $\mathbf{H}$ is feasible, -1 if $\mathbf{H}$ is infeasible.

1: **procedure** LIFTING-ALGORITHM($\mathbf{H}$)
2: Let $\mathbf{o}$ be an n vector.
3: Initialize every element of $\mathbf{o}$ to 0.
4: **if** ($\mathbf{b} \leq \mathbf{0}$) **then**
5:   **return 0**.
6: **for** ($r = 1$ **to** $(n-1)$) **do**
7:   **for** (each variable x_i) **do**
8:    $c := \max\{b_k : x_i \in P_k\}$.
9:    **if** ($c > 0$) **then**
10:    LIFT($\mathbf{H}$, $\mathbf{o}$, i, c).
11: **for** (each constraint $l_k \in \mathbf{H}$) **do**
12:   **if** ($b_k > 0$) **then**
13:    **return** -1.
14: **return o**.

5 A Certifying Version of the Lifting Algorithm

In this section, we describe a certifying version of the lifting algorithm and prove that it produces valid certificates. Let $\mathbf{H}$ be an HCS with m constraints over n variables. Throughout the remainder of this paper, we use $\mathbf{e}_i$ to denote a vector with i^{th} coordinate equal to 1, and all other coordinates equal to 0.

We can make the Lifting Algorithm in [4] certifying at the cost of increasing the running time. Note that we can modify the Lifting Algorithm in [4] to maintain a vector $\mathbf{y}_i \geq \mathbf{0}$ for each variable x_i. This vector $\mathbf{y}_i$ will have the property that $\mathbf{y}_i \cdot \mathbf{A} \leq \mathbf{e}_i$. Additionally, $\mathbf{y}_i \cdot \mathbf{b} = d_i$, where d_i is the best known lower bound on the variable x_i. Observe that,

$$\mathbf{y}_i \cdot \mathbf{b} \leq \mathbf{y}_i \cdot \mathbf{A} \cdot \mathbf{x} \leq \mathbf{e}_i \cdot \mathbf{x} = x_i.$$

Thus, the vector $\mathbf{y}_i$ serves as a certificate that the constraint $x_i \geq d_i$ is a valid bound on the variable x_i.

Algorithm 3 uses the updates the matrix $\mathbf{Y}$ of Farkas vectors to account for one round of lifting.

Algorithm 4 finds a feasible solution to an HCS $\mathbf{H}$ or returns a certificate of infeasibility. Algorithm 4 repeatedly calls Algorithm 3 to maintain a Farkas vector for each variable x_i. If $\mathbf{H}$ is infeasible, then after $(n+1)$ calls to Algorithm 3, either the lower bounds derived for each variable violate an all-negative constraint, or there is a variable x_i such that the current bound on x_i was derived using a previous bound on x_i.

Algorithm 3. Algorithm for updating the matrix of Farkas variables

Input: HCS **H**, matrix **Y**, and vector **d**.
Output: Updated matrix **Y**′

1: **procedure** UPDATE-FARKAS(**H**, **Y**, **d**)
2:     Let **Y**′ := **Y**.
3:     **for** (each constraint $l_k \in \mathbf{H} \setminus N(\mathbf{H})$) **do**
4:        Let x_i be the variable in P_k.
5:        **if** ($\mathbf{y}'_i \cdot \mathbf{b} < b_k + \sum_{x_j \in N_k} d_j$) **then**
6:           $\mathbf{y}'_i := \mathbf{e}_k + \sum_{x_j \in N_k} \mathbf{y}_j$.
7:     **return Y**′.

The Certifying Lifting Algorithm can terminate in one of three possible cases:

1. Setting each $x_i = d_i$ results in a feasible solution to **H**.
2. There is a constraint l_k of the form: $-\sum_{x_i \in N_k} x_i \geq b_k$ that is violated by setting each $x_i = d_i$.
3. There is a variable x_i such that the current bound on x_i was derived using a previous bound on x_i.

With these changes, the Lifting Algorithm in [4] will either produce a certificate of infeasibility or certify the optimality of the solution produced with respect to the optimization function $\min \sum_{i=1}^{n} x_i$.

Example 3. Let **H** be the following HCS:

$$l_1 : x_1 - x_2 - x_3 \geq 1$$
$$l_2 : x_2 - x_3 \geq 1$$
$$l_3 : x_3 \geq 1$$
$$l_4 : -x_1 - x_2 - x_3 \geq -6$$

We now see what happens when we run Algorithm 4 on **H**.

Initially, $\mathbf{Y}^{[0]} = \begin{bmatrix} 0 & 0 & 0 & 0 \\ 0 & 0 & 0 & 0 \\ 0 & 0 & 1 & 0 \end{bmatrix}$ and $\mathbf{d} = (0, 0, 1)$.

In the first round of lifts:

The algorithm first creates $\mathbf{Y}^{[1]}$ as a copy of $\mathbf{Y}^{[0]}$. The algorithm then attempts to lift using constraint l_1. In this lift the value of $\mathbf{y}_1^{[1]} \cdot \mathbf{b} = 0$ is compared to $b_1 + \sum_{x_i \in N_1} d_i = b_1 + d_2 + d_3 = 1 + 0 + 1 = 2$. Since, $b_1 + \sum_{x_i \in N_1} d_i > \mathbf{y}_1^{[1]} \cdot \mathbf{b}$, the algorithm lifts using l_1. This sets

$$\mathbf{y}_1^{[1]} = \mathbf{e}_1 + \sum_{x_i \in N_1} \mathbf{y}_i^{[0]} = \mathbf{e}_1 + \mathbf{y}_2^{[0]} + \mathbf{y}_3^{[0]} = (1, 0, 1, 0).$$

Algorithm 4. Certifying Lifting Algorithm
   Input: HCS **H**
   Output: Feasible solution **x** if **H** is feasible, Farkas vector **y** if **H** is infeasible.

1: **procedure** CERT-LIFT(**H**)
2:   Let $\mathbf{Y}^{[0]}$ be an $n \times m$ matrix.
3:   Initialize every element of $\mathbf{Y}^{[0]}$ to 0.
4:   Let **d** be an n vector.
5:   Initialize every element of **d** to 0.
6:   **for** (each constraint $l_k \in P(\mathbf{H})$) **do**
7:     Let x_i be the variable in P_k.
8:     $\mathbf{y}_i^{[0]} := \mathbf{e}_k$.
9:     $d_i := b_k$.
10: **for** $r = 1$ to $(n+1)$ **do**
11:    $\mathbf{Y}^{[r]} :=$ UPDATE-FARKAS(**H**, $\mathbf{Y}^{[r-1]}$, **d**).
12:    **for** (each variable x_i in **H**) **do**
13:      $d_i := \mathbf{y}_i^{[r]} \cdot \mathbf{b}$
14:      **if** ($\sum_{l_k : x_i \in P_k} y_{i,k}^{[r]} > 1$) **then**
15:        Let r' be the value found in the proof of Lemma 4.
16:        **return** the Farkas vector $(\mathbf{y}_i^{[r]} - \mathbf{y}_i^{[r']})$.
17:    **for** (each constraint $l_k \in N(\mathbf{H})$) **do**
18:      **if** ($b_k + \sum_{x_i \in N_k} d_i > 0$) **then**
19:        **return** the Farkas vector $(\mathbf{e}_k + \sum_{x_i \in N_k} \mathbf{y}_i^{[r]})$.
20: **return** the assignment **d**, and the Farkas vector $(\sum_{i=1}^n \mathbf{y}_i^{[n+1]})$.

The algorithm then attempts to lift using constraint l_2. In this lift the value of $\mathbf{y}_2^{[1]} \cdot \mathbf{b} = 0$ is compared to $b_2 + \sum_{x_i \in N_2} d_i = b_2 + d_3 = 1 + 1 = 2$. Since, $b_2 + \sum_{x_i \in N_2} d_i > \mathbf{y}_2^{[1]} \cdot \mathbf{b}$, the algorithm lifts using l_2. This sets

$$\mathbf{y}_2^{[1]} = \mathbf{e}_2 + \sum_{x_i \in N_2} \mathbf{y}_i^{[0]} = \mathbf{e}_2 + \mathbf{y}_3^{[0]} = (0, 1, 1, 0).$$

The algorithm then attempts to lift using constraint l_3. In this lift the value of $\mathbf{y}_3^{[1]} \cdot \mathbf{b} = 1$ is compared to $b_3 + \sum_{x_i \in N_3} d_i = b_3 = 1$. Since, $b_3 + \sum_{x_i \in N_3} d_i \leq \mathbf{y}_3^{[1]} \cdot \mathbf{b}$, the algorithm does not lift using l_3. Thus, after the first round of lifts,

$$\mathbf{Y}^{[1]} = \begin{bmatrix} 1 & 0 & 1 & 0 \\ 0 & 1 & 1 & 0 \\ 0 & 0 & 1 & 0 \end{bmatrix}.$$ For each variable x_i, the algorithm then sets $d_i = \mathbf{y}^{[1]} \cdot \mathbf{b}$.

Thus, $\mathbf{d} = (2, 2, 1)$.

At this point, $\sum_{l_k : x_1 \in P_k} y_{1,k}^{[1]} = 1$, $\sum_{l_k : x_2 \in P_k} y_{2,k}^{[1]} = 1$, and $\sum_{l_k : x_3 \in P_k} y_{3,k}^{[1]} = 1$. As a result, Algorithm 4 does not declare $\mathbf{H}$ infeasible for falling into Case 3. The only all-negative constraint in $\mathbf{H}$ is l_4. The algorithm then checks to see if this constraint is violated by setting $\mathbf{x} = \mathbf{d}$. Since $b_4 + \sum x_i \in N_4 d_i = b_4 d_1 + d_2 + d_3 = -6 + 2 + 2 + 1 = -1 \leq 0$, Algorithm 4 does not declare $\mathbf{H}$ infeasible for falling into Case 2.

In the second round of lifts:

The algorithm first creates $\mathbf{Y}^{[2]}$ as a copy of $\mathbf{Y}^{[1]}$. The algorithm then attempts to lift using constraint l_1. In this lift the value of $\mathbf{y}_1^{[2]} \cdot \mathbf{b} = 2$ is compared to $b_1 + \sum_{x_i \in N_1} d_i = b_1 + d_2 + d_3 = 1 + 2 + 1 = 4$. Since, $b_1 + \sum_{x_i \in N_1} d_i > \mathbf{y}_1^{[2]} \cdot \mathbf{b}$, the algorithm lifts using l_1. This sets

$$\mathbf{y}_1^{[2]} = \mathbf{e}_1 + \sum_{x_i \in N_1} \mathbf{y}_i^{[1]} = \mathbf{e}_1 + \mathbf{y}_2^{[1]} + \mathbf{y}_3^{[1]} = (1, 1, 2, 0).$$

The algorithm then attempts to lift using constraint l_2. In this lift the value of $\mathbf{y}_2^{[2]} \cdot \mathbf{b} = 2$ is compared to $b_2 + \sum_{x_i \in N_2} d_i = b_2 + d_3 = 1 + 1 = 2$. Since, $b_2 + \sum_{x_i \in N_2} d_i \leq \mathbf{y}_2^{[2]} \cdot \mathbf{b}$, the algorithm does not lift using l_2.

The algorithm then attempts to lift using constraint l_3. In this lift the value of $\mathbf{y}_3^{[2]} \cdot \mathbf{b} = 1$ is compared to $b_3 + \sum_{x_i \in N_3} d_i = b_3 = 1$. Since, $b_3 + \sum_{x_i \in N_3} d_i \leq \mathbf{y}_3^{[2]} \cdot \mathbf{b}$, the algorithm does not lift using l_3. Thus, after the second round of lifts,

$$\mathbf{Y}^{[2]} = \begin{bmatrix} 1 & 1 & 2 & 0 \\ 0 & 1 & 1 & 0 \\ 0 & 0 & 1 & 0 \end{bmatrix}.$$ For each variable x_i, the algorithm then sets $d_i = \mathbf{y}^{[2]} \cdot \mathbf{b}$.

Thus, $\mathbf{d} = (4, 2, 1)$. At this point, $\sum_{l_k : x_1 \in P_k} y_{1,k}^{[2]} = 1$, $\sum_{l_k : x_2 \in P_k} y_{2,k}^{[2]} = 1$, and $\sum_{l_k : x_3 \in P_k} y_{3,k}^{[2]} = 1$. As a result, Algorithm 4 does not declare $\mathbf{H}$ infeasible for falling into Case 3. The only all-negative constraint in $\mathbf{H}$ is l_4. The algorithm then checks to see if this constraint is violated by setting $\mathbf{x} = \mathbf{d}$. Since $b_4 + \sum x_i \in N_4 d_i = b_4 + d_1 + d_2 + d_3 = -6 + 4 + 2 + 1 = 1 > 0$, Algorithm 4 declares $\mathbf{H}$ infeasible for falling into Case 2.

In this case, the algorithm returns $\mathbf{y} = \mathbf{e}_4 + \sum_{x_i \in N_4} \mathbf{y}_i^{[2]} = \mathbf{e}_4 + \mathbf{y}_1^{[2]} + \mathbf{y}_2^{[2]} + \mathbf{y}_3^{[2]} = (1, 2, 4, 1)$ as a certificate of infeasibility. Observe that $\mathbf{y} \cdot \mathbf{A} = \mathbf{0}$ and $\mathbf{y} \cdot \mathbf{b} = 1 > 0$. Thus, $\mathbf{y}$ does indeed certify the infeasibility of $\mathbf{H}$.

Note that Algorithm 4, may not return a read-once refutation. Additionally, the algorithm may not return an optimal length tree-like or dag-like refutation. (A proof will be provided in the journal version of this paper).

5.1 Resource Analysis

We now analyze the resources taken by Algorithms 3 and 4.

Lemma 2. *Let $\mathbf{H}$ be an HCS with m constraints over n variables, let $\mathbf{Y}$ be an $n \times m$ matrix, and let $\mathbf{d}$ be an n-vector. Algorithm* UPDATE-FARKAS($\mathbf{H}$, $\mathbf{Y}$, $\mathbf{d}$) *runs in time $O(m^2 \cdot n)$.*

Proof. Consider the **for** loop on line 3 of Algorithm 3. This loop iterates at most m times. In each iteration, we update the values in $\mathbf{y}'_i$ for some variable x_i. Note that $\mathbf{y}'_i$ has m values and updating each value takes $O(n)$ time. Thus, each iteration of the **for** loop takes $O(m \cdot n)$ time. Thus, Algorithm 3 takes $O(m^2 \cdot n)$ time.

Theorem 1. *Let $\mathbf{H}$ be an infeasible HCS with m constraints over n variables. Algorithm* CERT-LIFT($\mathbf{H}$) *runs in $O(m^2 \cdot n^2)$ time.*

Proof. We first construct and initialize, the matrix $\mathbf{Y}^{[0]}$ and the vector $\mathbf{d}$. This takes $O(m \cdot n)$ time.

Now let us examine the **for** loop on line 10 of Algorithm 4. This loop iterates at most $(n+1)$ times. In each iteration, we call Algorithm 3. From Lemma 2, this takes $O(m^2 \cdot n)$ time. We then perform a $O(m)$ check for each variable in $\mathbf{H}$ and a $O(n)$ check for each constraint in $N(\mathbf{H})$. These checks take a total of $O(m \cdot n)$ time. Thus, each iteration of the **for** loop takes $O(m^2 \cdot n)$ time. Consequently, Algorithm 4 runs in $O(m^2 \cdot n^2)$ time.

5.2 Certifying the Lifting Algorithm

We now show that the modified Lifting Algorithm for Horn constraints is certifying.

Lemma 3. *Let $\mathbf{H}$ be an HCS with m constraints over n variables and let $\mathbf{Y}^{[r]}$, $r = 0 \ldots (n+1)$ be the $n \times m$ matrix of Farkas vectors maintained by the modified Lifting Algorithm. For $r = 0 \ldots (n+1)$ and each variable x_i, $\mathbf{y}_i \cdot \mathbf{A} \leq \mathbf{e}_i$.*

Proof. We will prove this by induction on r.

For each variable x_i, if $\mathbf{H}$ contains an absolute constraint l_k of the form $x_i \geq b_k$, then $\mathbf{y}_i^{[0]} = \mathbf{e}_k$. In this case, $\mathbf{y}_i^{[0]} \cdot \mathbf{A} = \mathbf{e}_k \cdot \mathbf{A} = \mathbf{e}_i$. If $\mathbf{H}$ contains no absolute constraints of the form $x_i \geq b_k$, then $\mathbf{y}_i^{[0]} = \mathbf{0}$ and $\mathbf{y}_i^{[0]} \cdot \mathbf{A} = \mathbf{0} \leq \mathbf{e}_i$.

Now assume that the desired property holds for $\mathbf{Y}^{[r-1]}$. Note that $\mathbf{Y}^{[r]}$ is created by lifting a variable in $\mathbf{H}$. For each variable x_i, there are two cases we need to consider, either $\mathbf{y}_i^{[r]} = \mathbf{y}_i^{[r-1]}$ or $\mathbf{y}_i^{[r]}$ was updated when the variable x_i was lifted.

In the first case, $\mathbf{y}_i^{[r]} = \mathbf{y}_i^{[r-1]}$. By the inductive hypothesis, $\mathbf{y}_i^{[r]} \cdot \mathbf{A} = \mathbf{y}_i^{[r-1]} \cdot \mathbf{A} \leq \mathbf{e}_i$ as desired.

In the second case, there exists a constraint l_k of the form: $x_i - \sum_{x_j \in N_k} x_j \geq b_k$ such that $\mathbf{y}_i^{[r]} = \mathbf{e}_k + \sum_{x_j \in N_k} \mathbf{y}_j^{[r-1]}$. By the inductive hypothesis,

$$\mathbf{y}_i^{[r]} \cdot \mathbf{A} = \mathbf{e}_k \cdot \mathbf{A} + \sum_{x_j \in N_k} \mathbf{y}_j^{[r-1]} \cdot \mathbf{A} = \mathbf{e}_i - \sum_{x_j \in N_k} \mathbf{e}_j + \sum_{x_j \in N_k} \mathbf{y}_j^{[r-1]} \cdot \mathbf{A}$$

$$\leq \mathbf{e}_i - \sum_{x_j \in N_k} \mathbf{e}_j + \sum_{x_j \in N_k} \mathbf{e}_j = \mathbf{e}_i$$

as desired.

Lemma 4. *Let* **H** *be an HCS. If* CERT-LIFT(**H**) *returns a Farkas vector* **y** *on line 16, then* **H** *is infeasible.*

Proof. Since CERT-LIFT(**H**) returned a vector **y** on line 16, there must be a variable x_i and $1 \leq r \leq n+1$, such that $\sum_{l_k : x_i \in P_k} y_{i,k}^{[r]} > 1$. Let l_{k_r} be the constraint used to obtain $\mathbf{y}_i^{[r]}$. If for each $x_j \in N_{k_r}$ it is the case that $\sum_{l_k : x_i \in P_k} y_{j,k}^{[r-1]} = 0$, then

$$\sum_{l_k : x_i \in P_k} y_{i,k}^{[r]} = \sum_{l_k : x_i \in P_k} (e_{k_r, k} + \sum_{x_j \in N_{k_r}} y_{j,k}^{[r-1]}) = 1 + \sum_{x_j \in N_{k_r}} \sum_{l_k : x_i \in P_k} y_{j,k}^{[r-1]} = 1.$$

Thus, for some $x_{j_{r-1}} \in N_{k_r}$, we must have that $\sum_{l_k : x_i \in P_k} y_{j_{r-1}, k}^{[r-1]} \geq 1$.

If $x_{j_{r-1}} \neq x_i$, let $l_{k_{r-1}}$ be the constraint used to obtain the current value of $\mathbf{y}_{j_{r-1}}^{[r-1]}$. Thus, for some $x_{j_{r-2}} \in N_{k_{r-1}}$, we must have that $\sum_{l_k : x_i \in P_k} y_{j_{r-2}, k}^{[r-2]} \geq 1$. We can continue this process until for some r' $x_{j_{r'}} = x_i$. When the variable x_i is lifted, both $\mathbf{y}_i^{[r]}$ and $\mathbf{y}_i^{[r]} \cdot \mathbf{b}$ increase. Thus, $\mathbf{y}_i^{[r]} \cdot \mathbf{b} > \mathbf{y}_i^{[r']} \cdot \mathbf{b}$ and $\mathbf{y}_i^{[r]} \geq \mathbf{y}_i^{[r']}$.

Observe the following:

$$\mathbf{y}_i^{[r]} \cdot \mathbf{A} = (\mathbf{e}_{k_r} + \sum_{x_j \in N_{k_r}} \mathbf{y}_j^{[r-1]}) \cdot \mathbf{A}$$

$$= \mathbf{e}_{k_r} \cdot \mathbf{A} + \sum_{x_j \in N_{k_r} \setminus \{j_{r-1}\}} \mathbf{y}_j^{[r-1]} \cdot \mathbf{A} + \mathbf{y}_{j_{r-1}}^{[r-1]} \cdot \mathbf{A}$$

$$= (\mathbf{e}_i - \sum_{j \in N_{k_r}} \mathbf{e}_j) + \sum_{x_j \in N_{k_r} \setminus \{j_{r-1}\}} \mathbf{y}_j^{[r-1]} \cdot \mathbf{A} + \mathbf{y}_{j_{r-1}}^{[r-1]} \cdot \mathbf{A}$$

$$\leq (\mathbf{e}_i - \sum_{j \in N_{k_r}} \mathbf{e}_j) + \sum_{x_j \in N_{k_r} \setminus \{j_{r-1}\}} \mathbf{e}_j + \mathbf{y}_{j_{r-1}}^{[r-1]} \cdot \mathbf{A}$$

$$= (\mathbf{e}_i - \mathbf{e}_{j_{r-1}}) + \mathbf{y}_{j_{r-1}}^{[r-1]} \cdot \mathbf{A}$$

$$\vdots$$

$$\leq (\mathbf{e}_i - \mathbf{e}_{j_{r-1}}) + (\mathbf{e}_{j_{r-1}} - \mathbf{e}_{j_{r-2}}) + \ldots + (\mathbf{e}_{j_{r'+1}} - \mathbf{e}_{j_{r'}}) + \mathbf{y}_{j_{r'}}^{[r']} \cdot \mathbf{A}$$

$$= (\mathbf{e}_i - \mathbf{e}_{j_{r'}}) + \mathbf{y}_{j_{r'}}^{[r']} \cdot \mathbf{A}$$

$$= (\mathbf{e}_i - \mathbf{e}_i) + \mathbf{y}_i^{[r']} \cdot \mathbf{A}$$

$$= \mathbf{y}_i^{[r']} \cdot \mathbf{A}.$$

Let $\mathbf{y} = \mathbf{y}_i^{[r]} - \mathbf{y}_i^{[r']} \geq \mathbf{0}$. Note that $\mathbf{y} \cdot \mathbf{b} = \mathbf{y}_i^{[r]} \cdot \mathbf{b} - \mathbf{y}_i^{[r']} \cdot \mathbf{b} > 0$. Additionally, $\mathbf{y} \cdot \mathbf{A} = \mathbf{y}_i^{[r]} \cdot \mathbf{A} - \mathbf{y}_i^{[r']} \cdot \mathbf{A} \leq \mathbf{0}$. Thus, from Farkas' Lemma, the vector $\mathbf{y}$ certifies that $\mathbf{H}$ is infeasible.

Lemma 5. *Let $\mathbf{H}$ be an HCS. If CERT-LIFT($\mathbf{H}$) returns a Farkas vector $\mathbf{y}$ on line 19, then $\mathbf{H}$ is infeasible.*

Proof. Since CERT-LIFT($\mathbf{H}$) returned a vector $\mathbf{y}$ on line 19, there must be an all-negative constraint l_k of the form: $-\sum_{x_i \in N_k} x_i \geq b_k$ such that, for some r, $b_k + \sum_{x_i \in N_k} \mathbf{y}_i^{[r]} \cdot \mathbf{b} > 0$.

Note that $\mathbf{y} = \mathbf{e}_k + \sum_{x_i \in N_k} \mathbf{y}_i^{[r]} \geq \mathbf{0}$. Observe that

$$\mathbf{y} \cdot \mathbf{A} = \mathbf{e}_k \cdot \mathbf{A} + \sum_{x_i \in N_k} \mathbf{y}_i^{[r]} \cdot \mathbf{A} \leq \sum_{x_i \in N_k} -\mathbf{e}_i + \sum_{x_i \in N_k} \mathbf{e}_i = \mathbf{0}.$$

Additionally,

$$\mathbf{y} \cdot \mathbf{b} = b_k + \sum_{x_i \in N_k} \mathbf{y}_i^{[r]} \cdot \mathbf{b} > 0.$$

Thus, from Farkas' Lemma, the vector $\mathbf{y}$ certifies that $\mathbf{H}$ is infeasible.

Lemma 6. *Let $\mathbf{H}$ be an HCS. If CERT-LIFT($\mathbf{H}$) returns an assignment $\mathbf{x}$ and a Farkas vector $\mathbf{y}$, then $\mathbf{H}$ is feasible and $\mathbf{x}$ optimizes the objective function $\min \sum_{i=1}^n x_i$.*

Proof. Let us assume that for some x_{j_1}, $\mathbf{y}_{j_1}^{[n+1]} \neq \mathbf{y}_{j_1}^{[n]}$. Thus, $\mathbf{y}_{j_1}^{[n+1]} \cdot \mathbf{b} > \mathbf{y}_{j_1}^{[n]} \cdot \mathbf{b}$. Let l_k be the constraint used to update $\mathbf{y}_{j_1}^{[n+1]}$. If for each $x_j \in N_k$, $\mathbf{y}_j^{[n]} = \mathbf{y}_j^{[n-1]}$, then

$$b_k + \sum_{x_j \in N_k} \mathbf{y}_j^{[n-1]} \cdot \mathbf{b} = b_k + \sum_{x_j \in N_k} \mathbf{y}_j^{[n]} \cdot \mathbf{b} = \mathbf{y}_{j_1}^{[n+1]} \cdot \mathbf{b} > \mathbf{y}_{j_1}^{[n]} \cdot \mathbf{b}.$$

Thus, Algorithm 3, would have set $\mathbf{y}_{j_1}^{[n]} = \mathbf{y}_{j_1}^{[n+1]}$. This contradicts our assumption. Thus, for some $x_{j_2} \in N_k$, we must have that $\mathbf{y}_{j_2}^{[n]} \neq \mathbf{y}_{j_2}^{[n-1]}$. Since the bound: $x_{j_2} \geq \mathbf{y}_{j_2}^{[n]} \cdot \mathbf{b}$ was used to derive the bound: $x_{j_1} \geq \mathbf{y}_{j_1}^{[n+1]} \cdot \mathbf{b}$, $\sum_{l_k : x_{j_2} \in P_k} y_{j_2,k}^{[n+1]} \geq 1$.

Now, let $l_{k'}$ be the constraint used to update $\mathbf{y}_{j_2}^{[n]}$. As before, there must exist a variable $x_{j_3} \in N_{k'}$ such that $\mathbf{y}_{j_3}^{[n-1]} \neq \mathbf{y}_{j_3}^{[n-2]}$. Additionally, $\sum_{l_k : x_{j_3} \in P_k} y_{j_1,k}^{[n+1]} \geq 1$ and $\sum_{l_k : x_{j_3} \in P_k} y_{j_2,k}^{[n]} \geq 1$.

Consider the sequence $j_1, j_2, j_3, \ldots, j_{n+1}$ constructed as above. Note that for each $p > q$, $\sum_{l_k : x_{j_p} \in P_k} y_{j_q,k}^{[n+1-q]} \geq 1$. Since each element of this sequence belongs to the set $1 \ldots n$, we must have that for some $p > q$, $j_p = j_q$. Thus, $\sum_{l_k : x_{j_q} \in P_k} y_{j_q,k}^{[n-q]} > 1$, since we need to account for the constraint used to derive $x_{j_q} \geq \mathbf{y}_{j_q}^{[n-q]} \cdot \mathbf{b}$ and the constraint used to derive $x_{j_q} \geq \mathbf{y}_{j_q}^{[n-p]} \cdot \mathbf{b}$. In this case, Algorithm 4 would have returned a Farkas vector on line 16. Thus, by Lemma 4, $\mathbf{H}$ is infeasible.

Thus, for each x_i, $\mathbf{y}_i^{[n+1]} = \mathbf{y}_i^{[n]}$. Consider a constraint $l_k \in \mathbf{H} \setminus N(\mathbf{H})$. Let x_i be the variable in P_k. We have that $\mathbf{y}_i^{[n+1]} \cdot \mathbf{b} \geq b_k + \sum_{x_j \in N_k} \mathbf{y}_j^{[n]} \cdot \mathbf{b}$. Thus, $\mathbf{y}_i^{[n+1]} \cdot \mathbf{b} - \sum_{x_j \in N_k} \mathbf{y}_j^{[n+1]} \cdot \mathbf{b} \geq b_k$. This means that setting $x_i = \mathbf{y}_i^{[n+1]} \cdot \mathbf{b}$ satisfies every constraint in $\mathbf{H} \setminus N(\mathbf{H})$.

Consider a constraint $l_k \in N(\mathbf{H})$. Since Algorithm 4 returned an assignment $\mathbf{x}$, then it could not have returned a Farkas vector on line 19. Thus, $b_k + \sum_{x_i \in N_k} \mathbf{y}_i^{[n+1]} \cdot \mathbf{b} \leq 0$. This means that $-\sum_{x_i \in N_k} \mathbf{y}_i^{[n+1]} \cdot \mathbf{b} \geq b_k$. Thus, setting $x_i = \mathbf{y}_i^{[n+1]} \cdot \mathbf{b}$ satisfies every constraint in $N(\mathbf{H})$.

Consequently, setting $x_i = \mathbf{y}_i^{[n+1]} \cdot \mathbf{b}$ satisfies every constraint in $\mathbf{H}$. Thus, $\mathbf{H}$ is feasible.

For each x_i, we have that $\mathbf{y}_i^{[n+1]} \cdot \mathbf{A} \leq \mathbf{e}_i$. Let $\mathbf{y} = \sum_{i=1}^n \mathbf{y}_i^{[n+1]}$. Observe that

$$\mathbf{y} \cdot \mathbf{A} = \sum_{i=1}^n \mathbf{y}_i^{[n+1]} \cdot \mathbf{A} \leq \sum_{i=1}^n \mathbf{e}_i = \mathbf{1}.$$

Additionally,

$$\mathbf{y} \cdot \mathbf{b} = \sum_{i=1}^n \mathbf{y}_i^{[n+1]} \cdot \mathbf{b}.$$

Thus, from the principle of weak duality, setting each $x_i = \mathbf{y}_i^{[n+1]} \cdot \mathbf{b}$ results in a solution to $\mathbf{H}$ that minimizes the objective function $\sum_{i=1}^n x_i$.

Theorem 2. *Algorithm 4 is a certifying algorithm from the linear feasibility problem in Horn constraints.*

Proof. From Lemmas 4 and 5, we know that if Algorithm 4 returns a Farkas vector $\mathbf{y}$ then the HCS $\mathbf{H}$ is infeasible, and the vector $\mathbf{y}$ is the certificate of infeasibility. From Lemma 6, we know that if Algorithm 4 returns an assignment $\mathbf{x}$ along with the Farkas vector $\mathbf{y}$, then the HCS $\mathbf{H}$ is feasible and $\mathbf{x}$ is the certificate of feasibility. Thus, Algorithm 4 is a certifying algorithm from the linear feasibility problem in Horn constraints.

6 Conclusion

In this paper, we developed a certifying version of the Lifting Algorithm [4]. Given an HCS $\mathbf{H}$ with m constraints over n variables, our algorithm returns a certificate of infeasibility if $\mathbf{H}$ is infeasible, and an optimal solution along with a certificate of optimality if $\mathbf{H}$ is feasible. The computation of these certificates increases the running time to $O(m^2 \cdot n^2)$.

HCSs are closely related to Leontief substitution systems. In particular, the transpose of the matrix defining an HCS is Leontief. Additionally, the dual of an optimization problem over an HCSs is an optimization problem over a Leontief substitution system.

The Certifying Lifting Algorithm (Algorithm 4) differs from the Lifting Algorithm (Algorithm 2) in three key factors:

1. Algorithm 4 keeps track of the Farkas vector corresponding to each bound. This allows the certifying algorithm to return a certificate of infeasibility if the input HCS is infeasible or a certificate certifying the optimality of a feasible solution. Maintaining these vectors increases the running time of the algorithm.
2. Algorithm 4 lifts every constraint in each round of lifts instead of only performing the greedy lifts. This has no impact on the correctness of the algorithm. The increase in running time caused by the extra lifts does not impact the asymptotic running time of the algorithm.
3. Algorithm 4 does not adjust the right-hand-sides of the constraints during the lifting procedure. Instead, whenever the right-hand-side of a constraint is needed, the value is adjusted using the currently known bounds. This has no impact on the correctness of the algorithm. The increase in running time caused by the extra computation does not impact the asymptotic running time of the algorithm.

References

1. Alekhnovich, M., Buss, S., Moran, S., Pitassi, T.: Minimum propositional proof length is NP-hard to linearly approximate. In: Brim, L., Gruska, J., Zlatuška, J. (eds.) MFCS 1998. LNCS, vol. 1450, pp. 176–184. Springer, Heidelberg (1998). https://doi.org/10.1007/BFb0055766
2. Bakhirkin, A., Monniaux, D.: Combining forward and backward abstract interpretation of horn clauses. In: Ranzato, F. (ed.) SAS 2017. LNCS, vol. 10422, pp. 23–45. Springer, Cham (2017). https://doi.org/10.1007/978-3-319-66706-5_2
3. Bjørner, N., Gurfinkel, A., McMillan, K., Rybalchenko, A.: Horn clause solvers for program verification. In: Beklemishev, L.D., Blass, A., Dershowitz, N., Finkbeiner, B., Schulte, W. (eds.) Fields of Logic and Computation II. LNCS, vol. 9300, pp. 24–51. Springer, Cham (2015). https://doi.org/10.1007/978-3-319-23534-9_2
4. Chandrasekaran, R., Subramani, K.: A combinatorial algorithm for Horn programs. Discret. Optim. **10**, 85–101 (2013)
5. Cormen, T.H., Leiserson, C.E., Rivest, R.L., Stein, C.: Introduction to Algorithms, 3rd edn. The MIT Press, Cambridge (2009)
6. Cousot, P., Cousot, R.: Abstract interpretation: a unified lattice model for static analysis of programs by construction or approximation of fixpoints. In: POPL, pp. 238–252 (1977)
7. Dantzig, G.B.: Optimal solution of a dynamic Leontief model with substitution. Econometrica **23**, 295 (1955)
8. de Moura, L., Owre, S., Rueß, H., Rushby, J., Shankar, N.: The ICS decision procedures for embedded deduction. In: Basin, D., Rusinowitch, M. (eds.) IJCAR 2004. LNCS (LNAI), vol. 3097, pp. 218–222. Springer, Heidelberg (2004). https://doi.org/10.1007/978-3-540-25984-8_14
9. Dutertre, B., de Moura, L.: The YICES SMT solver. Technical report, SRI International (2006)
10. Farkas, G.: Über die Theorie der Einfachen Ungleichungen. J. für die Reine und Angewandte Mathematik **124**(124), 1–27 (1902)

11. Ford, J., Shankar, N.: Formal verification of a combination decision procedure. In: Voronkov, A. (ed.) CADE 2002. LNCS (LNAI), vol. 2392, pp. 347–362. Springer, Heidelberg (2002). https://doi.org/10.1007/3-540-45620-1_29
12. Iwama, K., Miyano, E.: Intractability of read-once resolution. In: Proceedings of the 10th Annual Conference on Structure in Complexity Theory (SCTC 1995), Los Alamitos, CA, USA, June 1995, pp. 29–36. IEEE Computer Society Press (1995)
13. Jaffar, J., Maher, M.: Constraint logic programming: a survey. J. Logic Program. **503–581**(10), 19–20 (1994)
14. Jeroslow, R.G., Martin, R.K., Rardin, R.L., Wang, J.: Gainfree Leontief substitution flow problems. Math. Program. **57**, 375–414 (1992)
15. Kimura, K., Makino, K.: A combinatorial certifying algorithm for linear programming problems with gainfree Leontief substitution systems. In: Iwata, S., Kakimura, N. (eds.) 34th International Symposium on Algorithms and Computation, ISAAC 2023, Kyoto, Japan, 3–6 December 2023, volume 283 of *LIPIcs*, pp. 47:1–47:17. Schloss Dagstuhl - Leibniz-Zentrum für Informatik 2023
16. Komuravelli, A., Bjørner, N., Gurfinkel, A., McMillan, K.L.: Compositional verification of procedural programs using Horn clauses over integers and arrays. In: Formal Methods in Computer-Aided Design, FMCAD 2015, Austin, Texas, USA, 27–30 September 2015, pp. 89–96 (2015)
17. Lau, K.-K., Ornaghi, M.: Specifying compositional units for correct program development in computational logic. In: Bruynooghe, M., Lau, K.-K. (eds.) Program Development in Computational Logic. LNCS, vol. 3049, pp. 1–29. Springer, Heidelberg (2004). https://doi.org/10.1007/978-3-540-25951-0_1
18. Matouek, J., Gärtner, B.: Understanding and Using Linear Programming (Universitext). Springer, Heidelberg (2006). https://doi.org/10.1007/978-3-540-30717-4
19. McConnell, R.M., Mehlhorn, K., Näher, S., Schweitzer, P.: Certifying algorithms. Comput. Sci. Rev. **5**(2), 119–161 (2011)
20. Mehlhorn, K., Näher, S.: The LEDA Platform of Combinatorial and Geometric Computing. Cambridge University Press, Cambridge (1999)
21. Schrijver, A.: Theory of Linear and Integer Programming. Wiley, New York (1987)
22. Subramani, K.: Optimal length resolution refutations of difference constraint systems. J. Autom. Reason. (JAR) **43**(2), 121–137 (2009)
23. Subramani, K., Williamson, M., Gu, X.: Improved algorithms for optimal length resolution refutation in difference constraint systems. Formal Aspects Comput. **25**(2), 319–341 (2013)
24. Subramani, K., Wojciechowski, P.J.: A combinatorial certifying algorithm for linear feasibility in UTVPI constraints. Algorithmica **78**(1), 166–208 (2017)
25. Subramani, K., Wojciechowski, P.J.: A certifying algorithm for lattice point feasibility in a system of UTVPI constraints. J. Comb. Optim. **35**(2), 389–408 (2018)
26. Subramani, K., Wojciechowski, P.: Tree-like unit refutations in horn constraint systems. In: Leporati, A., Martín-Vide, C., Shapira, D., Zandron, C. (eds.) LATA 2021. LNCS, vol. 12638, pp. 226–237. Springer, Cham (2021). https://doi.org/10.1007/978-3-030-68195-1_18
27. Veinott, A.F., Dantzig, G.B.: Integral extreme points. SIAM Rev. **10**, 371–372 (1968)
28. Wojciechowski, P., Subramani, K., Chandrasekaran, R.: Analyzing read-once cutting plane proofs in Horn systems. J. Autom. Reason. (JAR) **66**, 239–274 (2022)
29. Wojciechowski, P., Subramani, K.: On the lengths of tree-like and dag-like cutting plane refutations of Horn constraint systems. Ann. Math. Artif. Intell. **90**, 979–995 (2022)

Deployment

Pick a Flavour: Towards Sustainable Deployment of Cloud-Edge Applications

Roberto Amadini[1,4], Simone Gazza[1,4], Jacopo Soldani[2],
Monica Vitali[3], Antonio Brogi[2], Stefano Forti[2], Saverio Giallorenzo[1],
Pierluigi Plebani[3], Francisco Ponce[2], and Gianluigi Zavattaro[1]

[1] University of Bologna, Bologna, Italy
{roberto.amadini,simone.gazza,saverio.giallorenzo,
gianluigi.zavattaro}@unibo.it
[2] University of Pisa, Pisa, Italy
{jacopo.soldani,antonio.brogi,stefano.forti,francisco.ponce}@unipi.it
[3] Politecnico di Milano, Milan, Italy
{monica.vitali,pierluigi.plebani}@polimi.it
[4] OPTIMA ARC Centre, Melbourne, Australia

Abstract. Multi-component Cloud-Edge applications, which rely on the Internet of Things, call for suitable and sustainable management techniques. On one hand, they have to guarantee their hardware and Quality of Service requirements over large-scale Cloud-Edge infrastructures. On the other hand, there is an urge to reduce carbon emissions related to the software life cycle at all phases. To this end, this work-in-progress article introduces a novel constraint optimisation approach to adapt multi-flavoured applications and their placement, while determining optimal trade-offs between Quality of Service, operational costs, and carbon emissions. To showcase the practical feasibility of the proposed approach, we provide an open-source implementation and run it over a lifelike example.

Keywords: cloud-edge continuum · carbon-awareness · constraint solving and optimisation

1 Introduction

The considerable growth of the Internet of Things (IoT) and IoT-based applications underscored the need of effectively managing multi-component applications over computing, storage, and networking resources along a seamless Cloud-Edge continuum [4]. Indeed, many of those applications (e.g., augmented reality, remote surgery, online gaming) feature stringent Quality of Service (QoS) requirements, e.g., low latency or high bandwidth between deployed distributed components, which, if unmet, can cause significant performance degradation.

Much literature (e.g., surveyed in [4,8,17]) has therefore tackled the problems related to suitably placing and orchestrating multi-component applications in Cloud-Edge settings while considering (and possibly optimising) various QoS aspects, along with hardware, software, and cost requirements, in the attempt to tame the complexity of the available infrastructure in terms of size, geographic distribution, dynamic nature and resource heterogeneity. Finally, with the increasing energy consumption and carbon footprint of the ICT sector, there is an urgent need to consider those in all phases of the software lifecycle to contribute to reducing its negative effects on the Planet [12,18].

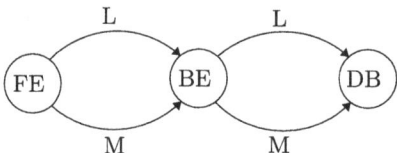

Fig. 1. Example 1's dependency graph. Node BE has no outgoing edge labeled T because a backend in flavour 'tiny' does not require any additional component.

Most existing works, like [1–3,11,14,23], targeted application placement with the goal of reducing energy consumption or carbon emissions. However, they do not envision the possibility to adapt applications (e.g., changing the application components to be deployed) based on predefined target objectives. Only few works indeed considered adapting applications and their placement to changing contextual settings, target objectives, or volume of incoming requests [17]. For instance, [10] considered the possibility to deploy each component of an application in different functionally-equivalent flavours (i.e., versions) depending on application operator's preferences and target operational costs, trying to minimise this latter metric through a greedy approach. However, [10] did not account for energy nor sustainability constraints. More recently, [9] addressed the problem of adapting intent-based applications to the number of users requiring a service, in an attempt to balance profit, carbon emissions, and overall hardware usage. They, however, focused only on chained application topologies, narrowing the applicability of their approach to networking applications.

This work-in-progress article proposes an initial approach enabling the adaptation of both component flavours and application topologies (i.e., by adding or removing components and dependencies between them) when deciding on application placement depending on hardware resource availability and given cost and carbon budgets (i.e., upper bounds). More precisely, we propose a constraint optimisation model to jointly decide (*a*) which application topology to deploy and (*b*) where to deploy application components, without exceeding budget constraints and optimising the number of components deployed in their *preferred* flavour – as specified by the application owner. A MiniZinc implementation of the proposed model and optimisation framework is open-sourced [13] and run over the following example, also relying on various state-of-the-art solvers.

Example 1. Consider a web application with 3 components: a frontend FE, deployable either in flavour *large* (L) or *medium*(M); a backend BE in either flavour *tiny* (T), *medium* (M), or *large* (L); a database DB always coming in flavour *large* (L). These components must be deployed on a distributed architecture of 3 interconnected nodes: n_1, n_2 and n_3. The frontend always needs a backend, regardless of the flavour. However, the backend only needs a database if deployed in flavour medium or large: no need for it if BE is deployed in flavour T. We can represent these relations through the *dependency graph* of Fig. 1.

2 Model

In this section, we formalize the main ingredients of our model: the input parameters, the variables, the constraints, and the objective function. The model is *parametric* in the sense that variables, constraints, and objective function are defined in terms of the input parameters.

2.1 Parameters

The input parameters are those *constant* values shaping a particular *instance* of the parametric model. We formalize them as follows.

Components and Flavours. Let Comps be the set of the components of our application and Flavs the set of the components' flavours. We assume these sets to be non-empty, finite and, w.l.o.g., subsets of $\{1, 2, 3, \dots\}$. Let MustComps $\subseteq$ Comps be the components that *must* always be deployed. This is an important parameter because it defines the '*entry-point*' components that must be deployed to allow the client to interact with the application.

Let $Flav : \mathsf{Comps} \to \mathcal{P}(\mathsf{Flavs})$ be the function returning the set $Flav(c)$ of all flavours offered by component c, while $Uses : \mathsf{Comps} \times \mathsf{Flavs} \to \mathcal{P}(\mathsf{Comps})$ returns the set $Uses(c, f)$ of all the components that *must* be used by c when deployed in flavour f. The corresponding *dependency (multi-)graph* has Comps as set of nodes, and one edge $c \xrightarrow{f} c'$ for each $f \in Flav(c)$ such that $c' \in Uses(c, f)$. We assume this graph to be *topological*, i.e., we do not allow circular dependencies. The introduction of different flavours for an application component increases the flexibility of the deployment. However, not all the flavours have the same quality. For this reason, we formalize the concept of *importance* with a function $imp : \mathsf{Comps} \times \mathsf{Flavs} \to \mathbb{N}$ such that $imp(c, f)$ is the importance of deploying c in flavour f (w.l.o.g. we assume the higher value of imp, the better).

In Example 1, we have Comps = $\{\text{FE}, \text{BE}, \text{DB}\}$ and Flavs = $\{\text{T}, \text{M}, \text{L}\}$. The flavours are $Flav(\text{FE}) = \{\text{M}, \text{L}\}$, $Flav(\text{BE}) = \{\text{T}, \text{M}, \text{L}\}$, and $Flav(\text{DB}) = \{\text{L}\}$. The dependency graph in Fig. 1 is defined by $Uses(\text{FE}, \text{M}) = Uses(\text{FE}, \text{L}) = \{\text{BE}\}$, $Uses(\text{BE}, \text{M}) = Uses(\text{BE}, \text{L}) = \{\text{DB}\}$, and $Uses(\text{BE}, \text{T}) = Uses(\text{DB}, \text{L}) = \emptyset$. Reasonably, we can expect that MustComps = $\{\text{FE}\}$.

Resources. We denote with $\mathsf{Res} = \mathsf{CRes} \cup \mathsf{NRes}$ the finite set of *resources*, defined by the disjoint union between its *consumable* (CRes) and *non-consumable* (NRes) resources. For example, resources like RAM, CPU, or storage are consumable because the resource requirements across all the components deployed on a node must not exceed the node's capacity for that resource.

Non-consumable resources are instead 'immaterial' resources denoting the quality of a service. For example, we may require that a component is deployed on a node available 99% of the time. In this case, we clearly cannot require that the sum of the percentages of the availability required by all components deployed on a node does not exceed the availability percentage promised by that node. Instead, the node's availability must serve as an *upper bound* for the availability of each component deployed on it.

The auxiliary function $Res : \mathsf{Comps} \times \mathsf{Flavs} \to \mathcal{P}(\mathsf{Res})$ returns the resources $Res(c, f)$ required by c when deployed in flavour f. We overload this definition with $Res : \mathsf{Comps} \times \mathsf{Comps} \to \mathcal{P}(\mathsf{Res})$, denoting the set $Res(c, c')$ of resources required for 'connecting' the source component c to the target component c'.

A *component requirement* is a function $comReq : \mathsf{Comps} \times \mathsf{Flavs} \times \mathsf{Res} \to \mathbb{R}$ such that $comReq(c, f, r)$ is the *minimum* amount of resource $r \in Res(c, f)$ required by c to be executed. A *dependency requirement* is instead a function $depReq : \mathsf{Comps} \times \mathsf{Comps} \times \mathsf{Res} \to \mathbb{R}$ such that $depReq(c, c', r)$ is the *minimum* amount of resource $r \in Res(c, c')$ required by the link connecting the "source" component c to the "target" component c' used by c.[1]

In Example 1, the required units of consumable resource $r = \mathrm{CPU}$ likely vary across the different components and their respective flavours. It could be e.g., $comReq(\mathrm{FE, M, CPU}) = comReq(\mathrm{BE, T, CPU}) = comReq(\mathrm{DB, L, CPU}) = 1$; $comReq(\mathrm{FE, L, CPU}) = comReq(\mathrm{BE, M, CPU}) = 2$; $comReq(\mathrm{BE, L, CPU}) = 4$. The availability required by a link connecting two components may also differ, e.g., $depReq(\mathrm{FE, BE, avail}) = 98$ and $depReq(\mathrm{BE, DB, avail}) = 99$.

Capacity and Budget. Let Nodes be the set of nodes composing the Cloud-Edge infrastructure. We represent the *node capacity* with a function $nodeCap$: $\mathsf{Nodes} \times \mathsf{Res} \to \mathbb{R}$ such that $nodeCap(n, r)$ is the maximum amount of resource r available in node n. Similarly, $linkCap : \mathsf{Nodes} \times \mathsf{Nodes} \times \mathsf{Res} \to \mathbb{R}$ models the *link capacity*, i.e., $linkCap(r, n, n')$ is the maximum amount of r that the link between n and n' can handle. Again, these capacities are upper bounds, but the case for lower bounds is symmetrical.

In Example 1, let's consider $\mathsf{Nodes} = \{n_1, n_2, n_3\}$ and $nodeCap(n_1, \mathrm{CPU}) = 2$. If $comReq(\mathrm{BE, L, CPU}) = 4$ then the backend cannot be deployed in flavour large on node n_1. If $linkCap(n_1, n_2, \mathrm{avail}) = 97.5$ and $depReq(\mathrm{FE, BE, avail}) = 98$, then surely we cannot deploy FE and BE over n_1 and n_2 respectively, even

---

[1] Note that both $comReq$ and $depReq$ define *lower* bounds (e.g., *at least* 4GB of RAM needed). However, we may also require *upper* bounds (e.g., *at most* some latency time). For simplicity, in the following, we shall only consider lower bounds, as the case for upper bounds is symmetrical.

if the individual nodes have enough capacity for FE and BE, because the availability required by this dependency exceeds the availability of the link $n_1 \to n_2$.

Finally, we have the *budget* requirements. We define $cost$: Nodes × Res → $\mathbb{R}$ returning the unit cost, in some currency, of using a resource deployed on a given node. Moreover, being one of our goals to be energy-aware, we introduce a function $carb$: Nodes × Res → $\mathbb{R}$ estimating the carbon emission per unit of a resource on a given node.

2.2 Variables

We now explain the "core" of our model. Different, equivalent ways of modeling the deployment are of course possible. We adopted *binary* decision variables to determine if a component is deployed in some flavour on some node. Precisely, we define |Comps| matrices $\mathcal{D}^c$ of $|Flav(c)| \times |$Nodes$|$ binary variables such that, for each $c \in$ Comps, the rows of $\mathcal{D}^c$ represent the flavours of c, and the columns of $\mathcal{D}^c$ the infrastructure nodes, such that for each $i \in Flav(c)$ and $j \in$ Nodes:

$$\mathcal{D}^c_{i,j} = \begin{cases} 1 & \text{if component } c \text{ is deployed in flavour } i \text{ on node } j \\ 0 & \text{otherwise.} \end{cases}$$

This approach facilitates the adoption of different solving technologies, e.g., we can use these variables as is in a mixed-integer linear programming (MIP) model or consider them as Boolean variables for a SAT problem.

2.3 Constraints and Objective

Constraints are relations over the variables, defining the feasible solutions and encoding the deployment requirements. First of all, we must enforce that each component $c \in$ Comps is deployed in at most one flavour, on at most one node:

$$\forall c \in \text{Comps} : \sum_{i \in Flav(c),\ j \in \text{Nodes}} \mathcal{D}^c_{i,j} \leq 1 \qquad (1)$$

Thanks to this constraint, for each $c \in$ Comps we can define an auxiliary variable $node_c = \sum_{i \in Flav(c), j \in \text{Nodes}} j \cdot \mathcal{D}^c_{i,j}$ representing the node where c is possibly deployed, i.e., $node_c > 0$ if and only if c is deployed on $node_c$. So, we can impose that each component $m \in$ MustComps must be deployed by enforcing:

$$\forall m \in \text{MustComps} : node_m > 0 \qquad (2)$$

Note that $node_m > 0$ if and only if $\sum \mathcal{D}^m_{i,j} = 1$.

Another requirement is that if component c is deployed in some flavour i, each component c' used by c in that flavour must be deployed:

$$\forall c \in \text{Comps}, \forall i \in Flav(c), \forall c' \in Uses(c,i) :$$

$$\sum_{j \in \text{Nodes}} \mathcal{D}^c_{i,j} \leq \sum_{i' \in Flav(c'),\ j' \in \text{Nodes}} \mathcal{D}^{c'}_{i',j'} \qquad (3)$$

In Example 1, we have $\mathcal{D}^{\mathrm{FE}} \in \{0,1\}^{2\times 3}$, $\mathcal{D}^{\mathrm{BE}} \in \{0,1\}^{3\times 3}$, $\mathcal{D}^{\mathrm{DB}} \in \{0,1\}^{1\times 3}$. From (1), $\sum \mathcal{D}^c_{i,j} \leq 1$ for each $c \in \{\mathrm{FE}, \mathrm{BE}, \mathrm{DB}\}$. Assuming MustComps = $\{\mathrm{FE}\}$, from (2) we have $\sum \mathcal{D}^{\mathrm{FE}}_{i,j} = 1$ and $node_{\mathrm{FE}} > 0$. Moreover, from $node_{\mathrm{FE}} > 0$ and (3) we can infer that $node_{\mathrm{BE}} > 0$ because $Uses(\mathrm{FE}, \mathrm{M}) = Uses(\mathrm{FE}, \mathrm{L}) = \{\mathrm{BE}\}$. Hence, the constraints in (3) become: $\mathcal{D}^{BE}_{\mathrm{M},n_1} + \mathcal{D}^{BE}_{\mathrm{M},n_2} + \mathcal{D}^{BE}_{\mathrm{M},n_3} \leq \mathcal{D}^{DB}_{\mathrm{L},n_1} + \mathcal{D}^{DB}_{\mathrm{L},n_2} + \mathcal{D}^{DB}_{\mathrm{L},n_3}$, and $\mathcal{D}^{BE}_{\mathrm{L},n_1} + \mathcal{D}^{BE}_{\mathrm{L},n_2} + \mathcal{D}^{BE}_{\mathrm{L},n_3} \leq \mathcal{D}^{DB}_{\mathrm{L},n_1} + \mathcal{D}^{DB}_{\mathrm{L},n_2} + \mathcal{D}^{DB}_{\mathrm{L},n_3}$.

Component Requirements. If a component c deployed in flavour i requires a certain amount of resource r, then $node_c$ must have capacity for r:

$$\forall c \in \mathsf{Comps}, \forall i \in Flav(c), \forall r \in Res(c,i):$$
$$node_c > 0 \implies comReq(c,i,r) \cdot \mathcal{D}^c_{i,node_c} \leq nodeCap(node_c, r) \quad (4)$$

Note that, unlike (1)–(3), this is not a linear formulation because $node_c$ is a variable whose value is generally unknown *a priori*. However, paradigms like Constraint Programming (CP) or Satisfiability Modulo Theory (SMT) can easily handle this formulation. For example, in CP, the *element* global constraint [5] allows the array's indices to be both numbers and variables. Moreover, constraint (4) can be easily *linearised* at the expense of adding more inequalities.

Furthermore, for *consumable* resources only, we must guarantee that a node fulfills the resource requirements for all the components deployed on it. Let CR_j be the set of all the consumable resources available on node j. We impose that:

$$\forall j \in \mathsf{Nodes}, \forall r \in \mathsf{CR}_j:$$
$$\sum_{\substack{c \in \mathsf{Comps}, \\ i \in Flav(c):\, r \in Res(c,i)}} comReq(c,i,r) \cdot \mathcal{D}^c_{i,j} \leq nodeCap(j,r) \quad (5)$$

In this way, we ensure that each node j has a sufficient quantity of a certain resource r to meet the demands of all the components deployed on it. For example, suppose that the parameters of Example 1 are those of Sect. 2.1. Then the constraint (4) for $c = \mathrm{DB}$ and $r = \mathrm{CPU}$ becomes $node_{\mathrm{DB}} > 0 \Rightarrow \mathcal{D}^{\mathrm{DB}}_{\mathrm{L},node_{\mathrm{DB}}} \leq nodeCap(node_{\mathrm{DB}}, \mathrm{CPU})$. Because CPU is consumable, we must ensure that a node has enough CPUs for all the components it hosts, i.e., for $j \in \{n_1, n_2, n_3\}$, $\mathcal{D}^{FE}_{\mathrm{M},j} + 2\mathcal{D}^{FE}_{\mathrm{L},j} + \mathcal{D}^{BE}_{\mathrm{T},j} + 2\mathcal{D}^{BE}_{\mathrm{M},j} + 4\mathcal{D}^{BE}_{\mathrm{L},j} + \mathcal{D}^{DB}_{\mathrm{L},j} \leq nodeCap(j, \mathrm{CPU})$.

Dependency and Budget Requirements. To ensure the satisfaction of the dependency requirements, we impose the following:[2]

$$\forall c \in \mathsf{Comps}, \forall i \in Flav(c), \forall c' \in Uses(c,i), \forall r \in Res(c,c'): \quad (6)$$
$$node_c > 0 \wedge node_{c'} > 0 \implies depReq(c,c',r) \leq linkCap(node_c, node_{c'}, r)$$

Like (4), this is a non-linear constraint that can, however, be linearized – even though in this case, the linearization is not as straightforward.

---

[2] Here we assume, for better readability, only non-consumable resources. For consumable resources, we only need to apply the same reasoning used in (5).

In Example 1 we would have, e.g., a constraint like $node_{BE} > 0 \land node_{DB} > 0 \Rightarrow depReq(BE, DB, avail) \leq linkCap(node_{BE}, node_{DB}, avail)$. Both preconditions $node_{BE} > 0$ and $node_{DB} > 0$ are necessary because a backend deployed in flavour tiny implies $node_{BE} > 0$, but not necessarily $node_{DB} > 0$.

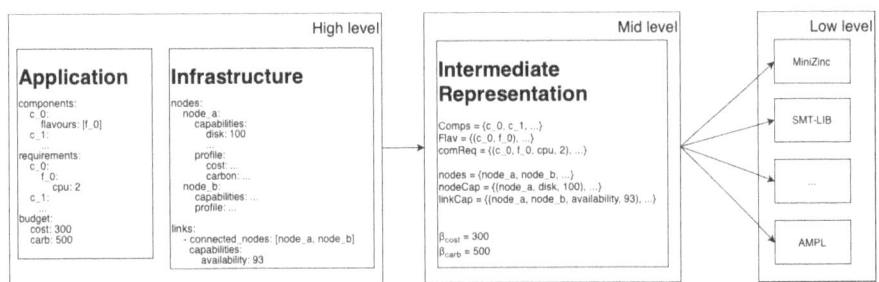

Fig. 2. Hierarchical structure of our architecture.

The following variables denote, respectively, the total cost and carbon emission of the deployment:

$$tot_{cost} = \sum_{c \in \text{Comps}, i \in Flav(c), r \in Res(c,i), j \in \text{Nodes}} comReq(c, i, r) \cdot cost(j, r) \cdot \mathcal{D}_{i,j}^c \quad (7)$$

$$tot_{carb} = \sum_{c \in \text{Comps}, i \in Flav(c), r \in Res(c,i), j \in \text{Nodes}} comReq(c, i, r) \cdot carb(j, r) \cdot \mathcal{D}_{i,j}^c \quad (8)$$

If β_{cost} is our money budget, and β_{carb} is our carbon budget, then we enforce $tot_{cost} \leq \beta_{cost}$ and $tot_{carb} \leq \beta_{carb}$.

Objective Function. To fulfil the definition of optimal deployment, we need to define what 'optimal' means for us. Multiple objective functions can be formulated. For example, a *'cost-oriented'* deployment, aiming to minimize tot_{cost}, an *'environmental-friendly'* deployment, minimizing tot_{carb}, or a *hybrid* approach that minimizes $\gamma \cdot tot_{cost} + \varepsilon \cdot tot_{carb}$ with γ and ε user-defined parameters. While the latter is more flexible, it is crucial to carefully tune γ and ε to reflect the relative importance of cost and carbon factors, typically involving different units of measurement. Furthermore, minimizing tot_{cost} and/or tot_{carb} will likely result in a *minimal* deployment with the least number of components in their "less powerful" flavour (e.g., for Example 1: FE in flavour M, BE in flavour T, and no DB deployed). To avoid the above issues, in this work we focus on the importance of the flavours of a component. The objective function is:

$$\max \left(\sum_{c \in \text{Comps}, i \in Flav(c)} imp(c, i) \cdot \left(\sum_{j \in \text{Nodes}} \mathcal{D}_{i,j}^c \right) \right) \quad (9)$$

3 Implementation

To facilitate flexibility and modularity, our idea is to define a hierarchical structure as schematized in Fig. 2. At the top, a *high-level* model encapsulates the deployment architecture, including its components, requirements, and infrastructure. This model, encoded, e.g., in YAML or JSON format, is then compiled into a *mid-level* model. This process encompasses syntactical and semantic checks, as well as pre-processing steps such as early inconsistency detection and code optimization, to return an encoding, in some format, of the parameters formalized in Sect. 2.1. Then, we can derive different *low-level*, executable models. For example, (4) and (6) can be kept as is, or linearised to better accommodate a MIP solver. Also, the same model can be tackled by solvers employing disparate technologies, necessitating translation into different languages. Furthermore, if pre-processing finds the model unsatisfiable, a proper theorem prover can be invoked to get a (possibly minimal) *explanation*.

Table 1. Runtimes over 100 instances. Best results in bold font.

Solvers	avg	min	max	flat	first	last	Borda
OR Tools	**0.58 s**	**0.42 s**	**1.05 s**	**0.35 s**	0.50 s	**0.55 s**	**173.99**
Gecode	0.81 s	0.44 s	5.69 s	**0.35 s**	**0.48 s**	0.63 s	162.23
Gurobi	0.75 s	0.45 s	2.08 s	0.43 s	0.56 s	0.69 s	149.76
Chuffed	8.81 s	0.79 s	226.34 s	0.35 s	0.88 s	1.27 s	114.02

As a proof of concept, we encoded Example 1 with the *MiniZinc* language [19]. Specifically, due to MiniZinc's capability to separate model and data, the infrastructure parameters outlined in Sect. 2.1 are encoded in .dzn format, while variables, constraints and objective function in .mzn format. We enriched the problem by randomly generating multiple components (from 2 to 3 frontends and backends and 2 databases), resource requirements (2, 4 or 8 CPUs, 3 different security levels, RAM and storage between 200 and 500 MBs, in/out bandwidth from 200 MB to 1 GB), dependency requirements (availability 95–97%), infrastructure (4 or 5 nodes having 8, 16 or 32 CPUs), link capacity (98–99%), tot_{carb} and tot_{cost} (5000–10000 times the number of components).

We maximize the importance of flavours as detailed in (9). Specifically, we randomly generated imp values in [1..10] by preserving for each $c \in$ Comps the invariant $imp(c, \text{T}) < imp(c, \text{M}) < imp(c, \text{L})$. The rationale is that the user usually prefers to deploy larger components, arguably offering a better performance. We leave the study of alternative objective functions as future work.

Table 1 summarizes the results using different solvers: Gecode [20] (a CP solver on finite domains), Chuffed [7] (a lazy clause generation CP solver), Gurobi [16] (a MIP solver), and OR-Tools [15] (its CP-SAT version). We generated 100 instances: 61 satisfiable, 39 unsatisfiable. Every solver concluded the search within 226 s.[3] The 'avg', 'min', and 'max' columns of Table 1 show, respectively, the average, minimum, and maximum solving time. The 'flat' column shows the average time to compile MiniZinc to Flatzinc, a step needed to convert the model in the format required by a solver – note that we consider 'flat' time as part of the solving time. Columns 'first' and 'last' show, respectively, the average time to find the first and last solution (i.e., the optimal one) for satisfiable problems. The former measures the "response time" of a solver, the latter gives an idea of how quickly a solver proves that the incumbent solution is optimal. The last column is inspired by the MiniZinc Challenge score [22]: for every instance, if solvers s, s' conclude the search in time t, t' respectively, t scores $\frac{t'}{t+t'}$ and t' scores $\frac{t}{t+t'}$ points (if $t = t' = 0$ they both score 0.5). The 'Borda' column shows the cumulative score of these pairwise comparisons.

All solvers provide a good performance, with the surprising exception of Chuffed, sometimes struggling to conclude the search. However, it is surely possible that using alternative models and/or search heuristics might improve its performance. This also holds for Gurobi, which would likely benefit more from a specialized linear model, obtained from the mid-level representation, instead of using the built-in linearisation of (4) and (6) provided by MiniZinc libraries [6].

4 Concluding Remarks

In this work-in-progress article, we have proposed a constraint optimisation model to jointly decide which application topology to deploy and where to deploy application components, without exceeding the available cost/carbon budget, and optimising the number of components deployed in their preferred flavour. We have also showcased the practical feasibility of the proposed approach, by introducing an open-source MiniZinc implementation of the model and by running it over a lifelike example – by relying on state-of-the-art solvers.

For future work, we plan to plug the proposed optimisation model into an end-to-end toolchain for enabling the sustainable and failure-resilient deployment of Cloud-Edge applications, as envisioned by the FREEDA project [21].

Acknowledgments. This work is mainly supported by the project FREEDA (CUP: I53D23003550006), funded by the frameworks PRIN (MUR, Italy) and Next Generation EU. The work is also partly supported by: PNRR - M4C2 - Investimento 1.3, Partenariato Esteso PE00000013 - "FAIR - Future Artificial Intelligence Research" - Spoke 8 "Pervasive AI"; "hOlistic Sustainable Management of distributed softWARE systems" (OSMWARE, PRA_2022_64) funded by the University of Pisa, Italy; *Energy-aware management of software applications in Cloud-IoT ecosystems* (RIC2021_PON_A18) funded by the Italian MUR over ESF REACT-EU resources through *PON Ricerca e Innovazione 2014–20*.

---

[3] We ran the experiments on a 2020 M1 MacBook Air with 8 GB of RAM and MacOS Sonoma 14.4.1 using the free search for each solver. A full description of the experiments, together with the source code, is publicly available at [13].

References

1. Abbasi-khazaei, T., Rezvani, M.H.: Energy-aware and carbon-efficient VM placement optimization in cloud datacenters using evolutionary computing methods. Soft. Comput. **26**(18), 9287–9322 (2022)
2. Ahvar, E., et al.: DECA: a dynamic energy cost and carbon emission-efficient application placement method for edge clouds. IEEE Access **9**, 70192–70213 (2021)
3. Aldossary, M., Alharbi, H.A.: Towards a green approach for minimizing carbon emissions in fog-cloud architecture. IEEE Access **9**, 131720–131732 (2021)
4. Apat, H.K., et al.: A comprehensive review on internet of things application placement in fog computing environment. Internet Things **23**, 100866 (2023)
5. Beldiceanu, N., et al.: Global constraint catalogue: past, present and future. Constraints **12**(1), 21–62 (2007). http://sofdem.github.io/gccat/
6. Belov, G., Stuckey, P.J., Tack, G., Wallace, M.: Improved linearization of constraint programming models. In: Rueher, M. (ed.) CP 2016. LNCS, vol. 9892, pp. 49–65. Springer, Cham (2016). https://doi.org/10.1007/978-3-319-44953-1_4
7. Chu, G., et al.: Chuffed, a lazy clause generation solver. https://github.com/chuffed/chuffed. Accessed May 2024
8. Costa, B.G.S., et al.: Orchestration in fog computing: a comprehensive survey. ACM Comput. Surv. **55**(2), 29:1–29:34 (2023)
9. Di Riccio, T., et al.: Sustainable placement of VNF chains in Intent-based Networking. In: IEEE/ACM UCC (2024)
10. Forti, S., Brogi, A.: Declarative osmotic application placement. In: Polyvyanyy, A., Rinderle-Ma, S. (eds.) CAiSE 2021. LNBIP, vol. 423, pp. 177–190. Springer, Cham (2021). https://doi.org/10.1007/978-3-030-79022-6_15
11. Forti, S., Brogi, A.: Green application placement in the cloud-IoT continuum. In: Cheney, J., Perri, S. (eds.) PADL 2022. LNCS, vol. 13165, pp. 208–217. Springer, Cham (2022). https://doi.org/10.1007/978-3-030-94479-7_14
12. Gaglianese, M., et al.: Green orchestration of cloud-edge applications: state of the art and open challenges. In: SOSE, pp. 250–261 (2023)
13. Gazza, S., Amadini, R.: Source code (2024). https://github.com/simonegazza/mzn-test-suite
14. Gnibga, W.E., et al.: Latency, energy and carbon aware collaborative resource allocation with consolidation and QoS degradation strategies in edge computing. In: ICPADS (2023)
15. Google AI: OR-Tools. https://developers.google.com/optimization. Accessed May 2024
16. Gurobi Optimization: Gurobi Optimizer. https://www.gurobi.com/. Accessed May 2024
17. Islam, M.M., et al.: Optimal placement of applications in the fog environment: a systematic literature review. J. Parallel Distrib. Comput. **174**, 46–69 (2023)
18. Manner, J.: Black software - the energy unsustainability of software systems in the 21st century. Oxford Open Energy **2** (2022)
19. Nethercote, N., Stuckey, P.J., Becket, R., Brand, S., Duck, G.J., Tack, G.: MiniZinc: towards a standard CP modelling language. In: Bessière, C. (ed.) CP 2007. LNCS, vol. 4741, pp. 529–543. Springer, Heidelberg (2007). https://doi.org/10.1007/978-3-540-74970-7_38
20. Schulte, C., Tack, G.: Gecode: Generic COnstraint Development Environment. http://www.gecode.org. Accessed May 2024

21. Soldani, J., et al.: Towards sustainable deployment of microservices over the cloud-edge continuum, with FREEDA. In: Workshop on Flexible Resource and Application Management on the Edge (FRAME 2024). ACM (2024). https://doi.org/10.1145/3659994.3660311
22. Stuckey, P.J., et al.: The MiniZinc challenge 2008–2013. AI Mag. **35**(2), 55–60 (2014)
23. Yu, Z., et al.: Less carbon footprint in edge computing by joint task offloading and energy sharing. IEEE Netw. Lett. **5**(4), 245–249 (2023)

Specification, Refactoring and Testing

An Axiomatic Category-Based Access Control Model for Smart Homes

Clara Bertolissi[1], Maribel Fernández[2](✉), and Bhavani Thuraisingham[3]

[1] Aix-Marseille Université, CNRS LIS UMR, 7020, Marseille, France
Clara.Bertolissi@univ-amu.fr
[2] King's College London, London, UK
maribel.fernandez@kcl.ac.uk
[3] University of Texas at Dallas, Richardson, USA
bxt043000@utdallas.edu

Abstract. Internet of Things (IoT) refers to a system of devices that send and receive data via the internet. In the Smart Home IoT, appropriate access controls are key for the security of the household. Popular access control models, such as RBAC and ABAC, have been adapted to this context, but user studies show that new hybrid models are required. We propose a logic-based access control model that is highly expressive: it subsumes the RBAC and ABAC models as well as a whole spectrum of hybrid models. Policies are specified via categorisation of users and devices (a natural mechanism for smart home owners) and have a logic semantics that facilitates policy verification. We have identified a simple yet expressive submodel that satisfies the criteria highlighted in user studies for smart home access control.

Keywords: Access Control · Internet of Things · Smart Home · Category-Based Access Control

1 Introduction

The Smart Home Internet of Things (IoT) consists of a network of devices (home appliances equipped with sensors and software, such as smart locks, TV, playstation, etc.) that can be controlled remotely via web and mobile applications. Access to devices is usually granted to household members (which may include children, babysitters, visiting relatives, etc.) under certain conditions, specified via access control policies. Popular access control models, such as Role-Based Access Control (RBAC) and Attribute-Based Access Control (ABAC), have been adapted to the Smart Home IoT scenario [2,3], however, user studies [17] show that the distinctive features of smart homes require new models combining the benefits of RBAC and ABAC. Indeed, RBAC policies, where users rights are defined on the basis of their role in the household (e.g., parent, child, babysitter) are easy to specify and review, but are not sufficiently flexible to take into account dynamic features (such as whether the babysitter is actually at home or not). ABAC policies assign users rights based on attribute values of users, devices and the environment, and permissions can automatically change when attribute value changes, making it more suitable to the dynamic smart home environment. However, ABAC policies are notoriously difficult to manage. Recent research [3,4] suggests that a hybrid model combining the benefits of RBAC and ABAC is required for smart homes.

Two hybrid models, one based on RBAC and the other on ABAC, have been proposed to address this need: $HyBAC_{RC}$ and $HyBAC_{AC}$ [4]. However, as stated by the authors [4], the "ultimate goal is to have a family of access control models ranging from relatively simple to more sophisticated [...] to provide policy designers with a range of models to choose from according to the environment requirements and the business needs".

In this paper, we propose to achieve this goal via Category-Based Access Control (CBAC). We define a general model, called *SHoCBAC*, from which a family of instances can be obtained ranging from simple RBAC-style models to sophisticated ABAC-style ones and covering the whole spectrum in between. An advantage of this unified approach, where all the models are obtained as instances of *SHoCBAC* by defining appropriate notions of categories, is that policy languages, enforcement mechanisms and policy analysis techniques can be defined once for *SHoCBAC* and then shared across all the instances, thus saving effort. Moreover, cognitive science indicates that categorisation is one of the main mechanisms to organise knowledge [6,13] and user studies confirm that category-based specifications are easier to manage than rule-based ones [19], making *SHoCBAC* well-suited for smart homes.

SHoCBAC is an axiomatic model with a formal semantics that makes it possible to reason about policies. As all CBAC models, it can be equipped with a rewriting-based operational semantics, which facilitates the analysis of policies (e.g., to verify consistency) [1,10].

Since administrative policies can also be defined in an instance of the CBAC metamodel, we are also able to define administrative policies for the Smart Home IoT in the same framework.

With the aim of simplifying policy creation and maintenance tasks for smart home owners, we have identified a core policy template with one administrative category and three basic user categories based on notions of trust, which can be parameterised to take into account environmental attributes. Actions for devices are categorised as safe or unsafe (e.g., opening a door lock or changing the oven temperature are unsafe actions, whereas turning off a light is safe) and assigned to user categories. We show that the core template covers most of the access control requirements for smart homes [17], while the general model, *SHoCBAC*, is able to express all the policies that can be expressed in previous models for Smart Home IoT [4].

The paper is organised as follows: Sect. 2 provides preliminary notions on category-based access control, Sect. 3 introduces *SHoCBAC* and Sect. 4 shows its expressive power by providing encodings of existing models. Section 5 discusses related work and Sect. 6 concludes.

2 Preliminaries: The CBAC Metamodel

In this section we recall the main concepts underlying the category-based access control metamodel [7]. We assume familiarity with first-order logic.

The CBAC metamodel aims at facilitating the definition of access control models. It consist of a family $\mathcal{E}$ of sets of **entities**, which are classified into **categories**, a family $\mathcal{R}el$ of **relationships between entities**, and a set $\mathcal{A}x$ of **axioms** that specify the properties that the model must satisfy.

The classification of entities into categories can be static (e.g., categories can be defined in terms of roles, which can only be updated by the administrator) or can be defined in terms of dynamic parameters (e.g., an airline may categorise clients in terms of the miles travelled, which can be automatically updated each time a user validates new air miles). Since categories can be application dependent, CBAC is extremely expressive: most of the existing access control models can be defined as instances of CBAC by selecting appropriate sets of entities and relationships and specifying adequate notions of categories (see [7,9]).

The metamodel includes the following generic sets of entities in addition to application-dependent entities: a countable set $\mathcal{C}$ of categories, denoted c_0, c_1, ...; a countable set $\mathcal{P}$ of principals, denoted p_0, p_1, ... (we assume that principals that request access to resources are pre-authenticated); a countable set $\mathcal{A}$ of named *actions*, denoted a_0, a_1, ...; a countable set $\mathcal{R}$ of *resource identifiers*, denoted r_0, r_1, ...; a finite set $\mathcal{A}uth$ of possible *answers* to access requests (e.g., {grant, deny, undetermined}).

The metamodel includes the following generic relationships:

- *Principal-Category Assignment*, $\mathcal{PCA} \subseteq \mathcal{P} \times \mathcal{C}$, which assigns categories to principals: $(p, c) \in \mathcal{PCA}$ iff the principal $p \in \mathcal{P}$ is in the category $c \in \mathcal{C}$;
- *Resource-Category Assignment*, $\mathcal{RCA} \subseteq \mathcal{R} \times \mathcal{C}$, which assigns categories to resources: $(r, c) \in \mathcal{RCA}$ iff the resource $r \in \mathcal{R}$ is in the category $c \in \mathcal{C}$;
- *Permissions*, $\mathcal{ARCA} \subseteq \mathcal{A} \times \mathcal{C} \times \mathcal{C}$, which assigns permitted actions to categories of principals and resources; such that $(a, c_r, c_p) \in \mathcal{ARCA}$ iff the action $a \in \mathcal{A}$ on resource category $c_r \in \mathcal{C}$ can be performed by principals assigned to the category $c_p \in \mathcal{C}$.
  Similarly, banned actions are specified via the relation $\mathcal{BARCA}$.
- *Authorisations*, $\mathcal{PAR} \subseteq \mathcal{P} \times \mathcal{A} \times \mathcal{R}$, such that $(p, a, r) \in \mathcal{PAR}$ iff the principal $p \in \mathcal{P}$ is allowed to perform the action $a \in \mathcal{A}$ on the resource $r \in \mathcal{R}$.

Similarly, a relation $\mathcal{BAR}$ defining prohibitions is also included.

Additional application-dependent relations can also be included. In this paper, we consider also categories of actions (safe/unsafe), and a relation $\mathcal{ACA} \subseteq \mathcal{A} \times \mathcal{C}$ (action-category assignment). In addition, the metamodel includes a reflexive-transitive relation $\subseteq$ between categories to specify a hierarchy of categories and permission inheritance.

Authorisations are directly deduced from the previous relations using the **core axiom** defined below (more axioms are added to specify how prohibitions are deduced from $\mathcal{BARCA}$ and to ensure that no inconsistencies arise; we refer to [10] for details).

(a1) $\forall p \in \mathcal{P}, \forall a \in \mathcal{A}, \forall r \in \mathcal{R}, (\exists c_p \in \mathcal{C}, \exists c'_p \in \mathcal{C}, \exists c_r \in \mathcal{C}, \exists c'_r \in \mathcal{C}, \exists c_a \in \mathcal{C},$
$(p, c_p) \in \mathcal{PCA} \wedge (r, c_r) \in \mathcal{RCA} \wedge (a, c_a) \in \mathcal{ACA}$
$\wedge\ c_p \subseteq c'_p \wedge c_r \subseteq c'_r \wedge (c_a, c'_r, c'_p) \in \mathcal{ARCA}) \Leftrightarrow (p, a, r) \in \mathcal{PAR}$

Based on this axiom, we can deduce that if a principal p is in the category c_p (that is, $(p, c_p) \in \mathcal{PCA}$), a resource r is in the category c_r (that is, $(r, c_r) \in \mathcal{RCA}$), and the category c_p is permitted to perform actions from category c_a on resource category c_r (that is, $(c_a, c_r, c_p) \in \mathcal{ARCA}$) then p is authorised to perform a on r (that is, $(p, a, r) \in \mathcal{PAR}$).

Definition 1. *Given a specification of $\mathcal{E}$ and $\mathcal{Rel}$, and a set $\mathcal{Ax}$ of axioms, a CBAC **policy** is a tuple $\langle E, Rel \rangle$ that defines the contents of $\mathcal{E}$ and $\mathcal{Rel}$ such that E, Rel satisfy all the axioms in $\mathcal{Ax}$.*

Sessions are not included in the core CBAC model but if needed they could be added as another set of entities, or they could be specified via user attributes.

RBAC and ABAC as Instances of CBAC. If categories of principals are defined using roles and resources are not categorised (i.e., each resource is in its own individual category) then we obtain the standard RBAC model. ABAC policies can be obtained by using a notion of category based on user, resource and environment attributes. C-ABAC [14] is a formal specification of the ABAC model in the CBAC metamodel. In C-ABAC, the set $\mathcal{E}$ of entities includes, in addition to the generic sets $\mathcal{P}, \mathcal{A}, \mathcal{R}, \mathcal{C}$, a countable set $\mathcal{E}nv$ of environment entities (e.g., networks, clock, etc.) and a set $\mathcal{A}t$ of *attributes*, ranging over *values $\mathcal{V}$*. A set $\mathcal{C}ond$ of Boolean expressions involving attributes and values is used to define categories: the expressions specify the acceptable ranges of attribute values required by each category. The set $\mathcal{Rel}$ of relationships in C-ABAC includes:

- *Principal-Attribute Assignment*, $\mathcal{PAtA} \subseteq \mathcal{P} \times (\mathcal{A}t \times \mathcal{V})^*$, which defines the attributes of each principal and their current values;
- *Resource-Attribute Assignment*, $\mathcal{RAtA} \subseteq \mathcal{R} \times (\mathcal{A}t \times \mathcal{V})^*$, which defines the attributes of each resource and their current values;
- *Environment-Attribute Assignment*, $\mathcal{EAtA} \subseteq \mathcal{E}nv \times (\mathcal{A}t \times \mathcal{V})^*$, which defines the relevant environmental attributes and their values;
- *Category-Attribute Assignment*, $\mathcal{CAtA} \subseteq \mathcal{C} \times \mathcal{C}ond$, which defines a condition on attribute values for each category. We write $\Gamma \vdash cond$ if Γ specifies attributes and values that satisfy $cond$. C-ABAC includes, in addition to the core axiom (a1), also categorisation axioms (see [14] for more details).

(c1) $\forall p \in \mathcal{P}, \forall c \in \mathcal{C}, ((p, c) \in \mathcal{PCA} \Leftrightarrow$
$\exists l_p, (p, l_p) \in \mathcal{PAtA}, \exists cond, (c, cond) \in \mathcal{CAtA}, (l_p, \mathcal{RAtA}, \mathcal{EAtA} \vdash cond))$
(c2) $\forall r \in \mathcal{R}, \forall c \in \mathcal{C}, ((r, c) \in \mathcal{RCA} \Leftrightarrow$
$\exists l_r, (r, l_r) \in \mathcal{RAtA}, \exists cond, (c, cond) \in \mathcal{CAtA}, (l_r, \mathcal{PAtA}, \mathcal{EAtA} \vdash cond))$

An administrative model for access control, *Admin-CBAC*, has also been defined as an instance of the CBAC metamodel (see [11]). For this, the sets of entities and relationships are split into standard and administrative subsets, and administration axioms are included to ensure safety.

3 Smart Home Instance of CBAC: *SHoCBAC*

Authorisation and privacy issues in the smart home domain have attracted attention recently. The threat model includes two major kinds of adversaries: external third parties that try to access devices or the data produced by the devices and insiders that misuse the devices or their generated data, with the intent of causing physical, financial or privacy-related damage.

To avoid device misuse, access control policies for the smart home need to be fine grained (to specify rights at the level of device actions rather than the device as a whole) and need to be dynamic to take into account environmental conditions, such as time and location. CBAC permits the definition of fine-grained, dynamic authorisation at the level of actions, has a formal definition and several proof of concept implementations exist (see [11]). Additionally, its category-based foundations make it easy to combine it with data sharing models [15], to offer not only access control but also privacy guarantees as required. These features make CBAC a suitable approach for the smart home domain.

In this section we define an instance of CBAC, *SHoCBAC*, for the Smart Home IoT domain. Categories will be specified using attributes, so *SHoCBAC* is actually an instance of C-ABAC, as specified in Sect. 2. We start by specifying the sets of entities and relationships considered.

SHoCBAC Entities: The set $\mathcal{P}$ of principals in *SHoCBAC* includes smart home users as well as services that require access to data generated by smart home devices. The set $\mathcal{R}$ of resources include devices and their generated data. The set $\mathcal{A}$ of actions includes actions on devices (such as play for entertainment devices, on/off for lights and locks, etc.) as well as actions that services can request on data, such as read (electricity consumption data or lock log status, for example). The set of attributes $\mathcal{A}t$ includes *role* for principal (e.g., owner, child, babysitter), and *type* for resources (e.g., entertainment device, security device). We consider environmental attributes in $\mathcal{E}\mathcal{A}t\mathcal{A}$: *date*, *time*, and additionally *weekDay*, *weekEnd*, *nightTime* and *dayTime* that can be customised by the owner to denote dates and times in certain ranges. For example, *nightTime* can be defined to be $\{21.00 - 7.00\}$, and *weekDay* to be *{Monday, Tuesday, Wednesday, Thursday, Friday}*. Similarly, we use the predefined Boolean attribute *holiday*, which is also customisable and it has the value true when the owners are on holidays. We also consider an additional attribute *onwer@Home*, which is a Boolean and has the value true if one of the owners is at home.

Policies in *SHoCBAC* include predefined sets of categories:

- Principal Categories: There are two core categories of principals: *Users*, which may have subcategories such as home-owner, partner, child, babysitter, neigh-

bour, visitor, and *Services* which may have subcategories such as health, insurance, energy-company, etc.
- Resource Categories: There are two core categories of resources: *Data*, which may have subcategories such as log-files, usage time, etc., and *Devices*, which has subcategories such as entertainment, cooking, etc.
- Action Categories: There are two core categories of actions: *Safe* and *Unsafe*. Even though in CBAC actions are not usually categorised, in the smart home domain it is convenient to distinguish categories of actions to facilitate the definition of policies (see examples below).

Although we assume two core categories for principals and two for resources (following a separation of concerns principle), it is possible for a household to define only a policy to control access to devices without any restriction on the use of data and vice-versa only a policy to restrict access to data but no restriction on the use of devices. In the rest of the paper we focus on the access control sub-model, that is we will not consider the subcategories Data and Services.

SHoCBAC Relationships and Axioms: The sets of relationships and axioms in *SHoCBAC* are the ones defined for C-ABAC.

SHoCBAC may be seen also as an instance of *Admin-CBAC*, including an administrative level where there is only one category: *owner-admin*. If administrative policies are required, the additional relationships and axioms in *Admin-CBAC* should be included.

We do not consider sessions in this paper, because they are not considered an essential feature of Smart Home IoT systems in user studies (see e.g. [17]). If needed, sessions can be modelled as an attribute of users in C-ABAC.

3.1 Use Case

We illustrate the use of *SHoCBAC* with an example, inspired from the user study in [17]. In this work, participants give their access control preferences for a list of pre-identified capabilities, according to the following characteristics: teenager, child, visiting family, babysitter, neighbour. We describe below some examples and show how they can be modelled using *SHoCBAC*.

Example 1 (Static Policy). Consider a family composed of Omar, who is the owner of the smart devices and administrator of the smart hub used to access the devices, and partner Oprah, who hired a babysitter (Bea) to take care of two children, Clohe aged 8 and Tom aged 14. They have several neighbours, of which Neil is trusted to take care of the house when the family is on holidays. They also host some other members of the family (e.g. aunt Vicky) who come to visit from time to time.

This can be specified in *SHoCBAC* by including in the set of principals all the members of the family, assigning them a role (e.g., owner, child, ...) and defining a set of (static) categories based on their role $\mathcal{C} =$ {owner, teenager, child, supervisedChild, houseKeeper, visitor, babysitter, neighbour}. Thus, we obtain the following $\mathcal{PCA}$ relation:

$\mathcal{PCA}$ ={(Omar, owner), (Oprah, owner), (Clohe, child), (Bea, Babysitter), (Tom, teenager), (Neil, neighbour), (Vicky, visitor)}.

We can suppose for now that principal categories are statically assigned by Omar, acting as the administrator of the policy.

The house is equipped with several IoT connected devices, such as a voice assistant, a smart TV, a playstation, smart heating, lights and cameras, smart oven, etc. These devices form the set $\mathcal{R}$ of resources of the *SHoCBAC* policy. Each device has a type attribute and a standard set of actions: open/close, lock/unlock, turn on/off, play, etc., which are categorised as safe or unsafe (e.g., turning the oven on, changing its temperature are considered unsafe, whereas turning the oven off is safe). Different device categories are defined on the basis of the device type: entertainment, cooking, lights, heating, etc.

Again, we can assume that device categories are static, based on the device type, which is assigned by the administrator. Thus, the $\mathcal{RCA}$ relation includes for instance: $\mathcal{RCA}$ ={ (oven, Cooking), (TV, Entertainment), ... }. The administrator of the policy (i.e. Omar) can then define permissions over categories of users and devices (the $\mathcal{ARCA}$ relation of the *SHoCBAC* policy). The use of categories of actions simplifies the definition of permissions, since it is possible for example to assign to the Child category the permission to perform any safe action on the cooking devices: $\mathcal{ARCA}$ = { (SafeCooking, Cooking, Child), (UnsafeHeat, Heating, Owner),... }.

To take into account the fact that authorisations may change depending on factors such as whether the family is on holidays or not, we can use dynamic attributes, such as time and location, in the definition of categories. Attributes are seen as functions returning a value at query evaluation time. In the following, we use the notation *entity.attribute* to denote principal, resource and environment attributes. Recall that in C-ABAC, category definitions consist of Boolean expressions. We give an example next.

Example 2 (Dynamic Policy). Categories as defined in Example 1 can be refined by using other attributes in addition to *role* for principals and *type* for resources, in such a way that authorisations adapt to changes in the environment without the need of manual changes by the administrator. For instance, we can refine the static category Child defined in Example 1, by defining a sub-category *SupervisedChild* that includes a condition about the presence of one of the parents at home, which may imply more permissions for the child. For this, we use the environment attribute *owner@Home*, which is true if one of the owners is at home (i.e., at a GPS location corresponding to the house).

$$SupervisedChild.cond = (p.role = Child \text{ and } env.owner@Home)$$

In this way, the $\mathcal{PCA}$ relation will change dynamically (unlike the $\mathcal{PCA}$ relation specified in Example 1, which is static):

$$\mathcal{PCA}(p) = \text{if } p.role = Child \text{ then (if } env.owner@Home \text{ then} \\ SupervisedChild \text{ else } p.role)$$

Similarly, we may want to define permissions according to specific intervals of time. To this end, we can use the environmental attributes defined above to specify, e.g. that children can turn on the TV only on weekends, or that neighbours can have access to the house when the owner is on holidays. For example:

$$\mathcal{PCA}(p) = \text{if } p.role = Neighbour \text{ then (if } env.holiday \text{ then} \\ HouseKeeper \text{ else ...}$$

where the category *HouseKeeper* has additional permissions w.r.t. the category *Neighbour*, such as access to Security devices (alarm, cameras, door lock, etc.).

We choose to specify restrictions using conditions in the definition of the category, rather than adding an ad-hoc condition in the definition of the permissions (i.e. $\mathcal{ARCA}$ in our case, or role-permission assignment in a RBAC style model [2]). In this way, we can obtain a structured policy where principals (resp. resources) that share the same permissions are grouped in a category.

Remark: Administrative categories could be similarly defined. For example, we could have several administrators (Ad1, Ad2,...) associated with different administrative categories (Entertainment-mngmnt, Ownership-control,...). We could also include prohibitions (e.g., to forbid administrators from making certain permission assignments) by using the relation $\mathcal{BARCA}$. In this way we could for example specify that the administrator Ad1 cannot add a permission for the Child category to access the Entertainment device category on school days.

3.2 Core *SHoCBAC* Policy Template

The above example highlights the expressive power of *SHoCBAC*. In this section we focus on another important property of access control models: usability. User studies show that access control models should align with natural cognitive processes [19]. Moreover, in the case of smart homes, simplicity is key [17]. We will define a simple *SHoCBAC* policy template, *Core*, that has both these features and is sufficiently expressive for Smart Home applications.

Minimal Template. Cognitive science studies [6,13] confirm that categorisation is one of the main mechanisms underlying cognition, and categorisation is easier if there is only a small number of groups. Taking this into account, we propose a default policy template with only one administrative category (owner) and three standard principal categories defined on the basis of trust levels. More precisely, each of the three principal categories, LT, MT, FT, are associated with a trust level, namely *low*, *medium* and *full* trust. Trust level is an attribute of principals (i.e., it will be recorded in $\mathcal{PAtA}$). Typically, the owners have full trust level, external people coming regularly into the house (e.g., for working reasons, such as babysitters and house cleaners, or to visit, such as grand parents), have medium trust level, and children have low trust level. Principals also have a dynamic attribute *location*: its value corresponds to the principal's location as defined by their mobile phone GPS. The conditions defining the three *Core* categories are:

$$LT.cond = (p.level = low \text{ and } p.location = atHome)$$
$$MT.cond = (p.level = medium \text{ and } p.location = atHome)$$
$$FT.cond = (p.level = full)$$

In *Core*, there is a hierarchical relation between categories: $FT \subseteq MT \subseteq LT$. Intuitively, LT is very restrictive (i.e., there are very few tuples in $\mathcal{ARCA}$ for LT), MT involves some restrictions and principals in FT will be granted full permissions (FT inherits the permissions of MT, which inherits those of LT).

Extended Template. The three *Core* categories can be refined: sub-categories can be defined by using environmental attributes in $\mathcal{EAtA}$. We consider an extended template with the following hierarchy of principal categories:

$$FT \subseteq FMT \subseteq MT \subseteq Supervised \subseteq LT$$

The principal category $Supervised \subseteq LT$ is predefined as:

$$Supervised.cond = (p.level = low \text{ and } p.location = atHome$$
$$\text{and } env.owner@Home)$$

Thus *Supervised* inherits all the permissions of LT according to the CBAC axiom ($a1$) and it could have more permissions. For example a child in the *Supervised* category might be able to access the TV even if the LT category does not have this permission.

Similarly, a category $FMT \subseteq MT$ is included with the following condition:

$$FMT.cond = (p.level = medium \text{ and } p.location = atHome \text{ and } env.holiday)$$

Thus, FMT inherits the permissions of MT and can have more permissions. In this way, permissions can be delegated from the FT category to FMT household members during holidays.

Summarising, *Core* includes the following pre-defined categorisation for principals:

$$\mathcal{PCA}(p) = \text{if } p.level = low \text{ and } p.location = atHome \text{ then}$$
$$\quad (\text{if } env.owner@Home \text{ then } Supervised \text{ else } LT)$$
$$\quad \text{else}$$
$$\quad \text{if } p.level = medium \text{ and } p.location = atHome \text{ then}$$
$$\quad (\text{if } env.holiday \text{ then } FMT \text{ else } MT)$$
$$\quad \text{else}$$
$$\quad \text{if } p.level = full \text{ then } FT$$
$$\quad \text{else } error$$

For users with a more technical background or in cases where more fine-grained policies are required (for example, families that prefer to differentiate young children from teenagers), the full *SHoCBAC* model can be used.

For devices, categories in *Core* are associated with the device *type* (an attribute stored in $\mathcal{RAtA}$) and are also hierarchically organised. The main device types are *Cooking* (e.g., smart oven), *Heating*, *Entertainment* (e.g., TV, playstation), *Lights*, *Security* (including devices such as alarm and door lock) and *Cameras*. The user can add, delete or refine these categories. Any number of devices can be assigned to a category, however, all the devices in a category must have the same actions.

Category Refinement. To refine the above-mentioned static device categories, environmental attributes can be used (e.g., date, time, weekDay/weekEnd, dayTime/nightTime). For instance, we may need to specify a category ensuring that unsupervised children cannot access the TV. In *Core*, the category *EntRestricted* $\subseteq$ *Entertainment* can be used to restrict access to entertainment devices. The idea is that $\mathcal{ARCA}$ includes permissions for principals in *LT* to access devices in the category *Entertainment*, but devices in *EntRestricted* can only be accessed by principals in *Supervised*. The hierarchical relation ensures that unsupervised children (in the category *LT*) cannot access the TV. Similarly, the device category Security can be restricted during the weekend:

$EntRestricted.cond = (r.type = Entertainment$ and $env.date \in env.weekDay$
$\qquad\qquad\qquad\qquad$ and $env.time \in env.nightTime)$
$SecRestricted.cond = (r.type = Security$ and $env.date \in env.weekEnd)$

Given the above definitions, the category of a device can be directly computed as specified in the C-ABAC axioms (i.e., by evaluating the conditions using the attribute values at the point of the access request). For example:

$\mathcal{RCA}(r) =$ if $r.type = Entertainment$ then
$\qquad\qquad$ (if $env.date \in env.weekDay$ and $env.time \in env.nightTime$
$\qquad\qquad$ then $EntRestricted$ else $Entertainment$)
$\qquad\quad$ else
$\qquad\quad$ if $r.type = Security$ then
$\qquad\qquad$ (if $env.date \in env.weekEnd$ then $SecRestricted$ else $Security$)
$\qquad\quad$ ...

There are two categories of actions for each category of device, namely *Safe* and *Unsafe*, as well as a category *All* which contains the union. For example, the actions On and changeTemperature are in *UnsafeCooking*, whereas Off is in *SafeCooking*. *Core* includes the assignment of safe and unsafe actions to categories of principals as follows: only safe actions in the categories Entertainment, Lights and Cooking are assigned to LT; Supervised and MT include all safe and unsafe actions in those categories, as well as safe and unsafe actions on the Heating and Security categories, the only restriction is the actions on Cameras (e.g., we may not want the house cleaner in MT to manipulate cameras, however this may be allowed for the house keeper in FMT); FT is the top of the hierarchy and has no restrictions. The tuples defining the core policy as well as the more refined policy, where the categories Supervised, FMT, EntRestricted and

SecRestricted could be used, are synthesised in Table 1. More general policies, where the categorisation of actions may depend for example on environmental attributes, could be defined using the general model.

Table 1. *SHoCBAC* policy templates

Minimal policy
Principal categories: LT, MT, FT.
Device categories: $Cooking, Lights, Entertainment, Heating, Security, Cameras$.
Permissions in $\mathcal{ARCA}$: (SafeCooking, Cooking, LT),
(SafeLights, Lights, LT),
(All, Entertainment, LT),
(All, Heating, MT),
(All, Security, MT),
(All, Cameras, FT).
Extended policy
Principal categories: $LT, Supervised, MT, FMT, FT$
Device categories: minimal categories plus $EntRestricted, SecRestricted$.
Permissions in $\mathcal{ARCA}$: *minimal policy permissions*,
(UnsafeCooking, Cooking, Supervised),
(All, EntRestricted, Supervised),
(All, SecRestricted, MT),
(All, Security, FMT),
(All, Cameras, FT).

Summarising, the administrator simply needs to assign a trust level (Full, Low or Medium) to each principal and a type to each device (Cooking, Entertainment, Security, etc.), then (optionally, if using the extended core template) parameterise environmental attributes such as holiday dates or Day/Night time and indicate whether any of the levels/types should be refined and whether the standard permissions associated to principal categories are appropriate, changing any assignments if required. The authorisations are then automatically derived.

Core includes an administrative level that contains only one category, Owner-Admin, the only category with access to the log files keeping track of the actions that have been carried out on all the devices.

Policy analysis techniques developed for CBAC are directly applicable to *SHoCBAC*, see for example [12, 14]).

3.3 Smart Home Policies in *Core*

In this section we consider some standard rules that user-studies [17] suggest for Smart Home IoT access control, and show how they can be modelled using *Core*.

- *Children should never be able to use certain capabilities without supervision.*
  To model this rule, it is sufficient to give children a low trust level. Then they will be categorised in LT, which does not have access to unsafe operations on devices. If using the Supervised category, they will automatically be assigned to *Supervised* when the parents are at home. For example, assume Chloe is a child and Chloe.level = low in $\mathcal{PAtA}$. Her category is computed using the following rule:

  $\mathcal{PCA}(Chloe) = $ if $Chloe.level = low$ and $Chloe.location = atHome$
  then if $env.owner@Home$ then $Supervised$ else LT

- *Babysitters, neighbours, and visitors should only be able to use any capabilities while in the house.*
  Assuming Babysitters, neighbours and visitors are assigned to the medium trust level, the condition $p.location = atHome$ in the definition of $\mathcal{PCA}$ ensures that they only have access to the devices while they are at home.
- *Any user who is currently at home should always be allowed to adjust lighting.*
  By assigning the actions to adjust lighting devices to the SafeLights category, we ensure that anyone at home can adjust lighting, since we have a tuple *(SafeLights, Lights, LT)* and all users belong to LT when they are at home. The latter is a consequence of the condition $p.location = atHome$ in the $\mathcal{PCA}$ rule that assigns the category LT to principals with low trust level, and the fact that $FT \subseteq MT \subseteq LT$, i.e., principals in the categories FT and MT inherit the permissions from LT.
- *Entertainment devices can be used freely by teenagers during the weekend, but need parental supervision on weekdays.*
  To specify this restriction for teenagers, we need a category with less permissions than MT but more than Supervised: If we assign teenagers medium trust level, then to ensure this restriction we will need to restrict all MT members, which might not be what is intended. If we assign teenagers low trust level then they will also need parental supervision on weekends (the Supervised category does not distinguish weekends from week days). This example shows the limits of *Core*, which is based on trust levels rather than specific characteristics of the principals. This restriction can be specified instead in the general *SHoCBAC* model.
- *Neighbours can have some additional capabilities when the owner is on vacation, e.g. the possibility to turn off the alarm.*
  By assigning neighbours medium trust level, they will be in the MT category if the owners are not on holidays, and in the FMT category when the owners are on holidays, which will give them the capability to perform all actions on security devices.
- *No one should be allowed to delete log files.*
  Since the action to delete log files has not been assigned to any category of users, no user is allowed to delete log files.
- *Spouses should have access to all capabilities, except for deleting log files.*
  It is sufficient to assign spouses the full trust level, which will categorise them as FT. All the capabilities are then available.

Core is sufficiently flexible to accommodate standard policies. As in RBAC, we can define static categories using trust levels, and as in ABAC, category membership can be refined dynamically without the intervention of an administrator. However, there may be situations where more involved policies are needed. This can be done by using the full power of *SHoCBAC*.

4 Expressive Power

In this section we show how to specify HyBAC policies [4] in *SHoCBAC*. Due to space limitations, we focus on $HyBAC_{RC}$, which follows a role-centric approach, and leave out the encoding of the attribute-centric $HyBAC_{AC}$ (which is closer to C-ABAC, and thus to *SHoCBAC*). As mentioned earlier, we do not consider sessions in this paper. If needed they can be represented as attributes of principals.

We start by briefly recalling the $HyBAC_{RC}$ model. We show next how CBAC policies can mimic $HyBAC_{RC}$ policies and produce the same result, while being more concise. The $HyBAC_{RC}$ definitions and examples are taken from [4].

$HyBAC_{RC}$ *Policy.* A $HyBAC_{RC}$ policy is composed of

- **users, roles and their relations**. Specifically, finite sets U of users and R of roles, and the user-role assignment relation $UA \subseteq U \times R$.
- **devices, operations and their relations**. Specifically, finite sets D of devices, DR of device roles, OP of operations, and a device-operation assignment relation $P \subseteq D \times OP$. A pair $(d \in D, op \in OP) \in P$ means that operation op is permitted on device d. For instance we may have $(Oven, Open_{oven}) \in P$. The relation PDRA is a many-to-many relation assigning permissions to device roles: $PDRA \subseteq P \times DR$. For instance, $((Oven, Open_{oven}),$
$Dangerous\_kitchen\_perm) \in PDRA$. For all permissions p, we denote $droles(p) = \{dr \mid (p, dr) \in PDRA\}$.
- **environment related information**. There is a finite set ER of environment roles, triggered by a set EC of environment conditions determined by sensors (e.g., in the kitchen, evening). They are related by a many-to-many environment role activation relation $EA \subseteq 2^{EC} \times ER$. For instance we may have $((weekend, evening), Kids\_Entertainment\_time) \in EA$.
- **dynamic attributes**. Specifically, finite sets of attributes associated with users ($DUSA$) and with devices (DDA), such that $DUSA \cap DDA = \emptyset$.
- **role pairs and their relations**. There is a relation $RP \subseteq R \times 2^{ER}$ pairing user roles and environment roles (e.g. $(kids, Kids\_Entertainment\_time) \in RP$) and derived relations $RPRA \subseteq RP \times R$ and $RPEA \subseteq RP \times 2^{ER}$.
The role-pair assignment relation $RPDRA \subseteq RP \times DR$ binds all these components together: for any role $rp \in RP$, the role r associated with it through $RPRA$ has access to all device roles assigned to it through $RPRA$, when the set of environment roles associated with it through $RPEA$ are active. Note that actions are not mentioned explicitly in this relation, the user roles are

linked with the device role which forces one to define different device roles, corresponding to the different capabilities of the device.

Evaluation of an Access Request. In $HyBAC_{RC}$, $CheckAccess(u, op, d, ec)$ is evaluated when a user u asks to perform an operation op on a device d when the environment condition $ec \in EC$ is active. The $CheckAccess$ predicate is composed of an authorisation function $Authorization(u, op, d)$ which evaluates to $True$ if u is allowed to perform op on d according to the current attributes values of u and d. The $CheckAccess$ predicate also includes a formula verifying

- role membership, i.e. there is an assignment role pair $(rp, dr) \in RPDRA$ such that dr is assigned the permission $(d, op) \in PDRA$
- role activation requirements: each environment role er in the role pair $rp = (r, ER_{rp})$ is activated by the currently active environment conditions ec.

Encoding. $HyBAC_{RC}$ policies are translated into $SHoCBAC$ as follows:

- **users, roles and their relations**. The set U of users is represented by the set $\mathcal{P}$ of principals in $SHoCBAC$, static attributes for users correspond to user categories $\mathcal{C}_P$ and their assignments $(u, a) \in UA$ are encoded by the relation $\mathcal{PCA}$.
- **devices, operations and their relations**. The set D of devices is represented by the set $\mathcal{R}$ of resources, device roles are resource categories $\mathcal{C}_R$ in $SHoCBAC$ and their assignments DR are encoded in the relation $\mathcal{RCA}$. Device operations OP are the actions $\mathcal{A}$ in $SHoCBAC$ and permissions P are assigned via the relation $\mathcal{ARCA}$ (we assume here each category action corresponds to a single action in $\mathcal{ACA}$). We don't need to create the relation P, as it generates redundancies. We assign via $\mathcal{ARCA}$ the actions to categories of resources instead of linking them to each resource. For instance, if $((Oven, Open_{oven}), Dangerous\_kitchen\_perm) \in PDRA$, then we may have $(Adult, Open, Dangerous\_kitchen\_perm) \in \mathcal{ARCA}$, with $Oven \in Dangerous\_kitchen\_perm \in \mathcal{RCA}$. More generally, for all $d \in D$, if $(d, op) \in P$ and $roles(d) = dr$ then dr is translated into a resource category c_r, op is translated into an action a to which the category c_r is related via the permission relation $\mathcal{ARCA}$.
- **environment related information**. Environmental conditions (possibly using sensors information) are built from environmental attributes in $\mathcal{EAtA}$ and are included in category-attribute assignment conditions in $\mathcal{CAtA}$. Environmental roles may thus be seen as principal or ressources (sub-)categories. Role activation is automatically ensured by (dynamic) category assignment $\mathcal{CAtA}$ in axioms (c1) and (c2).
 For instance $Kids\_Entertainment\_time \in ER$ can be seen as a subcategory of $Entertainment\_devices \in C_R$ when the condition $env.Evening$ and $env.weekEnd$ is satisfied (thus including the environmental attributes in EA, directly in the dynamic definition of ER, see Sect. 3.2).
- **dynamic attributes**. Attributes are assigned to users and resources using the relations $\mathcal{PAtA}$ and $\mathcal{RAtA}$, respectively.

- **role pairs and their relations**. We do not need a specific relation to pair environmental and user roles. They correspond to principal or device categories and are linked (with the corresponding actions) using the $\mathcal{ARCA}$ relation and using the hierarchical relation $\subseteq$ between categories.
  For instance, consider the pair
  $((kids, Kids\_Entertainment\_time), Entertainment\_devices)) \in RPDRA$,
  where $Entertainment\_devices$ is the role associated to the resources {TV, DVD, PlayStation} and possible actions are {On, Off}. This shows that kids can access $Entertainment\ Devices$ device role when the environment conditions $Kids\ Entertainment\ Time$ is active.
  This is encoded in $SHoCBAC$ with two categories $Kids\_Entertainment$ and $Entertainment\_devices$ such that $Entertainment\_devices \subseteq Kids\_Entertainment$. Both categories contain the resources {TV, DVD, PlayStation}. Assume there is a category of actions $Safe$ containing On and Off. To restrict kids access to Entertainment devices, we include only the tuple $(Safe, Kids\_Entertainment, kids)$ in $\mathcal{ARCA}$, whereas for other user categories we may have additional tuples granting access to Entertainment.

Proposition 1. *A user u is given permission to execute the operation op on a device d under certain environmental conditions ec in an $HyBAC_{RC}$ policy if and only if $(u, op, d) \in \mathcal{PAR}$ under the same environmental conditions in the translated SHoCBAC policy.*

Proof. (Sketch) $\Rightarrow$: A request by u to perform op on d is granted in $HyBAC_{RC}$ if $CheckAccess(s, op, d, ec)$ evaluates to $True$, that is, $Authorization(u, op, d)$ evaluates to $True$, i.e. the operation is allowed on the device for the user u considering the actual attribute values for the user, the device and their roles. In CBAC this is ensured by the fact that the $\mathcal{PCA}$ and $\mathcal{RCA}$ relations used to derive $\mathcal{PAR}$ are defined using the principal, resource (and environmental) attributes (see axioms c1 and c2). The user/device role is the principal/resource's category given by $\mathcal{PCA}$ ($\mathcal{RCA}$ respectively), and the permitted actions, i.e., the PDRA relation, are encoded in the $\mathcal{ARCA}$ relation.

Concerning the roles activation requirements, this is ensured by the definition of the relation $\mathcal{CAtA} \subseteq \mathcal{C} \times Cond$, which includes in its condition environmental attributes (relation $\mathcal{EAtA}$). This means that by definition a category (role) cannot be assigned in CBAC (i.e. activated) if the actual values of the environmental attributes are not satisfied (see axioms $c1$ and $c2$).

Role membership is ensured by construction. As explained, environmental conditions are included in the definition of principal or resource categories and the related permissions are given by the relation $\mathcal{ARCA}$, which is derived from $PDRA$.

$\Leftarrow$: If $\mathcal{PAR}(u, op, d)$ is granted in $SHoCBAC$, this means that there is a resource category $c_r \ni d$, a principal category $c_p \ni u$ and an action category $c_a \ni op$ such that $(c_a, c_r, c_p) \in \mathcal{ARCA}$. We need to define in the $HyBAC_{RC}$ policy a device role $c_r \in DR$, include $(d, op) \in P$ and include $((d, op), c_r) \in PDRA$. If environmental attributes in $\mathcal{EAtA}$ are used in the condition defining c_r (or

c_p), then the corresponding environmental role $er \in ER$ needs to be defined and paired with the device role c_r, i.e. $((c_p, er), c_r) \in RPDRA$. The function $Authorization(s, op, d)$ is defined as $c_p \in roles(u) \wedge c_r \in droles((op, d))$. □

The $HyBAC_{RC}$ model also includes some constraints such as PRConstraints, which prevent RPDRA assignments that would enable specific roles to access specifically prohibited permissions; Static Separation of Duty (SSD) and Dynamic Separation of Duty (DSD), which are the usual notions used e.g. in RBAC. PRConstraints can be modelled in CBAC using the relation $\mathcal{BARCA}$ (see [10]). SSD (as well as binding of duty constraints) can be specified by adding axioms at the metamodel level (see [11]). Verifying DSD can be done similarly to $HyBAC_{RC}$, if sessions are introduced.

The encoding of $HyBAC_{AC}$ is similar, but more direct since both $HyBAC_{AC}$ and $SHoCBAC$ are ABAC-style models.

5 Related Work

A variety of access control models have been proposed for the Internet of Things (e.g., [4,5,8,16,20]). Here we discuss the models that are closer to ours.

Following Kuhn et al. [18] RBAC and ABAC combinations can be classified as Attribute-centric, Role-centric, or using Dynamic roles. C-ABAC is in the third class: it is a dynamic version of RBAC, where roles (represented by categories), are defined by Boolean formulas on attribute values, and the same idea of categorisation is applied also to resources. Ameer et al. [4] follow the attribute-centric and role-centric approaches to define smart home access control models. Instead, here we follow the dynamic role approach, which was not considered in previous work due to the potentially large number of attributes in smart homes. However, even if smart home applications are rich in attributes, user studies [17] show that there is only a small set of meaningful attribute combinations, and CBAC offers a natural way to use them to define authorisations.

$SHoCBAC$ is able to express the policies defined in previous models, such as $HyBAC_{RC}$ and $HyBAC_{AC}$ (which were shown to generalise EGRBAC [2] and HABAC [3]) and can also express other models in between these two. $HyBAC_{RC}$ can specify role constraints to be checked at configuration time, whereas $HyBAC_{AC}$ can specify role constraints that are checked at execution time. The flexibility of CBAC allows us to define instances with both kinds of constraints, i.e., constraints checked at execution or configuration time.

Note that HABAC has been shown in previous work to be unable to express role constraints as defined in EGRBAC. Despite being an ABAC-based model, $SHoCBAC$ is able to express EGRBAC policies, including role constraints. This can be done using the administrative level (which is also an instance of the CBAC metamodel), i.e., we can ensure administrators are not allowed to include an $\mathcal{ARCA}$ tuple if there is a $\mathcal{BARCA}$ one. This would mimic EGRBAC behaviour.

While previous access control models for the Smart Home IoT need a separate administrative model, see for example [21], $SHoCBAC$ allows also the specification of administrative authorisations and prohibitions in a uniform way. In future work we will provide a detailed analysis of administration in *SHoCBAC*.

6 Conclusions and Future Work

SHoCBAC is a logic-based access control model for smart homes, which uses categorisation (a key mechanism underlying cognition). It is sufficiently expressive to satisfy the requirements of smart homes, where policies should be fine-grained and dynamic. From a usability point of view, we have also identified an instance of *SHoCBAC* based on trust levels, which is simple yet covers the standard Smart Home IoT requirements. *SHoCBAC* and its instances inherit the logic foundations of CBAC and its rewriting operational semantics, which provides a basis for the analysis of policy properties. Administrative policies can also be defined by using the CBAC administrative model. This will be the subject of future work.

References

1. Alves, S., Fernández, M.: A graph-based framework for the analysis of access control policies. Theor. Comput. Sci. **685**, 3–22 (2017)
2. Ameer, S., Benson, J.O., Sandhu, R.S.: The EGRBAC model for smart home IoT. In: 21st International Conference on Information Reuse and Integration for Data Science, IRI 2020, Las Vegas, NV, USA, 11–13 August 2020, pp. 457–462. IEEE (2020)
3. Ameer, S., Benson, J.O., Sandhu, R.S.: An attribute-based approach toward a secured smart-home IoT access control and a comparison with a role-based approach. Information **13**(2), 60 (2022)
4. Ameer, S., Benson, J.O., Sandhu, R.S.: Hybrid approaches (ABAC and RBAC) toward secure access control in smart home IoT. IEEE Trans. Dependable Secur. Comput. **20**(5), 4032–4051 (2023)
5. Ameer, S., Krishnan, R., Sandhu, R.S., Gupta, M.: Utilizing the DLBAC approach toward a ZT score-based authorization for IoT systems. In: Shehab, M., Fernández, M., Li, N. (eds.) Proceedings of the Thirteenth ACM Conference on Data and Application Security and Privacy, CODASPY 2023, Charlotte, NC, USA, April 24–26 2023, pp. 283–285. ACM (2023)
6. Augoustinos, M., Walker, I., Donaghue, N.: Social Cognition: An Integrated Introduction. Sage, Thousand Oaks (2014)
7. Barker, S.: The next 700 access control models or a unifying meta-model? In: SACMAT 2009, 14th ACM Symposium on Access Control Models and Technologies, Stresa, Italy, 3–5 June 2009, Proceedings, pp. 187–196. ACM Press, New York (2009)
8. Bertino, E., et al.: Security and privacy for emerging IoT and CPS domains. In: Joshi, A., Fernández, M., Verma, R.M. (eds.) CODASPY 2022: Twelveth ACM Conference on Data and Application Security and Privacy, Baltimore, MD, USA, 24–27 April 2022, pp. 336–337. ACM (2022)

9. Bertolissi, C., Fernández, M.: Category-based authorisation models: operational semantics and expressive power. In: Massacci, F., Wallach, D., Zannone, N. (eds.) ESSoS 2010. LNCS, vol. 5965, pp. 140–156. Springer, Heidelberg (2010). https://doi.org/10.1007/978-3-642-11747-3_11
10. Bertolissi, C., Fernández, M.: A metamodel of access control for distributed environments: applications and properties. Inf. Comput. **238**, 187–207 (2014)
11. Bertolissi, C., Fernández, M., Thuraisingham, B.: Admin-CBAC: an administration model for category-based access control. In: Roussev, V., Thuraisingham, B., Carminati, B., Kantarcioglu, M. (eds.) CODASPY 2020: Tenth ACM Conference on Data and Application Security and Privacy, New Orleans, LA, USA, 16–18 March 2020, pp. 73–84. ACM (2020)
12. Bertolissi, C., Fernández, M., Thuraisingham, B.: Graph-based specification of admin-CBAC policies. In: Joshi, A., Carminati, B., Verma, R.M. (eds.) CODASPY 2021: Eleventh ACM Conference on Data and Application Security and Privacy, Virtual Event, USA, 26–28 April 2021, pp. 173–184. ACM (2021)
13. Eysenck, M.W., Brysbaert, M.: Fundamentals of Cognition. Routledge, Milton Park (2018)
14. Fernández, M., Mackie, I., Thuraisingham, B.M.: Specification and analysis of ABAC policies via the category-based metamodel. In: Proceedings of the Ninth ACM Conference on Data and Application Security and Privacy, CODASPY 2019, Richardson, TX, USA, 25–27 March 2019, pp. 173–184. ACM Press, New York (2019)
15. Fernández, M., Jaimunk, J., Thuraisingham, B.: A privacy-preserving architecture and data-sharing model for cloud-IoT applications. IEEE Trans. Dependable Secure Comput. **20**(4), 3495–3507 (2023)
16. Gupta, M., Bhatt, S., Alshehri, A.H., Sandhu, R.S.: Access Control Models and Architectures For IoT and Cyber Physical Systems. Springer, Cham (2022). https://doi.org/10.1007/978-3-030-81089-4
17. He, W., et al.: Rethinking access control and authentication for the home internet of things (IoT). In: Enck, W., Felt, A.P. (eds.) 27th USENIX Security Symposium, USENIX Security 2018, Baltimore, MD, USA, 15–17 August 2018, pp. 255–272. USENIX Association (2018)
18. Kuhn, D.R., Coyne, E.J., Weil, T.R.: Adding attributes to role-based access control. IEEE Comput. **43**(6), 79–81 (2010)
19. Obrezkov, D., Sohr, K.: UCAT: the uniform categorization for access control. In: Mosbah, M., Sèdes, F., Tawbi, N., Ahmed, T., Boulahia-Cuppens, N., García-Alfaro, J. (eds.) FPS 2023, Part II. LNCS, vol. 14552, pp. 3–14. Springer, Cham (2023). https://doi.org/10.1007/978-3-031-57540-2_1
20. Ray, I., Abdunabi, R., Basnet, R.: Access control for internet of things applications. In: Proceedings of the 5th ACM on Cyber-Physical System Security Workshop, CPSS@AsiaCCS 2019, Auckland, New Zealand, 8 July 2019, pp. 35–36. ACM (2019)
21. Shakarami, M., Sandhu, R.S.: Role-based administration of role-based smart home IoT. In: Gupta, M., Abdelsalam, M., Mittal, S. (eds.) SAT-CPS@CODASPY 2021, Proceedings of the 2021 ACM Workshop on Secure and Trustworthy Cyber-Physical Systems, Virtual Event, USA, 28 April 2021, pp. 49–58. ACM (2021)

Towards Specification-Guarded Refactoring

Adam D. Barwell, Christopher Brown(✉), and Susmit Sarkar

School of Computer Science, University of St Andrews, St Andrews, Scotland, UK
{adb23,cmb21,susmit.sarkar}@st-andrews.ac.uk

Abstract. Refactoring is a common task in writing and maintaining software. Refactoring tools are available for a wide variety of programming languages, and they are intended to transform source code without altering the functional behaviour of the software being refactored. However, even though most refactoring tools are designed with this criterion in mind, and programmers expect it, their implementations do not provide formal guarantees of correctness. To address this limitation, we propose a methodology for extending refactoring implementations with correctness guarantees. Our approach leverages dependently-typed languages and techniques to *guard* refactoring implementations with proofs of conformance to formal specifications. We illustrate our proposed methodology using a renaming refactoring that includes correctness guarantees as an integral part of its implementation.

Keywords: Haskell · Refactoring · Dependent Types · Idris · Correctness · Soundness · Renaming · Equivalence · Semantics

1 Introduction

Refactoring [16] is the process of *changing the structure* of a program *without changing its behaviour*. It is a common exercise aimed at making a program more understandable, accessible, or amenable to further design alterations. Refactorings can be applied manually or via automated tooling. The (semi-)automatic approach is supported by many tools [24], and guards against common human mistakes, e.g. overlooking one file within hundreds, or one line in potentially millions [30]. Despite these advantages, automated tools have their own mode of failure, and could change the behaviour of the code silently and in unexpected ways. As refactoring tools grow in power and are applied to ever larger codebases, the key correctness criterion of behaviour preservation becomes critical.

Confidence of tool correctness frequently arises from writing, or generating, test cases to check that a refactoring has not obviously changed the semantics of the code. Whilst important, testing cannot catch every error, and bugs have been found in real-world tooling, e.g. the Eclipse refactoring tools for Java and C/C++ [17]. A complementary approach is to formally demonstrate correctness [22,27,28]. Akin to the problem of formally verifying a compiler [20], this is a significant undertaking, and thus usually approached by formally proving refactorings using an idealised algorithm over an idealised language. Consequently, the relationship of proofs to concrete implementations can be unclear. In order to address both the lack of formal correctness guarantees and to lessen the burden of developing verified refactoring tools, we propose a novel *specification-guarded refactoring* methodology.

Our proposed approach leverages dependently-typed languages and techniques in order to define *guard* functions that demonstrate conformance of a *core* refactoring implementation to formal refactoring *specifications*. Both specifications and guard functions are intrinsic components of the refactoring tool and can be used to extend existing implementations of that refactoring. Moreover, dependent types can be used to embed static semantics into the AST representation of a program. Consequently, we can ensure *for free* that well-formedness is invariant under refactoring: well-formed programs cannot be refactored into ill-formed programs.

We illustrate our methodology via the canonical renaming refactoring, included as part of a prototype refactoring tool, HESTER[1]. Renaming refers to the problem of substituting a function or variable name such that all occurrences, and only those occurrences, that are bound by a particular definition are transformed. The refactoring must ensure that it introduces no capturing or shadowing violations, i.e.where a new name that already exists within the target scope is introduced, and thus potentially altering the behaviour of the refactored code. The specific contributions of our paper are:

1. A novel methodology for introducing formal guarantees of correctness of refactoring implementations.
2. A prototype refactoring tool, HESTER, for a subset of Haskell 98, guaranteeing that well-formedness is invariant under refactoring.
3. An illustrative, specification-guarded implementation of the renaming refactoring, alongside a formal specification of renaming, included in HESTER.

2 Specification-Guarded Refactoring

Our proposed methodology aims to enrich refactoring tooling with correctness proofs by *guarding* refactorings. The concept extends from viewing refactorings as (endo)functions over a program's source code, which is usually represented by an Abstract Syntax Tree (AST). By leveraging dependently-typed languages and techniques, AST representations can be enriched with the static semantics of the target language and thereby restrict the set of programs that can be represented. A key consequence of such restrictions is the guarantee that no well-typed refactoring can produce an ill-formed program.

This enrichment provides a foundation that can be used to address the central correctness criterion of ensuring that a refactoring preserves the functional

---

[1] Named after Hester, Lee Scoresby's Dæmon Hare from *His Dark Materials*.

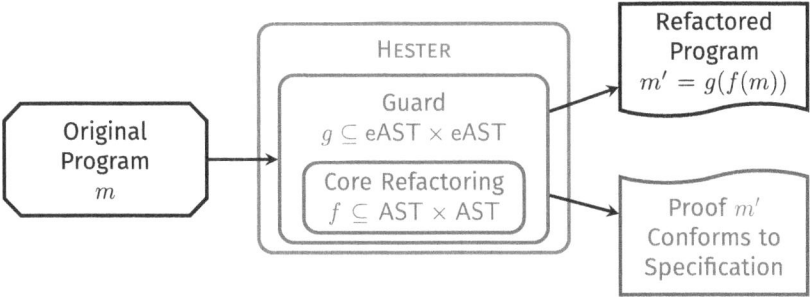

Fig. 1. HESTER: a prototype specification-guarded refactoring tool.

behaviour of the refactored program. Ensuring correctness is achieved by viewing the refactoring as a composition, $g \circ f$, of a *core* function, f, and a *guard* function, g. The core function, f, represents the standard refactoring implementation, which provides no correctness proof. Meanwhile, the guard function, g, serves two purposes: *i)* it acts as an interface for applying the refactoring; and *ii)* it guarantees that any resulting refactored program conforms to a given refactoring specification. Specifications are themselves represented within the tool's implementation, and g must derive the proofs that are required to demonstrate conformance.

To illustrate our methodology, we present HESTER, a prototype specification-guarded refactoring tool. HESTER represents a step towards extending the Haskell refactoring tool, HaRe [21], with intrinsic correctness proofs via specification-guarded refactorings. We illustrate the general idea for HESTER in Fig. 1, where the original program, represented by m, is refactored, resulting in a refactored program, m', that has been *guarded* (denoted by $g(f(m))$). HESTER itself is implemented by taking the core refactoring to be applied, which is an (endo)function from an AST representation to a (refactored) AST representation, and lifting the refactoring to an enriched AST representation (as defined in Sect. 4). The result of the guard is a transformation of this enriched AST, together with a correctness proof that the refactoring has conformed to its specification.

HESTER is implemented in Idris, a dependently-typed functional language, and supports a subset of Haskell 98 programs (Sect. 4). We demonstrate our approach using the canonical renaming refactoring (Sect. 3.2), providing both a formal specification and a guard function for renaming, alongside a core renaming function that is strongly inspired by HaRe's (Sect. 5). To simplify parsing, we use a standard Haskell parser, *haskell-src*[2]. An AST produced by *haskell-src* is converted into an equivalent representation in Idris, and enriched with proofs of conformity to a static semantics (Sect. 4). Our approach ensures that the AST resulting from an application of renaming, which can be pretty printed to reconstruct Haskell syntax, represents a Haskell module that has been successfully refactored without introducing errors.

---

[2] https://hackage.haskell.org/package/haskell-src.

```
1 data Elem : a -> List a -> Type where
2    Here  : Elem x (x :: xs)
3    There : Elem x xs -> Elem x (y : xs)
```

Fig. 2. Idris Elem definition

Although we reimplement the core function of renaming to simplify our presentation, we remark that, in principle, f can be the direct application of an *existing* refactoring. This reduces the implementation burden by leveraging existing tools. In such cases, the core and guard functions may operate over different AST representations, with the latter parsing both the original and refactored programs to check for conformance.

3 Background

3.1 Dependent Types

Dependently-typed languages, e.g. Idris [5], allow types to depend on any value. This enables properties to be expressed at the type level and proofs of properties as values of types, where both are verified by type-checking [6]. In this paper, we use Idris to: *i)* represent the AST and static semantics of programs in our Haskell 98 subset; *ii)* provide a soundness property for renaming over our AST; *iii)* implement a version of renaming; and *iv)* prove that this implementation is sound. We assume that the reader is familiar with Idris; for those unfamiliar, full texts are available elsewhere [7]. The advantage of dependent types, and the use of Idris, lies not only in the ability to prove that our implementation of renaming is sound but also that we can provide an *executable* tool, rather than providing proofs that are separate from the implementation.

To illustrate dependently-typed programming in Idris, consider the definition of Elem in Fig. 2, expressing a type that represents the membership $x \in xs$ (an element x is in the list xs). This membership relation is expressed by constructing an instance Elem x xs forming a proof that x is an element of xs. We form such a proof by constructing a term using the constructors of Elem, where Here represents that the element x is at the head of the list, x:xs, and There when x is in the tail of the list. Consider, for example, the list [1,2,3,4,5] and the following definition.

```
1 inList : Elem 3 [1,2,3,4,5]
2 inList = There (There Here)
```

Here, inList forms a proof of the proposition $3 \in [1,2,3,4,5]$. The proposition is represented by the type Elem 3 [1,2,3,4,5] and the proof is a value that inhabits that type, indicating the position of the element in the list, i.e. There (There Here).

```
1  module Sum where                      1  module Sum where
2                                        2
3  import Prelude hiding (sum)           3  import Prelude hiding (fold )
4                                        4
5  sum n []    = n                       5  fold  n []    = n
6  sum n (h:t) = (+) h rest              6  fold  n (h:t) = (+) h rest
7   where                                7   where
8    rest = (sumRest n t)                8    rest = (sumRest n t)
9    sumRest n t = sum n t               9    sumRest n t = fold  n t
10                                       10
11 main = sum 0 [1..4]                   11 main = fold  0 [1..4]
```

          (a) Before   (b) After

Fig. 3. Renaming sum; before and after (with differences highlighted) the application of the rename refactoring.

3.2 Renaming

Renaming is considered a canonical refactoring, with most refactoring tools providing an implementation. Renaming is also applicable as a refactoring for all programming languages. While we describe renaming here in terms of Haskell, it has strong analogues with α-conversion from the λ-calculus [2,27]. The concept of renaming variables, functions, and other programming constructs is a practice that developers do on a day-to-day basis. Requirements change, and variables and functions are renamed to better suit their purpose. Despite prior attempts at proving implementations of renaming for Haskell 98, e.g. the original implementation provided in the Haskell Refactorer (HaRe) [23], these implementations do not include their soundness proofs.

The refactoring renames all occurrences (and only those) of a variable or function name, including its free and bound occurrences, which may occur in function bindings, pattern bindings, free variables occurring in expressions, and calls to functions, recursive or otherwise. Performing these renamings must not alter the binding-structure of the program. No new names must be introduced into the binding-structure that result in capturing conflicts. To give an example, we define the Haskell module Sum in Fig. 3a, which defines two top-level functions: sum (Lines 5–9) and main (Line 11). The function sum is split across two clauses (Lines 5 & 6) with the second clause calling a local definition rest on its right-hand side as the second argument to (+). rest is defined in a where-block (Line 8) with sumRest (Line 9), which rest calls. As one step towards generalising sum, in order to promote code reuse, sum is renamed to fold. Figure 3b gives the result of the refactoring, where all occurrences and calls to sum have been transformed to calls to fold, including both the recursive call (Line 9) and the call in main (Line 11). In the case of attempting to rename a variable into one that causes a clash with another variable in scope, our refactoring implementation will fail to rename, returning the original program unmodified, and an error message.

This follows from the principle of the original renaming design of HaRe, where a refactoring fails, rather than compensating. For example, consider:

```
1 sum = ...
2 f :: [Int] → ...
3 f fold = ... sum fold ...
```

Here fold on the right-hand-side of the equation for f is bound in the argument position. Suppose the programmer wishes to rename the function sum (on Line 1) to fold to better reflect its purpose. A naïve implementation might result in an ill-formed program, as both occurrences of fold on the right-hand-side of f refer to the local bound variable in the argument position (and not the newly renamed definition on Line 1).

```
1 fold = ...
2 f :: [Int] → ...
3 f fold  = ... fold  fold   ...
```

A better solution would be to avoid capturing, and renaming the local argument to fold', resulting in a well-formed, and equivalent, program to the original:

```
1 fold = ...
2 f :: [Int] → ...
3 f fold'  = ... fold fold'   ...
```

Renaming is typically the first of many applied refactorings to generalise the definition of sum in order to promote code reuse. Employing additional refactorings, such as *generalisation* and *lifting of definitions* (similar to λ-lifting [19]), to end up with the following result:

```
1  module Sum where
2
3  import Prelude hiding (fold)
4
5  fold c n []    = n
6  fold c n (h:t) = c h (fold n t)
7
8  main = sum [1..4]
9
10 sum = fold (+) 0
```

Here, we have *unfolded* (in the transformational sense, à la Burstall and Darlington [13]) calls to the local functions rest and sumRest on the right-hand side of fold. The function fold has also been *generalised* (à la the original HaRe definition [21]) by adding an additional parameter c and changing all call sites to pass in (+) as an argument (Line 10). The call to fold has been renamed to a call to sum, specialising the instance of the original call to fold to better suit its original purpose: summing the integers [1..4].

$$p \in P := x_{\langle l,i \rangle} \quad \text{(Variable Patterns)}$$
$$\mid A_i \, \vec{p} \quad \text{(Constructor Patterns)}$$
$$e \in E := x_{\langle l,i \rangle} \quad \text{(Variable Exp.)}$$
$$\mid n \quad \text{(Integer Literal Exp.)}$$
$$\mid e_1 \, e_2 \quad \text{(Application Exp.)}$$
$$\mid \backslash \vec{p} \,\text{->}\, e \quad (\lambda\text{-Exp.})$$
$$c \in C := \vec{p} = e \,\, \text{where} \,\, \vec{d} \quad \text{(Function Clauses)}$$
$$d \in D := \text{def} \, f_{\langle l,i \rangle} \, \vec{c} \quad \text{(Function Decl.)}$$
$$\mid \text{def} \, p = e \,\, \text{where} \,\, \vec{d} \quad \text{(Pattern Decl.)}$$
$$m \in M := \text{mod} \, \vec{d} \quad \text{(Modules)}$$

Fig. 4. Hs98: a subset of Haskell 98.

```
1  data HsNameTy : (k : HsNameKind) → Type where
2    HsIdentVar : (lv : Nat) → (id : Nat) → (na : Name Variable)
3               → HsNameTy Variable
4    HsIdentCtr : (id : Nat) → (na : Name Constructor)
5               → HsNameTy Constructor
```

Listing 1.1. Definition for variables and constructors in Hs98.

4 A Subset of Haskell 98

We focus on the subset, Hs98, of the Haskell 98 standard given in Fig. 4. In Hs98, a valid program is a single module, $m \in M$, where *modules* are defined as a vector of declarations, $\vec{d}$. Modules implicitly import a Prelude, i.e. a set of constructors and variables that are valid in the module. *Declarations* are either pattern or function declarations. Pattern declarations consist of a pattern, p, an expression, e, and a where-block, $\vec{d}$. Function declarations comprise the function name, $f_{\langle l,i \rangle}$, and a vector of clauses, $\vec{c}$. Declarations within the same vector are assumed to be defined mutually. A *pattern* is either a variable, $x_{\langle l,i \rangle}$, or a constructor pattern comprising a constructor, A_i, and a vector of patterns, $\vec{p}$. An *expression* may be: a variable, $x_{\langle l,i \rangle}$; an integer literal, n; an application, $e_1 \, e_2$; or a λ, $\backslash \vec{p}$ -> e.

We define both the AST and static semantics for Hs98 via a series of types in Idris. This AST is the subject of the transformations effected by our renaming refactoring. While we elide some definitions here, the full implementation may be found in the supplementary material[3].

4.1 Variables and Constructors

For clarity of presentation we denote variables by $x_{\langle l,i \rangle}$ and constructors by A_i. We assume an infinite set of valid variable names, $f, x, y, \ldots$. A variable, $x_{\langle l,i \rangle}$, is a triple comprising a variable name, x, an identifier, $i \in \mathbb{N}$, and a level identifier,

---
[3] https://github.com/adbarwell/Hester.

$l \in \mathbb{N}$. Identifiers disambiguate variables since variable names may be shadowed by declarations in where-blocks. Similarly, level identifiers group variables in order to enforce linearity of variable names that are declared in the same vector of patterns or declarations. The linearity constraint represents a possible point of failure when renaming, e.g. when attempting to rename a function parameter y to x in f x y = ... Renaming does not allow capture violations as a precondition, of which this is an example; without this pre-condition, renaming can result in an ill-formed program. We additionally assume a set of valid constructor names, $A, B, \ldots$. A constructor, A_i, is a constructor name, indexed by an identifier, $i \in \mathbb{N}$. Identifiers are assumed to be unique within a given scope. Since type declarations are not included in Hs98, constructors are limited to those defined in the Prelude. All variables and constructors are *unqualified* since a program contains only a single module. Variables and constructors are represented by HsNameTy in Listing 1.1. Variables are constructed by HsIdentVar s.t. $x_{\langle l,i \rangle} =$ HsIdentVar l i x . Similarly, constructors are represented by HsIdentCtr, s.t. $A_i =$ HsIdentCtr i A . Both variable and constructor names are represented by Name, representing a restriction over strings. A value of type Name k, where k $\in$ {Variable, Constructor}, is a proof that the name conforms to the definitions of valid variable or constructor names in the Haskell 98 Report [25]. Symbols are not supported in our implementation but are represented by an equivalent textual name; e.g. Cons for (:).

4.2 An AST and Static Semantics for Hs98

We restrict the programs that are expressible in Hs98 to those that are well-formed. We thus define our AST to encode a subset of the static semantics of Haskell 98, but not including type information: we assume that all programs are well-typed and pass the type-checker.

Environments
We define an *environment*, $\sigma = (\overrightarrow{A_i}, \overrightarrow{\eta})$, to be a pair comprising a vector of constructors, $\overrightarrow{A_i}$, and a vector of *scope levels*, $\overrightarrow{\eta}$. Environments are defined by the type Env (Listing 1.2). In addition to $\overrightarrow{A_i}$ and $\overrightarrow{\eta}$, the sole constructor, (MkEnv $\overrightarrow{A_i} \overrightarrow{\eta}$ ok), takes a proof, ok, that all identifiers across $\overrightarrow{A_i}$ and $\overrightarrow{\eta}$ are indeed unique. Vectors allow the implicit ordering of scope levels.

A scope level, $\eta = (l, \overrightarrow{x_{\langle l,i \rangle}})$, is a pair comprising a level identifier, l, and a vector of variables, $\overrightarrow{x_{\langle l,i \rangle}}$, that are declared in the same pattern, vector of patterns, or where-block. For example, n, h, and t on Line 6 in Fig. 3a, form a scope level since they are all declared by the vector of patterns in the second clause of sum. All variables in $\overrightarrow{x_{\langle l,i \rangle}}$ must have the same level identifier. Scope levels are formally defined by the type ScopeLevel in Listing 1.2, such that η corresponds to (MkScopeLevel l $\overrightarrow{x_{\langle l,i \rangle}}$ lin) , where lin is a proof that all variable names in $\overrightarrow{x_{\langle l,i \rangle}}$ are distinct.

Environments are *extended*, denoted $\eta \cup \overrightarrow{x_{\langle l,i \rangle}}$, by a vector of variables, $\overrightarrow{x_{\langle l,i \rangle}}$, when all variables in $\overrightarrow{x_{\langle l,i \rangle}}$ have the same level identifier. Extending environments is defined by prepending a new scope level formed of $\overrightarrow{x_{\langle l,i \rangle}}$ to a given environment,

```
1 data EnvItem : HsNameKind → Type where
2   MkEnvItem : (id : Nat) → (na : Name k) → EnvItem k
3 data ScopeLevel : Type where
4   MkScopeLevel : (lvl : Nat)
5                → (vars : Vect nvars (EnvItem Variable))
6                → {lin : UniqueNames vars} → ScopeLevel
7 data Env : Type where
8   MkEnv : (ctrs : Vect nctrs (EnvItem Constructor))
9         → (lvls : Vect nlvls ScopeLevel)
10        → {ok : UniqueIds ctrs lvls} → Env
```

Listing 1.2. Definitions for environments and scope levels in Hs98.

```
1  (HsModule
2    σ₀
3    [sum⟨1,4⟩]
4    [HsFunBind {env=σ₁} sum⟨1,4⟩
5      [HsMatch {env=σ₁} {env_e=σ₂} [n⟨2,5⟩, Nil₁[]]
6        (HsVar n⟨2,5⟩) [],
7       HsMatch {env=σ₁} {env_e=σ₃} [n⟨2,5⟩, Cons₀[h⟨2,6⟩,t⟨2,7⟩]]
8        (HsApp (HsApp (HsVar (+)⟨0,2⟩) (HsVar h⟨2,6⟩))
9                (HsVar rest⟨3,11⟩))
10       [HsPatBind {env=σ₃} {env_e=σ₃} (rest⟨3,11⟩)
11         (HsApp (HsApp (HsVar succRest⟨3,10⟩) (HsVar n⟨2,5⟩))
12                 (HsVar t⟨2,7⟩)))) [],
13      HsFunBind {env=σ₃} sumRest⟨3,10⟩
14       [HsMatch {env=σ₃} {env_e=σ₄} [n⟨4,12⟩,t⟨4,13⟩]]
15         (HsApp (HsApp (HsVar sum⟨1,4⟩) (HsVar n⟨4,12⟩))
16                 (HsVar t⟨4,13⟩)) []]]])
```

Fig. 5. Abstract Syntax Tree representation of sum.

assuming that all variables in $\overrightarrow{x_{\langle l,i \rangle}}$ have the same scope level and that no variable breaks the linearity or uniqueness constraints.

Each pattern, p, expression, e, declaration, d, and function clause, c, is indexed by an environment, σ; denoted $\langle p \rangle^\sigma$, $\langle e \rangle^\sigma$, $\langle d \rangle^\sigma$, and $\langle c \rangle^\sigma$, respectively. Modules also have an environment, σ_0, representing the Prelude.

Each term propagates the environment by which it is indexed to its subterms. In the case of declarations and λ-expressions, the environment is extended by those variables declared in patterns and while-blocks, where appropriate.

To give an illustrative example of how environments are formed, consider the example in Fig. 5, which is an AST representation of the sum example from Fig. 3, with the following environments.

$$\sigma_0 := ([Cons_0, Nil_1], [(0, [(+)_{\langle 1,2 \rangle}, enumFromTo_{\langle 0,3 \rangle}])])$$
$$\sigma_1 := \sigma_0 \cup [sum_{\langle 1,4 \rangle}, main_{\langle 1,14 \rangle}]$$
$$\sigma_2 := \sigma_1 \cup [n_{\langle 2,5 \rangle}]$$
$$\sigma_3 := \sigma_2 \cup [n_{\langle 2,5 \rangle}, h_{\langle 2,6 \rangle}, t_{\langle 2,7 \rangle}, rest_{\langle 3,11 \rangle}, sumRest_{\langle 3,10 \rangle}]$$
$$\sigma_4 := \sigma_3 \cup [n_{\langle 4,12 \rangle}, t_{\langle 4,13 \rangle}]$$

```
 1  data HsPatTy : (env : Env) → Type where
 2    HsPat : (defs : Vect (S n) (HsPatDefTy (S n) env))
 3         → {ok : IsTree defs} → HsPatTy env
 4
 5  data HsExpTy : (env : Env) → Type where
 6    HsVar : (na : HsNameTy Variable) → {ok : ElemEnv env na}
 7         → HsExpTy env
 8
 9  data HsDeclTy : (env : Env) → (kind : HsDeclKindTy) → Type where
10    HsMatch : (ps : Vect (S nps) (HsPatTy env))
11         → {psec : HsPatSecTy env ps env_ws}
12    ...
```

Listing 1.3. Definitions for declarations and function clauses in Hs98.

```
1  HsMatch ((SrcLoc "<unknown>" 47 1)) (HsIdent "sum")
2    [HsPVar (HsIdent "n"),
3     HsPParen (HsPInfixApp (HsPVar (HsIdent "h"))
4         (Special HsCons)
5         (HsPVar (HsIdent "t")))]
```

Listing 1.4. AST fragment from *haskell-src*

Here, σ_0 is the Prelude environment, σ_1 is the environment for each declaration in the module, σ_2 is the environment for the RHS of the first clause of $sum_{\langle 1,4 \rangle}$, σ_3 is the environment for the RHS and where-block definitions of the second clause of $sum_{\langle 1,4 \rangle}$, and σ_4 is the environment for the RHS of $sumRest_{\langle 3,10 \rangle}$.

4.3 Patterns, Expressions, Functions and Modules

Representing the abstract syntax of Haskell 98 programs in Idris follows in the standard way, where language constructs are indexed by the environment. To give a flavour of this implementation, Listing 1.3 illustrates the high-level data type signatures for patterns (HsPatTy), expressions (HsExpTy), and declarations (HsDeclTy), together with an example constructor.

This representation is obtained via an enrichment process applied to an equivalent representation in *haskell-src* (Listing 1.4). Since Hs98 represents a more restrictive subset of the Haskell 98 standard when compared to *haskell-src*, enrichment is a partial mapping; cases for any Haskell 98 program that cannot currently be represented in our system remain undefined (e.g. modules with imports or functions with guards).

Our AST representation takes inspiration from the structure of *haskell-src* and hence our enrichment is mostly straightforward: constructors are mapped to like-constructors. Proofs are derived for cases where Hs98 requires them, e.g. environment membership for variables in expressions. Enrichment for *haskell-src* modules where such proofs cannot be derived is considered undefined. Some

```
1  data RenamePrf : (lv,id : Nat) → (oldN,newN : Name Variable)
2                → (m1 : HsModuleTy)
3                → (rn : Maybe
4                        (DecldHsModuleTy lvl id oldN m1,
5                         (m2 : HsModuleTy **
6                          (StructEq m1 m2,
7                           DecldHsModuleTy lvl id newN m2))))
8                → Type where
9    RenameFail : RenamePrf lvl id oldN newN m1 sPrf Nothing
10   RenameSucc : (oldD : DecldHsModuleTy lvl id oldN m1)
11              → (newD : DecldHsModuleTy lvl id newN m2)
12              → (eq : StructEq m1 m2)
13              → RenamePrf lvl id oldN newN m1 sPrf
14                          (Just (oldD, (m2 ** (eq,newD))))
```
Listing 1.5. Definition of soundness for renaming, over Hs98.

notable differences between *haskell-src* and Hs98 representations include declarations and patterns. For example, the function clause in Listing 1.4 is represented by a type, `HsMatch`, distinct from declarations, `HsDecl`. Conversely, in Hs98, and in order to satisfy totality constraints, these types are merged into `HsDeclTy`, with constructors' provenance indicated by an additional index (`kind` in Listing 1.3). Patterns similarly deviate from the *haskell-src* representation in order to satisfy totality. In Hs98, patterns are represented via a vector with pointers to successive elements that indicate sub-patterns. This contrasts with the more standard tree structure found in *haskell-src*. All functions and definitions that comprise the enrichment process can be found in the supplied repository.

5 Specification-Guarded Renaming

We define a specification for renaming that *guards* our implementation such that the result of the refactoring conforms to the specification. The implementation is thus divided into a standard *core* function, `renameModule`, and a *guard* function, `renamePrf`, where the former effects rename and the latter ensures conformity.

5.1 Renaming Specification

For $\langle m'\rangle^{\sigma_0}$ to be a valid result of renaming $x_{\langle l,i\rangle}$ to $y_{\langle l,i\rangle}$ in a module $\langle m\rangle^{\sigma_0}$, we require that $\langle m\rangle^{\sigma_0}$ and $\langle m'\rangle^{\sigma_0}$ be *structurally equivalent* and that both $x_{\langle l,i\rangle}$ and $y_{\langle l,i\rangle}$ are declared in $\langle m\rangle^{\sigma_0}$ and $\langle m'\rangle^{\sigma_0}$ respectively. The type `RenamePrf` in Listing 1.5 defines these requirements for our implementation. To ensure that the specification is not restricted to a proof of a specific implementation, `RenamePrf` does not specify *how* $\langle m'\rangle^{\sigma_0}$ is obtained. Unsuccessful refactorings of $\langle m\rangle^{\sigma_0}$ are

$$\begin{aligned}
\pi \in P' &:= i_l & \text{(Variable Patterns)} \\
&\mid i \, \vec{p} & \text{(Constructor Patterns)} \\
\varepsilon \in E' &:= i_l & \text{(Variable Exp.)} \\
&\mid n & \text{(Integer Literal Exp.)} \\
&\mid e_1 \, e_2 & \text{(Application Exp.)} \\
&\mid \backslash \vec{p} \texttt{ -> } e & \text{(λ-Exp.)} \\
\chi \in C' &:= \vec{p} = e \text{ where } \vec{d} & \text{(Function Clauses)} \\
\delta \in D' &:= \texttt{def } i_l \, \vec{c} & \text{(Function Decl.)} \\
&\mid \texttt{def } p = e \text{ where } \vec{d} & \text{(Pattern Decl.)} \\
\mu \in M' &:= \texttt{mod } \vec{d} & \text{(Modules)}
\end{aligned}$$

Fig. 6. Simplified syntax for Hs98.

```
1  data Struct : (m1 : HsModuleTy) → (m1s : SHsModuleTy) → Type where
2    MkStruct : Map StructDecl ds ds'
3             → Struct (HsModule env vars ds) (SHsModule ds')
4
5  transStructModule : (m  : HsModuleTy)
6                    → (m' : SHsModuleTy ** Struct m m')
```

Listing 1.6. Structural simplification definition and covering function in HESTER.

represented by `RenameFail`, where failures can arise in either the core function or as an inability to demonstrate conformance. A capture violation precipitated by a violation of linearity guarantees is one such example of a failure for renaming, either having been successfully identified by the core function or as a violation of well-formedness guarantees of the extended AST in the guard function. Conversely, refactorings successfully producing $\langle m' \rangle^{\sigma_0}$ must also provide proofs of structural equivalence, `eq`, and of the respective variable declarations, `oldD` and `newD`. Variable declaration proofs are analogous to `Elem` proofs of membership in vectors: a proof `oldD` : `DecldHsModuleTy` $l \, i \, x \, \langle m \rangle^{\sigma_0}$ is a pointer to the pattern or function declaring $x_{\langle l, i \rangle}$ in $\langle m \rangle^{\sigma_0}$.

Two modules $\langle m \rangle^{\sigma_0} \equiv \langle m' \rangle^{\sigma_0}$ are *structurally equivalent* when they result in the same structural simplification, $\langle m \rangle^{\sigma_0} \Downarrow \langle m' \rangle^{\sigma_0}$. The *structural simplification* of a module $\langle m \rangle^{\sigma_0} \Downarrow \mu$ is a surjection from $\langle m \rangle^{\sigma_0}$ to an equivalent, undecorated module $\mu \in M'$. The syntax of simplified programs in Fig. 6 is achieved by stripping our enriched AST of static semantic elements, and both variable and constructor names. The type `Struct` in Listing 1.6 defines this relation in our implementation, with the concomitant covering functions, including `transStructModule`, producing both μ and proof of $\langle m \rangle^{\sigma_0} \Downarrow \mu$. Note that this is not a de Bruijn representation since only the names of variables and constructors are removed; unique identifiers and scope levels remain in the form i_l for variables and i for constructors. Structural equivalence $\langle m \rangle^{\sigma_0} \equiv \langle m' \rangle^{\sigma_0}$ is defined as propositional equality over structural simplifications.

```
prfRename : (lv, id      : Nat)
          → (oldN, newN : Name Variable)
          → (m          : HsModuleTy)
          → RenamePrf lv id oldN newN m msP (g' lvl id oldN newN m)
prfRename lv id oldN newN m with (g' lvl id oldN newN m)
  | Nothing = RenameFail
  | Just (oldD, (m' ** (eq, newD))) = RenameSucc oldD newD eq
  where
    g' : (lv,id : Nat) → (oldN, newN : Name Variable)
       → (m : HsModuleTy)
       → Maybe (DecldHsModuleTy lv id oldN m,
                (m' : HsModuleTy **
                    (StructEq m m', DecldHsModuleTy lv id newN m')))
    ...
```

Listing 1.7. Guard function for renaming in HESTER.

5.2 Guarding Implementations

A refactoring is an (endo)function over the enriched AST of the form $g \circ f$, where f is the *core* function and g is the *guard* function. The core function f represents a standard implementation of the refactoring taken, in principle, from an existing refactoring tool: it transforms the given AST, can succeed or fail, and produces no proofs. For renaming in HESTER, f is renameModule, and is inspired by the HaRe implementation. In order to ensure that the program, $\langle m' \rangle^{\sigma_0}$, resulting from applying the refactoring to a given program, $\langle m \rangle^{\sigma_0}$, conforms to the corresponding specification, g derives the necessary proofs and acts as an interface to the refactoring as a whole.

For our renaming implementation, prfRename in Listing 1.7 acts as g. Here, prfRename returns the proof of conformance to RenamePrf. For clarity of presentation we elide the definition of g' but it is unsurprising: constructing the necessary proofs oldD, newD, and eq via their respective decision procedures. When any one decision procedure fails, g' returns Nothing.

Future work may extend the failure case to indicate the reason for failure; for renaming in particular, prfRename could be a full decision procedure, providing a proof of failure. Since renameModule is the only implementation of renaming in HESTER, and to facilitate use of prfRename as an interface function, renameModule is called directly from g'. In principle, our methodology allows for using an existing core function directly, which will not use our enriched AST. In such cases, the guard function converts both the original and refactored programs into our enriched AST, and thereby guard the refactoring.

6 Related Work

Refactoring has a long history, with the original refactoring work going back to the *fold/unfold* system from 1977 [13]. Since then, refactoring has been applied to a wide range of applications as an approach to program transformation (characterised by a survey by Mens [24]), and has been applied to a wide range of applications as an approach to program transformation, with refactoring tools a feature of popular IDEs including, e.g., Eclipse and Visual Studio.

The Haskell Refactorer, HaRe, is a refactoring tool for full Haskell 98 that is itself also implemented in Haskell [8,11,21,23]. The tool supports a wide number of Haskell refactorings, such as renaming, generalisation, lifting, folding, data type refactoring, clone detection [12] and others. The refactorings are described in terms of pre-conditions, with a set of unit tests given as part of the implementation. However, proofs of correctness of the refactorings are not given, apart from separate formalisation attempts of renaming for a small language based on a λ-calculus with recursion primitives [22]. A further mechanised attempt at proving soundness of extracting functions is given in [28] with mechanised proofs for Isabelle. However, the mechanised proofs are, again, for a small subset of a language based on a λ-calculus with recursion primitives.

Apart from Haskell, there has been some recent attempts at formalising refactoring for other functional languages, including OCaml [27], where the authors propose a set-theoretic framework of soundness, including a static semantics for renaming in OCaml and mechanised proofs of general soundness. This is perhaps the closest to our approach, with the complex features of a real-life language (OCaml modules) treated formally in a refactoring tool. The subset of Haskell we treat is arguably less complex semantically; for example, we do not deal with multiple modules. We do however have a fully verified refactoring tool using dependent types, where the proof is carried along with the refactoring implementation, rather than separate mechanised proofs of correctness.

Formalisation of refactoring for some other languages has followed similar lines: to our knowledge we are the first to produce an executable proof, leaving no gap between proof and implementation. A refactoring tool for Erlang was proven, using a mechanised proof system, for a small subset of the language [18]. Using a system of semantic rules, a formalisation of refactorings for a subset of Java including compositions of refactoring was proposed in [15]. Proving correctness properties of refactorings (including renaming) using correctness conditions and operational semantics is given in [1].

While we do not treat all of Haskell 98, our semantics deal with a substantial subset. Previous work on a static semantics for Haskell is rather limited, with the main prior work on static semantics for Haskell proposed in the 90s [26]. Work on semantics of functional languages also extends recently to Erlang, with formalisation of the semantics of core Erlang [4,14]. Finally, Idris has been used to formalise refactorings for side-channel attacks in [10] for C, and to model the semantics of imperative programming w.r.t. non-functional properties in [3,9].

7 Conclusions and Future Work

In this paper we introduced a new methodology that uses dependent types to define *guard* functions that demonstrate conformance of *core* refactoring implementations to formal refactoring specifications. This was achieved through a prototype refactoring tool, called HESTER, implemented in Idris using dependent types. Our dependently-typed implementation enabled us to embed static semantics into the AST representation of the programs, allowing *well-formedness* to be an invariant under the refactoring process. This invariant property means that well-formed programs cannot be refactored into ill-formed programs. We demonstrated our prototype on a renaming refactoring, illustrating conformance of the original renaming refactoring to its specification. Although we demonstrated our prototype tool on Haskell, the techniques demonstrated in this paper are completely general, and would easily transfer to other programming languages and paradigms, such as Java, C and Python. In the future, we will extend our prototype to include more refactorings from the original HaRe implementation, such as generalisation, folding/unfolding, lifting/narrow, adding/removing parameters and others. We also plan to extend our approach to allow compositions of refactorings. By composing refactorings together, and composing their proofs, we can show the refactoring composition is correct by virtue of the composition of its proofs. We will also extend our approach with more Haskell refactorings from the traditional set (lifting definitions, adding arguments, etc.) and to refactorings that allow us to model aspects of the type system. This will require extending our static semantics to model aspects of Haskell's type system. Doing so would allow both refactorings that transform Haskell programs in a type-safe way and those that transform aspects of a program's type structure, e.g. λ-lifting, where changing the scope of a definition (i.e. promoting a let declaration to a where; see [8] for details) could introduce monomorphism restrictions on definitions that are inferred to be polymorphic in some of their arguments; and refactorings that aim to generalise a function making it more polymorphic, and capturing this in its type signature. Allowing compositions of refactorings would also enable us to support a wider range of more complex refactorings. For example, in [29], Thompson et al. discuss how complex refactorings can be expressed as a composition of substitution and rewriting transformations, we will build on this technique. Finally, our prototype requires a specification of the refactoring to be provided as part of the implicit implementation of the refactoring. It would be better to separate out the production of the refactoring specification as a further step in the methodology, by requiring a specification as an additional parameter to be provided by the programmer. HESTER would then derive, automatically, the guarded procedure for the given specification. This would eliminate errors in the implementation of the refactorings and also their specifications.

References

1. Bannwart, F., Müller, P.: Changing programs correctly: refactoring with specifications. In: Misra, J., Nipkow, T., Sekerinski, E. (eds.) FM 2006. LNCS, vol. 4085, pp. 492–507. Springer, Heidelberg (2006). https://doi.org/10.1007/11813040_33
2. Barendregt, H.P.: The Lambda Calculus Its Syntax and Semantics, revised edn., vol. 103. North Holland (1984). http://www.cs.ru.nl/~henk/Personal Webpage
3. Barwell, A.D., Brown, C.: A Trustworthy framework for resource-aware embedded programming. In: Proceedings of the 31st Symposium on Implementation and Application of Functional Languages, IFL 2019. Association for Computing Machinery, New York (2021). https://doi.org/10.1145/3412932.3412944
4. Bereczky, P., Horpácsi, D., Thompson, S.: Machine-checked natural semantics for core erlang: exceptions and side effects. In: Proceedings of the 19th ACM SIGPLAN International Workshop on Erlang, pp. 1–13. ACM, August 2020. https://kar.kent.ac.uk/82515/
5. Brady, E.: Idris, a general-purpose dependently typed programming language: design and implementation. J. Funct. Program. **23**(5), 552–593 (2013)
6. Brady, E.: Type-driven development of concurrent communicating systems. Comput. Sci. **18**(3), 219–240 (2017). https://doi.org/10.7494/csci.2017.18.3.1413
7. Brady, E.: Type-driven Development with Idris. Manning Publications Co., Shelter Island (2017)
8. Brown, C.: Tool support for refactoring Haskell programs. Ph.D. thesis, Computing Laboratory, University of Kent, Canterbury, Kent, UK, September 2008
9. Brown, C., Barwell, A.D., Marquer, Y., Minh, C., Zendra, O.: Type-driven verification of non-functional properties. In: Proceedings of the 21st International Symposium on Principles and Practice of Declarative Programming, PPDP 2019. Association for Computing Machinery, New York (2019). https://doi.org/10.1145/3354166.3354171
10. Brown, C., Barwell, A.D., Marquer, Y., Zendra, O., Richmond, T., Gu, C.: Semi-automatic ladderisation: improving code security through rewriting and dependent types. In: Proceedings of the 2022 ACM SIGPLAN International Workshop on Partial Evaluation and Program Manipulation, PEPM 2022, pp. 14–27. Association for Computing Machinery, New York (2022). https://doi.org/10.1145/3498886.3502202
11. Brown, C., Li, H., Thompson, S.: An expression processor: a case study in refactoring Haskell programs. In: Page, R., Horváth, Z., Zsók, V. (eds.) TFP 2010. LNCS, vol. 6546, pp. 31–49. Springer, Heidelberg (2011). https://doi.org/10.1007/978-3-642-22941-1_3
12. Brown, C., Thompson, S.: Clone detection and elimination for Haskell. In: Proceedings of the 2010 ACM SIGPLAN Workshop on Partial Evaluation and Program Manipulation, PEPM 2010, pp. 111–120. Association for Computing Machinery, New York (2010). https://doi.org/10.1145/1706356.1706378
13. Burstall, R.M., Darlington, J.: A transformation system for developing recursive programs. J. ACM **24**(1), 44–67 (1977). https://doi.org/10.1145/321992.321996
14. Caballero, R., Martin-Martin, E., Riesco, A., Tamarit, S.: A core Erlang semantics for declarative debugging. J. Log. Algebraic Methods Program. **107**, 1–37 (2019). https://doi.org/10.1016/j.jlamp.2019.05.002
15. Cornelio, M., Cavalcanti, A., Sampaio, A.: Sound refactorings. Sci. Comput. Program. **75**(3), 106–133 (2010). https://doi.org/10.1016/j.scico.2009.10.001

16. Fowler, M.: Refactoring: Improving the Design of Existing Code. Addison-Wesley Longman Publishing Co., Inc., USA (1999)
17. Gligoric, M., Behrang, F., Li, Y., Overbey, J., Hafiz, M., Marinov, D.: Systematic testing of refactoring engines on real software projects. In: Castagna, G. (ed.) ECOOP 2013. LNCS, vol. 7920, pp. 629–653. Springer, Heidelberg (2013). https://doi.org/10.1007/978-3-642-39038-8_26
18. Horpácsi, D., Kőszegi, J., Thompson, S.: Towards trustworthy refactoring in erlang. Electron. Proc. Theoret. Comput. Sci. **216**, 83–103 (2016). https://doi.org/10.4204/eptcs.216.5
19. Johnsson, T.: Lambda lifting: transforming programs to recursive equations. In: Jouannaud, J.-P. (ed.) FPCA 1985. LNCS, vol. 201, pp. 190–203. Springer, Heidelberg (1985). https://doi.org/10.1007/3-540-15975-4_37
20. Leroy, X.: Formal verification of a realistic compiler. Commun. ACM **52**(7), 107–115 (2009). https://doi.org/10.1145/1538788.1538814
21. Li, H.: Refactoring Haskell Programs, September 2006. https://kar.kent.ac.uk/14425/
22. Li, H., Thompson, S.: Formalisation of Haskell refactorings. In: van Eekelen, M., Hammond, K. (eds.) Trends in Functional Programming, pp. 182–196, September 2005. http://www.cs.kent.ac.uk/pubs/2005/2250
23. Li, H., Thompson, S., Reinke, C.: The haskell refactorer: HaRe, and its API. In: Boyland, J., Hedin, G. (eds.) Proceedings of the 5th workshop on Language Descriptions, Tools and Applications (LDTA 2005), pp. 182–196, April 2005. http://www.cs.kent.ac.uk/pubs/2005/2158. published as Volume 141, Number 4 of Electronic Notes in Theoretical Computer Science, http://www.sciencedirect.com/science/journal/15710661
24. Mens, T., Tourwé, T.: A survey of software refactoring. IEEE Trans. Softw. Eng. **30**(2), 126–139 (2004). https://doi.org/10.1109/TSE.2004.1265817
25. Peyton Jones, S.L.: Haskell 98: lexical structure. J. Funct. Program. **13**(1), 7–16 (2003). https://doi.org/10.1017/S0956796803000418
26. Peyton Jones, S.L., Wadler, P.: A static semantics for Haskell, February 1992. https://www.microsoft.com/en-us/research/publication/a-static-semantics-for-haskell/
27. Rowe, R.N.S., Férée, H., Thompson, S.J., Owens, S.: characterising renaming within OCaml's module system: theory and implementation. In: Proceedings of the 40th ACM SIGPLAN Conference on Programming Language Design and Implementation, PLDI 2019, pp. 950–965. Association for Computing Machinery, New York (2019). https://doi.org/10.1145/3314221.3314600
28. Sultana, N., Thompson, S.: Mechanical verification of refactorings. In: Proceedings of the 2008 ACM SIGPLAN Symposium on Partial Evaluation and Semantics-Based Program Manipulation, PEPM 2008, pp. 51–60. Association for Computing Machinery, New York (2008). https://doi.org/10.1145/1328408.1328417
29. Thompson, S., Horpácsi, D.: Refactoring = Substitution + Rewriting (2023). https://arxiv.org/abs/2211.11550
30. Thompson, S.: Refactoring functional programs. In: Vene, V., Uustalu, T. (eds.) AFP 2004. LNCS, vol. 3622, pp. 331–357. Springer, Heidelberg (2005). https://doi.org/10.1007/11546382_9

Impact and Performance of Randomized Test-Generation Using Prolog

Marcus Gelderie[✉][iD], Maximilian Luff, and Maximilian Peltzer

Aalen University of Applied Sciences, Beethovenstr. 1, 73430 Aalen, Germany
{marcus.gelderie,maximilian.luff,maximilian.peltzer}@hs-aalen.de
https://www.hs-aalen.de

Abstract. We study randomized generation of sequences of test-inputs to a system using Prolog. Prolog is a natural fit to generate test-sequences that have complex logical inter-dependent structure. To counter the problems posed by a large (or infinite) set of possible tests, randomization is a natural choice. We study the impact that randomization in conjunction with SLD resolution have on the test performance. To this end, this paper proposes two strategies to add randomization to a test-generating program. One strategy works on top of standard Prolog semantics, whereas the other alters the SLD selection function. We analyze the mean time to reach a test-case, and the mean number of generated test-cases in the framework of Markov chains. Finally, we provide an additional empirical evaluation and comparison between both approaches.

Keywords: software testing · randomization · prolog

1 Introduction

The need for software testing is well established. The idea to auto-generate tests is a constant theme in the field of software testing, stretching back many decades (e.g. [11,13,15,17]). Automatically generating software tests can be done in a number of ways, depending on the specific test-goal: In the past, tests have been generated from UML specifications [12], based on natural language [21], and, more recently, using large-language models [8,19]. But tests have also been generated according to formal or semi-formal specifications [5,24]. Particularly when formal methods are used, one often has to deal with a very large, even infinite, number of test cases. Exploring such a large set of tests in a randomized fashion is a natural approach and has been used extensively in various different ways and contexts for a long time (see e.g. [3,6,14,14,18,18]).

Prolog is a natural fit to generate test-cases that follow a logical pattern (as opposed to unstructured testing, as is done, for example, in many forms of

---

This work was created as part of a project funded by the German Federal Ministry of Education and Research under grant number 16KIS193K. The authors are responsible for the content of this publication.

© The Author(s), under exclusive license to Springer Nature Switzerland AG 2024
J. Bowles and H. Søndergaard (Eds.): LOPSTR 2024, LNCS 14919, pp. 166–182, 2024.
https://doi.org/10.1007/978-3-031-71294-4_10

Fuzzing [14]). Generating test-cases using Prolog has been studied in the past [3, 4,9,17]. It has been applied to software-testing in general, but also to specialized areas, such as security testing [5,23]. Some approaches also use randomization to explore the space of test-cases [3]. Randomization solves some of the problems inherent in the SLD resolution algorithm—particularly the fact that it is not complete when the resolution works in a depth-first manner. It may also yield a more diverse set of test-cases, because it permits exploring distant parts of an infinite SLD tree. Randomization seems to be a logical fit in the context of test-case generation using Prolog.

In the light of its apparent utility, it is natural to study randomization itself and its properties. What are possible strategies to implement randomized search strategies for test-cases in Prolog running on current state-of-the-art implementations? What is the probability of a hitting a particular test case, and how long will it take? To our surprise, we only found very few papers dealing with the properties of randomization itself (see also *Related Work* below).

In this paper, we study randomized test-case generation using Prolog. Our main contributions are (i) We propose strategies to implement randomized search in both unmodified Prolog runtimes, and via specific modifications to the usual SLD implementations. (ii) We show how adding randomness naturally turns the SLD resolution into an infinite discrete-time Markov chain and propose to use this framework to study the runtime-effects. We do this for our proposed scheme and give tight asymptotic bounds on the expected time to hit a particular test-case. (iii) Finally, we study the effect that various Prolog implementations have on the efficiency of randomizing test-case generation.

We present two ways of adding randomization to Prolog programs. The first way works without altering the semantics of standard Prolog and thus works on existing implementations. It works by adding a predicate, called a *guard*, that randomly fails to every rule. Crucially, failure is determined by an independent event for every successive call to the same rule. In a second strategy, we propose a modification to the resolution algorithm: Given a goal and a set of matching rules, drop an indeterminate number of rules from the set and permute the remaining ones. Again, we do this in an independent fashion every time a goal is resolved with the input program. This second modification is reminiscent of that proposed in [3], but differs in that it also *drops* a random number of rules from the set. This, in effect, prevents an infinite recursion with probability 1.

In the following we study the effects on randomizing the resolution in this way. Due to space constraints, we focus on the first proposed randomization strategy (though the analysis is almost identical in both cases). We give a detailed description of the resulting Markov chain and analyze its probability structure. We show that, provided the guards are chosen appropriately, the number of test cases produced is finite and given by a simple equation in terms of the guard probabilities. We also show that, if we repeat the initial query infinitely many times, we will reach each test-case after a finite number of steps on average. This *hitting time* is a well-known concept in the study of Markov chains. We again give a closed formula representation and accompanying asymptotic bound in the depth of the given test case in the SLD resolution tree.

Finally, we study the randomization procedures from an empirical perspective and provide comparisons between the two aforementioned approaches. We implement the *guards approach* to randomization in SWI-Prolog [20] and the *drop-and-shuffle* approach in Go-Prolog [10]. We chose Go-Prolog for its accessible code-base, which lends itself to experimental modifications. We then compare the number of test-cases produced before a specific test-goal is seen and the number of iterations that were required to do so.

Related Work. Some early works on test-case generation using Prolog are [2,4,9,17]. Automated test-case generation in Prolog was described by Pesch and Schaller [17]. The authors state how to test individual syscalls with logic programming. The used specifications state a set of pre-conditions then the actual invocation of the respective syscall and afterwards what the expected post-conditions are. This paper demonstrates that test-case generation using Prolog is very beneficial to test systems in a structured manner. This problem domain does not deal with any problems of recursion since the authors only test input sequences of length one. This means that this paper does not deal with recursion problems that are witnessed for many other test scenarios.

Another approach showcasing Prolog's capabilities used in test case generation was shown by Hoffman and Strooper [9]. The authors automated the generation procedure of tests for modules written in C with Prolog.

Bougé et al. [2] start the testing procedure with the definition of a Σ-algebra and respective axioms. The aim of this testing procedure is based on the regularity and uniformity testing hypothesis. Prolog is used to generate test cases and to partition the test cases into test classes following the uniformity hypothesis. The authors also recognize the problems of recursion in Prolog test case generation and apply different search strategies to solve them. Since the paper enforces a length limit on the generated solution, it will not find any test case that exceeds that length.

Richard Denney [4] also researches test-case generation based on specifications written in Prolog. In his paper, he implements a meta-interpreter in Prolog to be able to track which rules, generated from the specification, were already applied. This is done by constructing a finite automaton. Each arc between states corresponds with respective rules in the Prolog database. Final states in this automaton are test cases produced in the test-case generation process. With this solution, he addresses the problems of recursion, evaluable predicates, and ordering, which are challenging aspects of test-case generation using Prolog. However, the recursion problem is only addressed heuristically, which means that a user has to specify a threshold of how often an arc can be traversed during the execution of test case generator. We argue that the estimation of the threshold is an error-prone task and, if not set correctly, could miss important test cases.

Gorlick et al. [7] also introduce a methodology for formal specifications. For this task, they use constraint logic programming to describe the system under test's behaviour. With this approach, the authors also recognize that they both have a test oracle and a test case generator at the same time. One challenge the authors addressed is, yet again, the recursion problem. To solve this challenge they used a randomization approach. This feature enables the proposed framework to pick probabilistically from the predicates. However, they do not provide any statements about test case duplication or infinite looping.

Casso et al. [3] approach assertion-based testing of Prolog programs with random search rules. They rely on the `Ciao` model and its capabilities to specify pre- and post-conditions for static analysis and the runtime checker. Further, the authors develop a test-case generator based on these conditions. For randomizing the test case search, Casso et. al. use a selection function that randomly chooses clauses to be resolved. The authors do not study the randomization itself, nor its properties. We will revisit this paper and its randomization strategy in Sect. 3, where we will also explain the differences from our approach in more detail.

Prolog was also used in security testing. For web applications, Zech et al. [22,23] first build an expert system to filter test cases according to some attack pattern and later apply this risk analysis to filter test cases in the generation process. Since the paper, yet again, only addresses single input sequences, it effectively circumvents the problem of recursion. Prolog as was also used in Fuzzing [5] by Dewey et al. to use CLP in order to produce fuzzing inputs to compilers.

2 Preliminaries

Given a (usually finite) set Σ of elements, we write Σ^* for the set of all finite length sequences $w_1 \cdots w_l$ with $w_i \in \Sigma$ and $l \in N_0$. The empty sequence is denoted by ε. We write $\Sigma^+ = \Sigma^* \setminus \{\varepsilon\}$. If $\Sigma = \{x\}$ is a singleton, we write x^* or x^+ instead of $\{x\}^*$. Concatenation is denoted by $(u_1 \cdots u_l) \cdot (v_1 \cdots v_r) = u_1 \cdots u_l v_1 \cdots v_r$. We write $|w| = |w_1 \cdots w_l| = l \in N_0$.

We use the theory of *Markov chains*. For a detailed introduction and proofs of the following claims, the reader is referred to standard literature on the subject, e.g. [16]. We revisit the concepts, notation, and central results from the theory of Markov chains that we will use throughout this paper for convenience.

We consider a countable set $\mathcal{S}$ of *states*, a mapping $p\colon \mathcal{S} \times \mathcal{S} \to [0,1]$ that assigns *transition probabilities* to pairs of states with the property that for all $s \in \mathcal{S}$ it holds that $\sum_{s' \in \mathcal{S}} p(s, s') = 1$, and an *initial state*[1] $\mathsf{Init} \in \mathcal{S}$. Let $(X_n)_{n \in N_0}$ be an infinite sequence of random variables $X_n \in \mathcal{S}$. The triple $((X_n)_{n \in N_0}, p, \mathsf{Init})$ is a *Markov chain*, if $\Pr[X_0 = \mathsf{Init}] = 1$ and for all $n \in N$ and all $s_1, \ldots, s_n \in \mathcal{S}$:

---

[1] In the literature one usually considers *initial distributions* to model uncertainty about the initial state. We do not need this capability in the present paper and consider initial distributions whose support is a single state of probability 1.

$$\Pr[X_n = s_n \mid X_0 = s_0 = \mathsf{Init}, \ldots, X_{n-1} = s_{n-1}]$$
$$= \Pr[X_n = s_n \mid X_{n-1} = s_{n-1}] = p(s_{n-1}, s_n)$$

For two states s, s' we write $s \rightsquigarrow s'$, if $\Pr[X_n = s'$ for some $n] > 0$ in the Markov chain $((X_n)_{n \in \mathbb{N}}, p, s)$. Intuitively, there is a way to get from s to s'. A set $A \subseteq \mathcal{S}$ is *absorbing*, if for every $s \in A$ and every $s' \in \mathcal{S}$ with $s \rightsquigarrow s'$ it holds that $s' \in A$. If $A = \{s\}$ is a singleton, the state s is said to be absorbing. If any two states are reachable from one-another ($s \rightsquigarrow s'$ for any $s, s' \in \mathcal{S}$), the Markov chain is *irreducible*.

Let $A \subseteq \mathcal{S}$ be a non-empty set of states and let $H^A = \inf\{n \in \mathbb{N}_0 \mid X_n \in A\} \in \mathbb{N}_0 \cup \{\infty\}$ denote the random variable such that $X_H \in A$ visits A for the first time. H^A is the *hitting time* of A. Then conditioned on $H^A < \infty$ and $X_H = s$, the sequence $(X_{H+n})_{n \in \mathbb{N}_0}$ is a Markov chain with initial state s and is independent of $X_0, \ldots, X_H$. This is called the *strong Markov property*. It is sometimes useful to consider the hitting times for initial states other than Init. Write H_s^A for the hitting time of A with starting state s.

The expected value $h^A \stackrel{\text{def}}{=} \mathsf{E}[H^A]$ is known as the *mean hitting time*. Given any state $s \in \mathcal{S}$, we define $h_s^A = \mathsf{E}[H_s^A]$ for the mean hitting time of A from initial state s. The mean hitting times are then the unique minimal solution to the equations

$$\begin{cases} h_s^A = 0 & \text{if } s \in A \\ h_s^A = 1 + \sum_{s' \in \mathcal{S}} p(s, s') h_{s'}^A & \text{if } s \notin A \end{cases} \quad (1)$$

A state $s \in \mathcal{S}$ is *recurrent*, if $\Pr[\sum_{n=0}^{\infty} \mathbb{1}_{X_n = s} = \infty] = 1$ (where $\mathbb{1}_A$ is the indicator random variable for event A). Otherwise, it is *transient*. It can be shown that a state is recurrent iff $\Pr[X_m = s$ for some $m \geq 1] = 1$ in the chain $((X_n)_{n \in \mathbb{N}_0}, p, s)$ (the probability of returning s, once visited, is 1). One can show that if a Markov chain is irreducible and contains one recurrent state, then all states are recurrent. In the case we call the chain itself *recurrent* (or *transient*).

3 Randomized Test Generation with Prolog

In this paper, we view a test as a sequence of inputs to a system. For example, given a web-application with a REST-interface, we could think of a test as a sequence of HTTP-Requests using various methods (`GET`, `POST` and so forth) against different API-endpoints (e.g. `/login`, `/items/{USERID}/list`). Since our focus is on randomization, we do not explicitly model a concept of "valid" test-cases. We also do not model the *test-oracle* which determines the success or failure of the test (e.g. "requests are processed in < 700 ms").

At a very abstract level, such a sequence of test-inputs could be generated with the Prolog program shown in Listing 1.1. All valid substitutions for `X` in the query `t(X)` are input sequences to our fictional system. Since this program will only ever output test sequences of the type `[command1, command1, command1, ...]`, a straightforward approach is to add *guard clauses* of the form

Listing 1.1. A program generating randomized sequences of test inputs.

```
1 t([]).
2 t([H|T]) :- command(H), t(T).
3 command(X) :- command1(X); /* ... */ ; commandr(X).
4 command1(X) :- /* ... */ .
5 %  ...
```

Listing 1.2. Guard clauses.

```
1 guard_t :- random(X), X < p_cont.
2 guard_1 :- random(X), X < p_1.
3 %  ...
4 t([H|T]) :- guard_t, command(H), t(T).
5 command1(X) :- guard_1, /* ... */ .
6 %  ...
```

shown in Listing 1.2. Note that the symbols p_cont, p_1,... are meant to represent float constants between 0 and 1, and can be adjusted as needed. In effect, some sub-trees of the SLD-tree are then randomly left unexplored.

A similar approach was proposed by Casso et. al. in [3]. Their randomization is presented as a modification to the Prolog interpreter; equivalently, it can be implemented using meta-predicates. Essentially, Casso et. al. shuffle the list of input clauses whose head unifies with the current goal, instead of iterating over it in the usual left-to-right fashion. They do not drop rules. The termination of the program is instead enforced via depth-control. It is thus not difficult to see that the random approach itself merely alters the order of test-cases, but not their number. As such, the questions concerning the number of test-cases (that we study here) do not make sense for their approach.

However, one can augment the shuffling approach due to Casso et. al. by *additionally* dropping several items from the set of unifying rules prior to shuffling. We do this with an independent Bernoulli trial for each rule (i.e. the number of dropped rules follows a Binomial). The resulting algorithm shares many properties with our scheme above (in particular, the results from the next section apply). Results regarding hitting times require a more involved variant of the analysis given in Sect. 3.2. Due to space constraints, we cannot elaborate here.

3.1 Number of Generated Tests

The program $\mathcal{P}$ shown in Listing 1.1 and 1.2 gives rise to a probabilistic number of test cases. We study the questions: Is this number finite? If so, what is the expected number of test-case?

The program $\mathcal{P}$ gives rise to an infinite Markov chain, which is based on the SLD-tree corresponding to $\mathcal{P}$. Recall that $\mathcal{P}$ is governed by some probabilities p_cont, p_1, ..., which we will denote by $p_c, p_1, \ldots, p_r$. The Markov chain is

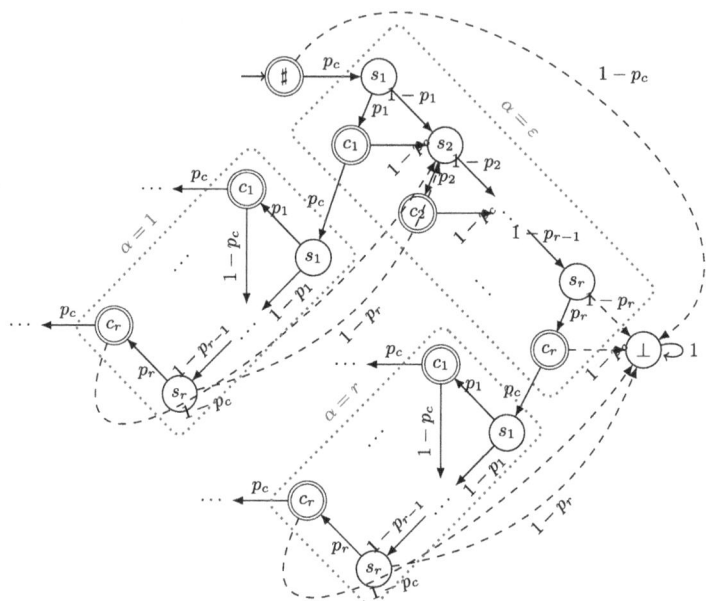

Fig. 1. The Markov chain corresponding to $\mathcal{P}$ with $\mathsf{Init} = \sharp$ and *blocks* $\alpha \in \{1,\ldots,r\}*$ surrounded by blue boxes. (Color figure online)

depicted in Fig. 1. Note that we model choice points via the states s_i. This is necessary, because $\mathcal{P}$ will backtrack when a call to command1 fails, and proceed to command2 with probability p_2. Double-circles denote *output states*—that is, whenever such a state is visited, a test case terminating in that state is generated. Node $\sharp$ corresponds to the empty list. $\bot$ is the only absorbing state. It corresponds a termination of the resolution algorithm.

The blue boxes denote areas that share a common structure. We call these areas *blocks*. We can uniquely identify each block by a finite sequence $\alpha \in \{1,\ldots,r\}*$. For any state s in the Markov chain, we denote by $\mathsf{Block}(x)$ the unique block that contains it. For any label occurring in a block ($s_1, s_2, \ldots, s_r$ and $c_1 \ldots, c_r$) and a block α, write s_1^α, c_1^α, ... for the unique state with that label in block α. In this way, we can identify any state in the Markov chain. Put differently, the Markov chain is given by a state space $\mathcal{S} = \{s_i^\alpha, c_i^\alpha \mid 1 \leq i \leq r,\ \alpha \in \{1,\ldots,r\}*\} \cup \{\bot, \sharp\}$ and transition probabilities $p(s, s')$ for $s, s' \in \mathcal{S}$:

$$p(\sharp, s_1^\varepsilon) = p_c$$
$$p(\sharp, \bot) = 1 - p_c$$
$$p(s_i^\alpha, s_{i+1}^\alpha) = 1 - p_i \qquad 1 \leq i < r$$
$$p(c_i^\alpha, s_{i+1}^\alpha) = 1 - p_c \qquad 1 \leq i < r$$
$$p(c_i^\alpha, s_1^{\alpha \cdot i}) = p_c \qquad 1 \leq i \leq r$$
$$p(s_i^\alpha, c_i^\alpha) = p_i \qquad 1 \leq i \leq r$$

The dashed *upward* arrows (which correspond to backtracking to a lower-recursion level) are somewhat more technical to define. Those arrows originate in states of the form s_r^α or c_r^α. There are several cases to consider:

a) $\alpha \in \{1,\ldots,r\}^* \cdot i$ for some $1 \leq i < r$
b) $\alpha \in \{1,\ldots,r\}^* \cdot i \cdot r^+$ for some $1 \leq i < r$
c) $\alpha \in r^*$

This motivates the following transition probabilities

$$p(s_r^{\alpha \cdot i}, s_{i+1}^\alpha) = 1 - p_r \qquad 1 \leq i < r \qquad \text{case a)}$$
$$p(c_r^{\alpha \cdot i}, s_{i+1}^\alpha) = 1 - p_c \qquad 1 \leq i < r \qquad \text{case a)}$$
$$p(s_r^{\alpha' \cdot i \cdot r \cdots r}, s_{i+1}^{\alpha'}) = 1 - p_r \qquad 1 \leq i < r \qquad \text{case b)}$$
$$p(c_r^{\alpha' \cdot i \cdot r \cdots r}, s_{i+1}^{\alpha'}) = 1 - p_c \qquad 1 \leq i < r \qquad \text{case b)}$$
$$p(s_r^{r \cdots r}, \bot) = 1 - p_r \qquad \text{case c)}$$
$$p(c_r^{r \cdots r}, \bot) = 1 - p_c \qquad \text{case c)}$$

We call edges from a block $\beta \cdot \alpha$ to a state in block β or from any block to $\bot$ an *upward* edge. They correspond precisely to the dashed arrows in Fig. 1. If the Markov chain follows such an edge, we say block $\beta \cdot \alpha$ is *left upward* or that the chain *traverses upward* at that point. A block that has been left upward, is never visited again.

It is immediate that every state is visited at most once. There are no two states that can be reached from one-another. Note further that if we omit the dashed arrows and the state $\bot$, the resulting graph structure is an infinite, finitely branching tree. Yet, it is conceivable that the terminal state $\bot$ is never reached, because the sequence of states visited from ♯ is infinite. The following proposition shows that this is not the case, provided $p_c < 1$.

Proposition 1. *Let $s \in \mathcal{S}$ be any state. If $p_c < 1$, then all sequences originating in s eventually leave* Block(s) *upward. In particular, $\bot$ is visited eventually.*

Proof. Let $\alpha =$ Block(s). It is sufficient to show the result for $s = s_1^\alpha$. We first study the special case that there is an infinite path that *never* traverses upward. Pick an infinite path $s_0 s_1 s_2 \cdots$ through the chain that never traverses upward. For every n, the prefix $s_0 \cdots s_n$ must traverse at least $t_n \stackrel{\text{def}}{=} 1 + \lfloor \frac{n}{2r} \rfloor$ edges of the form $(c_i^\beta, s_1^{\beta \cdot i})$ (for correspondingly many distinct blocks β). This is because inside a block, there are only $2r$ states and no cycles. Hence, the probability of such a prefix is at most $p_c^{t_n}$ which tends to 0 as $n \to \infty$. As a result, the probability of any path that never traverses upward is 0.

Now for any i, consider the subtree of nodes below c_i^α that are visited. Since every node in the Markov chain can be visited at most once, the only option to remain in this tree indefinitely is for the tree to be infinite. However, the Markov chain is finitely branching. Therefore, the subtree of visited nodes below c_i^α is finitely branching. By König's lemma, this tree contains an infinite path and hence has probability 0. □

Corollary 1. *Let α be any block. The probability of reaching α from s_1^ε (i.e. from the initial block) is:*

$$\Pr[H^\alpha < \infty] = p_c^{|\alpha|} \cdot \prod_{i=1}^{|\alpha|} p_{\alpha_i} < \infty$$

Consequently, the probability of reaching α from $\sharp$ is $p_c^{|\alpha|+1} \cdot \prod_{i=1}^{|\alpha|} p_{\alpha_i}$.

Let $s \in \mathcal{S}$. We denote by $N(s)$ the random variable that counts the total number of states visited from s (including those in downstream blocks), before Block(s) is left upward. A useful observation is that $N(s) = H_s^E$ can also be expressed as a hitting time, where $E = \{s_i^\beta \mid \beta \prec \mathsf{Block}(s),\ 1 \leq i \leq r\} \cup \{\bot\}$. Note that it would suffice to take the subset of E which contains $s_{\alpha_i+1}^\beta$ for any $\beta = \alpha_1 \cdots \alpha_{i-1} \prec \alpha$. To define this set, we would have to work around the case $\alpha_i = r$—indeed, if $\alpha \in r^*$, then $E = \{\bot\}$. So we define E as larger than needed purely to simplify notation. Note moreover that E depends on $\alpha = \mathsf{Block}(s)$. Since α is usually clear from context, we simply write E, but also use the notation E_α when needed.

Lemma 1. *Let $s \in \mathcal{S}$ and write $p_{\max} = \max\{p_1, \ldots, p_r\}$. If $p_c < 1$ and $\eta \stackrel{def}{=} r \cdot p_{\max} \cdot p_c < 1$, then $\mathsf{E}[N(s)]$ is finite.*

Proof. Let $\alpha = \mathsf{Block}(s)$. It is obvious that $N(x^\alpha) \leq N(s_1^\alpha)$ for all $x \in \mathcal{S}$ with $\mathsf{Block}(x) = \alpha$. It therefore suffices to show that $N(s_1^\alpha)$ is finite. In the remainder of this proof, we write $\hat{s} = s_1^\alpha$.

Let now β be any block and let M_β denote the number of states visited in block β from s_1^β. Clearly $M_\beta \leq 2r$. Let furthermore $I_\beta = \mathbb{1}_{H_{\hat{s}}^\beta < \infty}$ denote the indicator random-variable of the event that β is visited from $\hat{s}$. Note that both random-variables are independent because the underlying random events in $\mathcal{P}$ are independent and we count by M_β only states that are visited once β is entered. We have:

$$N(\hat{s}) = \sum_{\beta \in \alpha \cdot \{1,\ldots,r\}^*} I_\beta \cdot M_\beta$$

There are precisely r^l blocks that have distance $l \in \mathbb{N}$ from α. For each such block $\beta = \alpha \cdot \beta_1 \cdots \beta_l$, the probability of reaching it from α is $\Pr[I_\beta = 1] = p_c^l \cdot \prod_{i=1}^l p_{\beta_i}$ by Corollar 1 (if $l = 0$ then $\beta = \alpha$ and the probability is 1). Then $\Pr[I_\beta = 1] \leq (p_{\max} \cdot p_c)^l$ for all β. This gives (using linearity of expectation and that I_β is independent from X_β for all β):

$$\mathsf{E}[N(\hat{s})] = \sum_{\beta \in \alpha \cdot \{1,\ldots,r\}^*} \mathsf{E}[I_\beta] \cdot \mathsf{E}[X_\beta] \leq \sum_{l=0}^\infty r^l \cdot (p_c^l \cdot p_{\max}^l) \cdot 2r$$

$$= 2r \cdot \sum_{l=0}^\infty \eta^l = \frac{2r}{1-\eta}$$

$\square$

Let $s \in \mathcal{S}$. Denote by $O(s)$ the number of output states that are visited from s before $\mathsf{Block}(s)$ is left upward. Clearly $O(s) \le N(s)$.

Theorem 1. *Let $p_c < 1$ and $p_c \cdot r \cdot \max\{p_1, \ldots, p_r\} < 1$. Then for any block α*

$$\mathsf{E}[O(s_1^\alpha)] = \frac{\sum p_i}{1 - p_c \sum p_i} \quad \text{and} \quad \mathsf{E}[N(s_1^\alpha)] = \frac{r + \sum p_i}{1 - p_c \sum p_i}$$

Proof. $C \stackrel{\text{def}}{=} \mathsf{E}[N(s_1^\alpha)]$ is finite by Lemma 1. Note that C is independent of α by the strong Markov property. We recall that $N(s_1^\alpha) = H_{s_1^\alpha}^A$ is a hitting time, where $A = \{s_i^\beta \mid \beta \prec \alpha\} \cup \{\bot\}$. In the remainder of the proof, we drop the superscript Greek letter for all states in α; i.e. s_1 is understood to mean s_1^α.

Every path from s_1 to A must visit $s_2, \ldots, s_r$. Thus, by the strong Markov property, $N(s_1) = \left(\sum_{i=1}^{r-1} H_{s_i}^{s_{i+1}}\right) + H_{s_r}^A$. By linearity of expectation and Eq. (1)

$$C = \mathsf{E}[N(s_1)] = \sum_{i=1}^{r-1} h_{s_i}^{s_{i+1}} + h_{s_r}^A = \sum_{i=1}^{r-1} 1 + p_i \cdot h_{c_i}^{s_{i+1}} + (1 + p_r \cdot h_{c_r}^A)$$

$$= \sum_{i=1}^{r-1} 1 + p_i(1 + p_c \cdot \underbrace{h_{s_1^{\alpha \cdot i}}^{s_{i+1}}}_{C}) + (1 + p_r(1 + p_c \cdot \underbrace{h_{s_1^{\alpha \cdot r}}^A}_{C}))$$

$$= r + \sum_{i=1}^{r} p_i + C p_c \sum_{i=1}^{r} p_i$$

Solving for C proves the theorem. The proof for $\mathsf{E}[O(s_1)]$ is similar. □

Note that for any given state s_1^α, the mean hitting time $h_\sharp^{s_1^\alpha} = \sum_{n \ge 0} \Pr[H_\sharp^{s_1^\alpha} \ge n] \ge \sum_{n \ge 0} 1 - p_c - \infty$ (where we use $\mathsf{E}[X] = \sum_{n \ge 0} \Pr[X \ge n]$ for any random variable that only takes on positive integer values). So although we have a non-zero probability of selecting every test, we won't, informally speaking, do so on average. Naturally, this is solved by repeating the experiment a sufficient number of times. This is the content of the next section.

3.2 Infinite Looping and Time-to-Hit

As shown in Lemma 1, the program in Listing 1.1 terminates eventually. As a result, every state except $\bot$ in the Markov chain we studied above is transient and, moreover, the number of produced test-cases is always finite. In testing, one aims at a high test coverage, and the number of test cases we produce in this fashion, though free of duplicates, has a low chance of visiting tests in deep blocks. A natural approach is to loop on the predicate `t/1` like so:

```
main_loop(X) :- repeat, t(X).
```

With respect to our Markov chain this amounts to removing $\bot$ and to instead redirect any arc into $\bot$ to $\sharp$. The resulting chain is recurrent (indeed positive recurrent) and we compute the mean hitting time of any state. In what follows, we will assume $p_1 = p_2 = \cdots = p_r \stackrel{\text{def}}{=} p$ such that $r \cdot p \cdot p_c < 1$ (as in Theorem 1). Moreover, we assume that $p(\sharp, s_1^\varepsilon) = 1$, so that the empty list is never selected as an output. This simplifies the formulas below slightly, but has otherwise no effect on the line of reasoning we give here.

Given the conditions of Theorem 1, there is a constant $C = N(s_1^\alpha)$ that is independent of the value of α. As noted before, $C = h_{s_1^\alpha}^{E_\alpha}$ is a mean hitting time where $E_\alpha = \{s_i^\beta \mid \beta \prec \alpha, 1 \leq i \leq r\} \cup \{\sharp\}$ (note that we modified the definition of E used in the previous section by replacing $\bot$ by $\sharp$). Recall that we usually drop the subscript α, because it is clear from context.

If we hop from one state s_i^α to its neighbor s_{i+1}^α we might traverse the tree below $s_1^{\alpha \cdot i}$ with probability $p \cdot p_c$. That step will visit C states. This means (by Eq. (1)):

$$h_{s_i^\alpha}^{s_{i+1}^\alpha} = 1 + p(1 + p_c C) \stackrel{\text{def}}{=} \Delta$$

More generally the mean hitting time within a block is again independent of α and can be computed as:

$$h_{s_1^\alpha}^{s_i^\alpha} = h_{s_1^\alpha}^{s_2^\alpha} + h_{s_2^\alpha}^{s_3^\alpha} + \cdots + h_{s_{i-1}^\alpha}^{s_i^\alpha} = (i-1)\Delta \qquad (2)$$

We define the *leave upward time* $U_{s_i^\alpha} = h_{s_i^\alpha}^E$ where $E = E_\alpha$ as above. Note that the value $U_{s_i^\alpha} \in \mathbb{N} \cup \{\infty\}$ does not actually depend on α. This justifies writing $U_i = U_{s_i^\alpha}$. It is obvious that $C = U_1$. Moreover, by using the same derivation as that in Eq. (2):

$$U_i = (r - i + 1)\Delta \quad 1 \leq i \leq r \qquad (3)$$

We already noted that $C = U_1$. A related quantity is the hitting time of $\sharp$ from any s_i^α, $\alpha = \alpha_1 \cdots \alpha_t$, which we may compute using the intermediate leave upward times:

$$h_{s_i^\alpha}^\sharp = U_i + U_{\alpha_t + 1} + \cdots U_{\alpha_1 + 1}$$

Note that we abuse notation: Eq. 3 gives $U_{r+1} = 0$. While s_{r+1}^α does not exist and hence the corresponding hitting time is not defined, it is convenient to allow such terms and exploit that $U_{\alpha_j + 1} = 0$ whenever $\alpha_j = r$ ($1 \leq j \leq t$).

With this, we may compute:

$$h_{s_i^\alpha}^\sharp = U_i + \sum_{k=1}^t U_{\alpha_k+1} = \Delta \cdot \left((r - i + 1) + \sum_{k=1}^t (r - (\alpha_k + 1) + 1) \right)$$

$$= \Delta \cdot \left(1 + \sum_{k=0}^t r - \alpha_k \right) \qquad \text{where } \alpha_0 \stackrel{\text{def}}{=} i \qquad (4)$$

Note again that the formula works correctly, if $i = r + 1$: Say $\alpha = rrr$. Then we are in the process of falling back to $\sharp$ and the equation gives 0. While the hitting time is again not defined for the non-existent state s_{r+1}, we will sometimes have to compute the hitting time of $\sharp$ from the "right neighbor" of s_{i+1}. In these situations, abusing notation in this way is useful because we need not distinguish between cases where $i < r$ and those where $i = r$.

Finally, we may now compute the hitting time of an arbitrary state in terms of hitting times in intermediate blocks, again using Eq. (1). Let $\alpha = \beta \cdot j$.

$$\begin{aligned}h_\sharp^{s_i^\alpha} &= h_\sharp^{s_j^\beta} + 1 \\ &\quad + (1-p)(h_{s_{j+1}^\beta}^\sharp + h_\sharp^{s_i^\alpha}) &\text{(fall through to } s_{j+1}^\beta) \\ &\quad + p(1+(1-p_c))(h_{s_{j+1}^\beta}^\sharp + h_\sharp^{s_i^\alpha})) &\text{(no visit to next block } \alpha) \\ &= h_\sharp^{s_j^\beta} + 1 + p + (h_{s_{j+1}^\beta}^\sharp + h_\sharp^{s_i^\alpha})(1-pp_c)\end{aligned}$$

This gives

$$h_\sharp^{s_i^\alpha} = \frac{h_\sharp^{s_j^\beta} + 1 + p + (1-pp_c)h_{s_{j+1}^\beta}^\sharp}{pp_c}$$

and together with Eq. (2) and Eq. (4), recalling that $\alpha_{|\alpha|} = j$, we have:

$$\begin{aligned}h_\sharp^{s_i^\alpha} &= \frac{h_\sharp^{s_j^\beta} + 1 + p + (1-pp_c)h_{s_{j+1}^\beta}^\sharp}{pp_c} + (i-1)\Delta \\ &= \frac{h_\sharp^{s_j^\beta}}{pp_c} + \frac{1}{pp_c}\left(1 + p + (1-pp_c)(\sum_{k=1}^{|\alpha|} r - \alpha_k)\Delta\right) + (i-1)\Delta \quad (5)\end{aligned}$$

The following theorem gives a closed formula:

Theorem 2. *Let s_i^α for some $\alpha = \alpha_1 \cdots \alpha_t$. Let $\nu = pp_c$ and $\nu \cdot r < 1$. Then*

$$h_\sharp^{s_i^\alpha} = \nu^{-t} + (i-1)\Delta + \sum_{k=1}^{t} \frac{1 + p + (\alpha_k - 1)\Delta + (1-\nu)\sum_{s=1}^{k}(r-\alpha_s)\Delta}{\nu^{t+1-k}} \quad (6)$$

Proof. By induction on t. If $t = 0$, then $\alpha = \varepsilon$ and by Eq. (2), we have $h_\sharp^{s_i^\varepsilon} = 1 + (i-1)\Delta$. Moreover the empty sum in Eq. (6) equates to 0 establishing the induction base.

Now let $t > 0$ and assume the statement holds for $t - 1$. By induction, we may replace $h_{\#}^{s_j^\beta}$ in Eq. (5) with Eq. (6):

$$\frac{1}{\nu}\left(\nu^{-(t-1)} + (\alpha_t - 1)\Delta + \sum_{k=1}^{t-1} \frac{1 + p + (\alpha_k - 1)\Delta + (1 - \nu)\sum_{s=1}^{k}(r - \alpha_s)\Delta}{\nu^{t-k}}\right)$$

$$+ \frac{1}{\nu}\left(1 + p + (1 - \nu)(\sum_{s=1}^{t} r - \alpha_s)\Delta\right) + (i - 1)\Delta$$

$$= \nu^{-t} + \sum_{k=1}^{t-1} \frac{1 + p + (\alpha_k - 1)\Delta + (1 - \nu)\sum_{s=1}^{k}(r - \alpha_s)\Delta}{\nu^{t+1-k}}$$

$$+ \frac{(1 + p + (\alpha_t - 1)\Delta + (1 - \nu)(\sum_{s=1}^{t} r - \alpha_s)\Delta)}{\nu^{t+1-t}} + (i - 1)\Delta$$

$$= \nu^{-t} + (i - 1)\Delta + \sum_{k=1}^{t} \frac{1 + p + (\alpha_k - 1)\Delta + (1 - \nu)\sum_{s=1}^{k}(r - \alpha_s)\Delta}{\nu^{t+1-k}}$$

□

Corollary 2. *Let $\nu = p \cdot p_c$ with $\nu \cdot r < 1$. Then $h_{\#}^{s_i^\alpha} \in \Theta(\nu^{-t})$ for any $\alpha = \alpha_1 \cdots \alpha_t$.*

Proof. Write Eq. (6) as

$$\nu^{-t} + A + \nu^{-t-1} \sum_{k=1}^{t} \frac{B_k + \sum_{s=1}^{k} D_s}{\nu^{-k}}$$

for suitable constants (in t) $A \geq 0$, $B_k \geq 0$, and $D_s \geq 0$, whereby $h_{\#}^{s_i^\alpha} \in \Omega(\nu^{-t})$.

Choose suitable largest values $B \geq B_k$ for all $t \in \mathbb{N}$, $1 \leq k \leq t$, and $D \geq D_s$ for all $1 \leq s \leq t$. Bound Eq. (6) from above by

$$\nu^{-t} + A + \nu^{-t-1} \sum_{k=1}^{t} \frac{B + D \cdot k}{\nu^{-k}} \leq \nu^{-t} + A + \nu^{-t-1}(B + D) \cdot \sum_{k=1}^{t} \frac{k}{\nu^{-k}}$$

A well-known calculation via derivatives gives $\sum_{k=1}^{t} k \cdot \nu^k = \nu \sum_{k=1}^{t} k\nu^{k-1} \leq \nu \sum_{k=1}^{\infty} k\nu^{k-1} = \nu \cdot \frac{d}{d\nu} \sum_{k=0}^{\infty} \nu^k = \frac{\nu}{(1-\nu)^2}$. With that we have

$$\nu^{-t} + A + \nu^{-t-1} \sum_{k=1}^{t} \frac{B + D \cdot k}{\nu^{-k}} \leq \nu^{-t} + A + (B + D)\frac{\nu^{-t}}{(1-\nu)^2} \in \mathcal{O}(\nu^{-t})$$

□

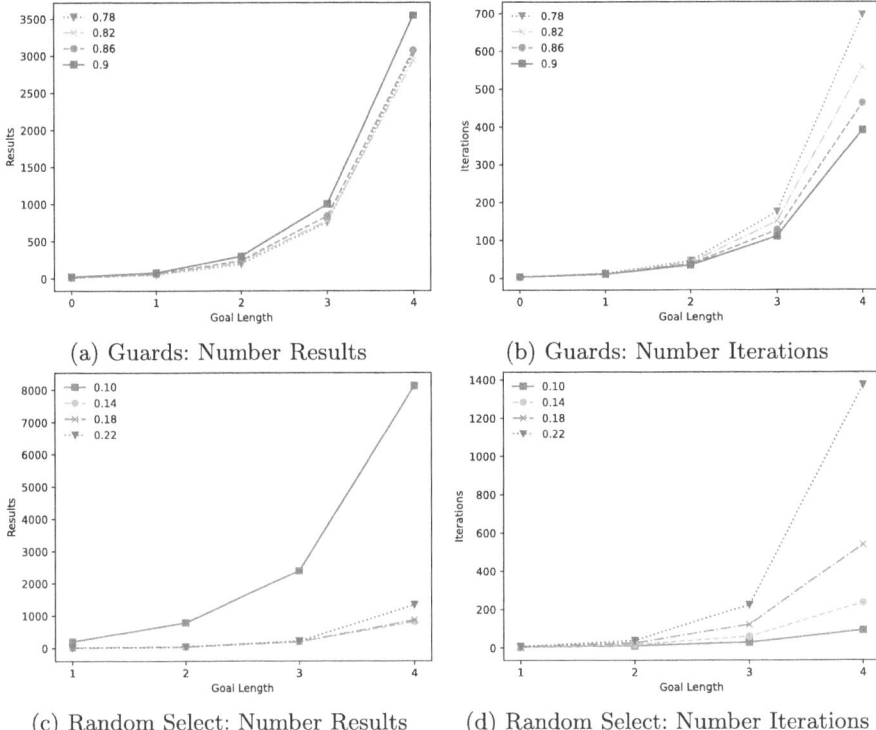

Fig. 2. Iterations and results until test-case [second, ..., second] is reached.

4 Evaluation

In the following section, we empirically evaluate the randomization approaches outlined in Sect. 3. We implement the strategy via *guards* using SWI-Prolog [20]. To modify the *shuffle-and-drop* strategy, we choose Go-Prolog [10] Go-Prolog has a small and easily modifiable code-base, which simplifies experiments of this kind. We benchmark these approaches with various choices for the configurable probabilities. All the benchmarks were done using a slightly altered version of the program from Listing 1.1: We limited the number of available commands to three. Moreover, each command/1 predicate simply unifies its argument with a corresponding constant (in our case first, second, and third). We executed each benchmarks 1000 times. The results are shown in Fig. 2.

The *goal length* each benchmark lists on its x-axis is the length of a list consisting solely of the constant symbol second the respective number of times. This guarantees that we would not find this test case with the standard depth-first, left-first search behaviour, but also that is not the path that would be picked last with depth-first search. For our implementation, we relied on Janus [1] for SWI as the Python-Prolog bridge to gather the results.

Guard-Approach Benchmarks: Every command/1 predicate had an equal probability of $\frac{1}{3}$ for the steady probability. The different plots mark different continuation probabilities p_c used for the respective queries.

Figure 2a shows the number of results until a specific target test-case is found. As expected, the number of results drastically increases with the target list size. Further, only a continuation probability of 0.9 has a higher number of results. Figure 2b shows how many iterations were necessary until the determined target was found. Similar to the number of results, the number of iterations also grows with increasing list size of the expected outcome. As the continuation probability increases the total number of iterations decreases.

Shuffle-and-Drop Benchmarks: As described above, for the Go-Prolog variant we implemented a drop-probability as discussed in Sect. 3. Otherwise, the benchmarks are still conducted using the same pattern as described above for increasing goal lengths. Figure 2c shows the number of produced results whereas Fig. 2d shows the number of iterations.

Note that dropping a clause with probability 0.1 is the same as proceeding to explore it with probability 0.9. Hence, the probabilities in Figs. 2c and 2d are dual to those above. However, the shuffle-and-drop randomization strategy is much more coarse than the guard strategy by design: The 0.1 drop probability applies to *both* the t predicate as well as the command{1,2,3} predicates. This is quite different from the previous scenario, where $p_c = 0.9 \gg p_1 = \cdots = p_r = 0.33$ were distinct. Consequently, both the number of iterations and the number of results are notably higher for the shuffle-and-drop approach: A probability of 0.1 to drop a clause produces significantly more solutions until a specified goal is found. The drop probabilities of 0.14 and 0.18 are rather similar for all specified goal lengths. On the other hand, the number of iterations signals that the number of iterations rises with a higher dropping probability. In Fig. 2c, the probability 0.14 outperforms both 0.10 and 0.18, hinting at an inflection point. This is likely due to the dual role of the dropping probability, which governs the exploration of a specific test *and* the probability of exploring the SLD-tree below it: Too high, and the number of tests-until-target increases; too low, and the target-test is skipped at the target depth.

5 Conclusion

We have presented two approaches to randomize the SLD derivation of test-cases in Prolog and studied their performance in terms of expected time to hit a test-case, and mean number of test-cases produced. To this end, we presented a detailed analysis of the random behavior of test-case generation using Prolog using Markov chains. Our theorems allow a precise calibration of the probabilities to adjust the expected number of test-cases per query. When looping on such a query, the rate of growth of the mean-hitting time for a given test-case is exponential in its depth, where the base is the product of the involved probabilities. We then compared both strategies and various sets of values for the involved probabilities empirically. We find that the guard approach that uses an unmodified Prolog implementation provides a very fine-grained control over the randomization and thus produces test-cases quicker.

In future work, we plan to study the semantics of this approach when negation-as-failure is involved. In particular, randomization may lead to a false refutation of q(t_1,...,t_k) in the goal \+ q(t_1,..., t_k). However, this may be acceptable, if it occurs with low probability. In a similar vein, the treatment of negation as failure might require entirely different randomization strategies than we have presented here, which is another interesting topic for future research.

References

1. Andersen, C., Swift, T.: The janus system: a bridge to new prolog applications. In: Warren, D.S., Dahl, V., Eiter, T., Hermenegildo, M.V., Kowalski, R., Rossi, F. (eds.) Prolog: The Next 50 Years, LNCS, vol. 13900, pp. 93–104. Springer, Cham (2023). https://doi.org/10.1007/978-3-031-35254-6_8
2. Bougé, L., Choquet, N., Fribourg, L., Gaudel, M.C.: Application of PROLOG to test sets generation from algebraic specifications. In: Ehrig, H., Floyd, C., Nivat, M., Thatcher, J. (eds.) TAPSOFT 1985. LNCS, vol. 186, pp. 261–275. Springer, Heidelberg (1985). https://doi.org/10.1007/3-540-15199-0_17
3. Casso, I., Morales, J.F., López-García, P., Hermenegildo, M.V.: An integrated approach to assertion-based random testing in prolog. In: Gabbrielli, M. (ed.) LOPSTR 2019. LNCS, vol. 12042, pp. 159–176. Springer, Cham (2020). https://doi.org/10.1007/978-3-030-45260-5_10
4. Denney, R.: Test-case generation from prolog-based specifications. IEEE Softw. **8**(2), 49–57 (1991)
5. Dewey, K., Roesch, J., Hardekopf, B.: Language fuzzing using constraint logic programming. In: Proceedings of the 29th ACM/IEEE International Conference on Automated Software Engineering, ASE 2014, pp. 725–730. Association for Computing Machinery, New York, NY, USA (2014). https://doi.org/10.1145/2642937.2642963
6. Duran, J.W., Ntafos, S.C.: An evaluation of random testing. IEEE Trans. Software Eng. **4**, 438–444 (1984)
7. Gorlick, M.M., Kesselman, C.F., Marotta, D.A., Parker, D.S.: Mockingbird: a logical methodology for testing. J. Logic Program. **8**(1–2), 95–119 (1990)
8. Gu, Q.: Llm-based code generation method for golang compiler testing. In: Proceedings of the 31st ACM Joint European Software Engineering Conference and Symposium on the Foundations of Software Engineering, pp. 2201–2203 (2023)
9. Hoffman, D.M., Strooper, P.: Automated module testing in prolog. IEEE Trans. Software Eng. **17**(9), 934 (1991)
10. ichiban/prolog: ichiban/prolog. https://github.com/ichiban/prolog, Accessed 03 May 2024
11. Ince, D.C.: The automatic generation of test data. Comput. J. **30**(1), 63–69 (1987). https://doi.org/10.1093/comjnl/30.1.63
12. Kim, Y.G., Hong, H.S., Bae, D.H., Cha, S.D.: Test cases generation from UML state diagrams. IEE Proc. Softw. **146**(4), 187–192 (1999)

13. Meyer, B., Ciupa, I., Leitner, A., Liu, L.L.: Automatic testing of object-oriented software. In: van Leeuwen, J., Italiano, G.F., van der Hoek, W., Meinel, C., Sack, H., Plášil, F. (eds.) SOFSEM 2007: Theory and Practice of Computer Science, pp. 114–129. Springer, Berlin Heidelberg, Berlin, Heidelberg (2007)
14. Miller, B.P., Fredriksen, L., So, B.: An empirical study of the reliability of unix utilities. Commun. ACM **33**(12), 32–44 (1990). https://doi.org/10.1145/96267.96279
15. Miller Jr, E.F., Melton, R.A.: Automated generation of testcase datasets. In: Proceedings of the International Conference on Reliable Software, pp. 51–58 (1975)
16. Norris, J.: Markov Chains. Cambridge Series in Statistical and Probabilistic Mathematics, Cambridge University Press, Cambridge (1998)
17. Pesch, H., Schnupp, P., Schaller, H., Spirk, A.P.: Test case generation using prolog. In: Proceedings of the 8th International Conference on Software Engineering, pp. 252–258 (1985)
18. Ramler, R., Winkler, D., Schmidt, M.: Random test case generation and manual unit testing: Substitute or complement in retrofitting tests for legacy code? In: 2012 38th Euromicro Conference on Software Engineering and Advanced Applications, pp. 286–293. IEEE (2012)
19. Siddiq, M.L., Santos, J., Tanvir, R.H., Ulfat, N., Rifat, F.A., Lopes, V.C.: Exploring the effectiveness of large language models in generating unit tests. arXiv preprint arXiv:2305.00418 (2023)
20. Wielemaker, J., Schrijvers, T., Triska, M., Lager, T.: SWI-Prolog. Theory Pract. Logic Program. **12**(1–2), 67–96 (2012)
21. Xu, F.F., Vasilescu, B., Neubig, G.: In-ide code generation from natural language: promise and challenges. ACM Trans. Softw. Eng. Methodol. **31**(2) (2022). https://doi.org/10.1145/3487569
22. Zech, P., Felderer, M., Breu, R.: Security risk analysis by logic programming. In: Bauer, T., Großmann, J., Seehusen, F., Stølen, K., Wendland, M.-F. (eds.) RISK 2013. LNCS, vol. 8418, pp. 38–48. Springer, Cham (2014). https://doi.org/10.1007/978-3-319-07076-6_3
23. Zech, P., Felderer, M., Breu, R.: Knowledge-based security testing of web applications by logic programming. Int. J. Softw. Tools Technol. Transfer **21**, 221–246 (2019)
24. Zeng, Z., Ciesielski, M., Rouzeyre, B.: Functional test generation using constraint logic programming. In: Robert, M., Rouzeyre, B., Piguet, C., Flottes, M.-L. (eds.) SOC Design Methodologies. ITIFIP, vol. 90, pp. 375–387. Springer, Boston, MA (2002). https://doi.org/10.1007/978-0-387-35597-9_32

Term and Graph Rewriting

Proving Uniqueness of Normal Forms w.r.t Reduction of Term Rewriting Systems

Takahito Aoto

Niigata University, Niigata, Japan
aoto@ie.niigata-u.ac.jp

Abstract. Most of the studies concerning uniqueness of normal forms of rewriting systems address confluence (CR); but other such properties are also known, such as unique normal forms w.r.t. conversion (UNC), etc. Among such properties, we address in this paper unique normal forms w.r.t. reduction (UNR), aiming for automated proof of it for first-order term rewriting systems (TRSs). UNR is less known compared to CR or UNC, but captures a most natural notion of "uniqueness of normal forms" for computational systems. Although CR (UNC) implies UNR, there have been few methods for showing UNR of TRSs not having CR (UNC). Furthermore, UNR is not closed under signature extensions in contrast to CR and UNC, and some subtleties are required to treat UNR. In this paper, we give some transformation methods for proving UNR that can be applied for TRSs without CR/UNC, and report on an implementation and experiments on automated verification of UNR of TRSs.

Keywords: Unique normal forms w.r.t. reduction · Unique normal form properties · Confluence · Term rewriting systems

1 Introduction

Automated proofs of properties of rewrite systems concerning uniqueness of normal forms have been getting addressed recently for various formats of rewrite systems. Some properties concerning uniqueness of normal forms have been known: most notably, *confluence (CR)*, *unique normal forms w.r.t. conversion (UNC)*, *unique normal forms w.r.t. reduction (UNR)*, and *the normal form property (NFP)*. These properties constitute a logical hierarchy, that is, the implications CR $\implies$ NFP $\implies$ UNC $\implies$ UNR hold, and these implications are proper, i.e. the inverse implications do not hold. These properties are sometimes referred as *unique normal form properties*.

Among these properties, confluence is most well-known and has been studied most extensively (e.g. [4,6,8,9,14,17,19]). Confluence naturally arises when proving uniqueness of normal forms and can be regarded as the central property of these properties. UNC is less known in comparison with confluence, but several important criteria are known; furthermore, UNC has played a substantial role in the study of rewriting (e.g. [9,15,18,20,21]). So far, NFP and UNR have been extensively explored in rewriting only for limited topics such as the decidability (in subclasses of TRSs) [6,7,16] and modularity [1,2,12,13]. Concerning the tools for automatically verifying these properties for rewrite systems, among participants of the confluence competitions[1], CR is the target of many tools, UNC for some, and there are only few tools that target NFP or UNR.

In this paper, we address the automated proofs of UNR for TRSs. UNR is the weakest property that constitutes the logical hierarchy of unique normal form properties, and it captures the most natural notion of "uniqueness of normal forms" of the computational models—a rewrite system is said to be UNR if each term has at most one normal form. So far, most of possible ways to prove UNR is by proving CR or UNC, via the implication mentioned above. In particular, up to our knowledge, there is no general method for showing UNR of non-UNC TRSs, apart from UNR decision procedures for some specific subclasses of TRSs. As well, currently, there is a quite limited ability for automatically proving UNR of the rewrite systems not having CR/UNC. An important fact, which one should be careful about in UNR proving in contrast to CR/UNC proving, is that UNR is not closed under signature extensions [10]. More specifically, there is a case where UNR is lost by addition of auxiliary function symbols. This fact adds some subtleties dealing with proofs of UNR, automatically as well as manually.

We present in this paper some basic techniques for proving UNR and some specific methods that can be also applied for proving UNR of TRSs without CR/UNC. Presented criteria are implemented in our tool ACP, and we report on an experimental evaluation of presented techniques. A preliminary version of our tool, with (preliminary versions of) techniques of this paper, has won in the UNR category of the 2023 edition of the Confluence Competition.

The rest of the paper is organized as follows. In the next section, we present some standard notions of term rewriting which will be used in this paper, and explain some basic properties and facts to give our UNR criteria. Here, we adapt some non-standard presentations, in order to take care of the subtleties arising from the fact that UNR is not closed under signature extensions. In Sect. 3, we give criteria of transformations of TRSs that preserves UNR, for transformations consisting of addition and elimination of rewrite rules. In Sect. 4, we give an abstract criteria for UNR-preserving transformations. Sections 5 and 6 deal with methods for proving UNR; more specifically, we give dedicated transformations that can be applied for proving UNR of TRSs without CR/UNC properties. In Sect. 7, we explain implementations of our UNR (dis)proving methods and report on some experiments to estimate the power of presented techniques. Section 8 concludes.

2 Preliminaries

We denote by $\mathcal{V}$ the set of *variables*. A *signature* is a set of arity-fixed function symbols. The set of *terms over a signature* $\mathcal{F}$ is denoted by $\mathrm{T}(\mathcal{F}, \mathcal{V})$. We will deal with multiple signatures, and assume that all function symbols are supplied from

---
[1] http://project-coco.uibk.ac.at/.

the *universe* $\mathcal{F}_{univ}$ *of function symbols*; unless specific signatures are mentioned, we consider the signature $\mathcal{F}_{univ}$. The set of *positions* of a term t is denoted by $\text{Pos}(t)$; we use ε to denote the *root* position. A *subterm* of t at a position $p \in \text{Pos}(t)$ is denoted by $t|_p$. We write $u \trianglelefteq t$ to denote that u is a subterm of the term t; u is a *proper* subterm if $u \neq t$. For a term t over the signature $\mathcal{F}$, the set of function symbols in t is denoted by $\mathcal{F}(t)$. The set of variables in a term t is denoted by $\mathcal{V}(t)$. A term t is said to be *ground* if $\mathcal{V}(t) = \emptyset$; t is said to be *linear* if no variable has multiple occurrences in t. A *substitution over a signature* $\mathcal{F}$ is a mapping $\sigma : \mathcal{V} \to \text{T}(\mathcal{F}, \mathcal{V})$ with a finite domain $\text{dom}(\sigma) = \{x \in \mathcal{V} \mid x \neq \sigma(x)\}$, where each substitution σ is identified with its homomorphic extension $\sigma : \text{T}(\mathcal{F}, \mathcal{V}) \to \text{T}(\mathcal{F}, \mathcal{V})$. We write $t\sigma$ for the shorthand of $\sigma(t)$. A binary relation R over the set $\text{T}(\mathcal{F}_{univ}, \mathcal{V})$ is *closed under substitutions* if $s\ R\ t$ implies $s\sigma\ R\ t\sigma$ for any substitution σ. A *context over a signature* $\mathcal{F}$ is a special 'term' over $\mathcal{F}$ containing precisely one occurrence of a special constant $\square$ (*hole*). We assume $\square \notin \mathcal{F}_{univ}$. For a context C, the term obtained by replacing the hole with a term t is denoted by $C[t]$; we also write $C[t]_p$ to specify that the hole occurs at the position p in C. A binary relation R over the set $\text{T}(\mathcal{F}_{univ}, \mathcal{V})$ is *closed under contexts* if $s\ R\ t$ implies $C[s]\ R\ C[t]$ for any context C. A binary relation over $\text{T}(\mathcal{F}_{univ}, \mathcal{V})$ is said to be a *rewrite relation* if it is closed under contexts and substitutions. A term l *encompasses* a term t (denoted by $l \trianglelefteq\!\!\!\cdot\, t$) if $l\sigma \trianglelefteq t$ for some substitution σ. Clearly, the encompassment relation is reflexive and transitive. A term t *(properly) overlaps on a term* l if there exists a non-variable (proper) subterm u of l such that t and u are unifiable. A *rewrite rule* is a pair $\langle l, r \rangle$ of terms satisfying $l \notin \mathcal{V}$ and $\mathcal{V}(r) \subseteq \mathcal{V}(l)$, which will be written as $l \to r$. A rewrite rule $l \to r$ is said to be *left-linear* (*linear, non-erasing*) if l is linear (resp. l, r are linear, $\mathcal{V}(l) \subseteq \mathcal{V}(r)$). Let $\mathcal{R}$ be a set of rewrite rules. We put $\text{LHS}(\mathcal{R}) = \{l \mid l \to r \in \mathcal{R}\}$.

We define sets $\text{NF}(\mathcal{R})$ and $\text{RED}(\mathcal{R})$ of terms as follows:

$$\text{RED}(\mathcal{R}) = \{t \in \text{T}(\mathcal{F}_{univ}, \mathcal{V}) \mid l \trianglelefteq\!\!\!\cdot\, t \text{ for some } l \to r \in \mathcal{R}\}$$
$$\text{NF}(\mathcal{R}) = \{t \in \text{T}(\mathcal{F}_{univ}, \mathcal{V}) \mid l \trianglelefteq\!\!\!\cdot\, t \text{ for no } l \to r \in \mathcal{R}\}$$

Terms in $\text{RED}(\mathcal{R})$ are *$\mathcal{R}$-reducible terms* and those in $\text{NF}(\mathcal{R})$ are *$\mathcal{R}$-normal forms*. Clearly, $t \in \text{RED}(\mathcal{R})$ iff $t \notin \text{NF}(\mathcal{R})$ for any $t \in \text{T}(\mathcal{F}_{univ}, \mathcal{V})$, and hence $\text{RED}(\mathcal{R}) \cap \text{NF}(\mathcal{R}) = \emptyset$. Also, for sets $\mathcal{R}, \mathcal{R}'$ of rewrite rules such that $\text{LHS}(\mathcal{R}) \subseteq \text{LHS}(\mathcal{R}')$, we have $\text{RED}(\mathcal{R}) \subseteq \text{RED}(\mathcal{R}')$ and $\text{NF}(\mathcal{R}') \subseteq \text{NF}(\mathcal{R})$.

For a set $\mathcal{R}$ of rewrite rules, we define a binary relation $\to_\mathcal{R}$ over $\text{T}(\mathcal{F}_{univ}, \mathcal{V})$ as follows: $s \to_\mathcal{R} t$ if $s = C[l\sigma]$ and $t = C[r\sigma]$ for some $l \to r \in \mathcal{R}$, context C and substitution σ. The symmetric (reflexive, reflexive transitive, equivalence) closure of $\to_\mathcal{R}$ is denoted by $\leftrightarrow_\mathcal{R}$ (resp. $\stackrel{=}{\to}_\mathcal{R}, \stackrel{*}{\to}_\mathcal{R}, \stackrel{*}{\leftrightarrow}_\mathcal{R}$). The inverse of $\to_\mathcal{R}$ ($\stackrel{=}{\to}_\mathcal{R}, \stackrel{*}{\to}_\mathcal{R}$) is denoted by $\leftarrow_\mathcal{R}$ (resp. $\stackrel{=}{\leftarrow}_\mathcal{R}$ and $\stackrel{*}{\leftarrow}_\mathcal{R}$). We note that all of these relations are rewrite relations. We will sometimes mention the notion of critical pairs and the set of critical pairs $\text{CP}(\mathcal{R})$ of rewrite rules $\mathcal{R}$—for which we refer to textbooks of term rewriting such as [5], but whose details will not be necessary. We here note the facts that $u \leftarrow_\mathcal{R} \circ \to_\mathcal{R} v$ for $\langle u, v \rangle \in \text{CP}(\mathcal{R})$, which fact will be implicitly used later. A set of rewrite rules $\mathcal{R}$ is said to be *strongly closed* if, for any $\langle u, v \rangle \in \mathrm{CP}(\mathcal{R})$, both $u \xrightarrow{*}_{\mathcal{R}} \circ \xleftarrow{=}_{\mathcal{R}} v$ and $u \xrightarrow{=}_{\mathcal{R}} \circ \xleftarrow{*}_{\mathcal{R}} v$ hold. Here $- \circ -$ stands for the composition of binary relations.

A triple $\langle s, t, \to_{\mathcal{R}} \rangle$ is a *rewrite step* if $s \to_{\mathcal{R}} t$ holds. For brevity, a rewrite step $\langle s, t, \to_{\mathcal{R}} \rangle$ is also written as $s \to_{\mathcal{R}} t$. To specify a rewrite step $s \to_{\mathcal{R}} t$, we also use a notation $s \to_{p, l \to r, \sigma} t$, $s \to_{l \to r, \sigma} t$, $s \to_{p, l \to r} t$, etc. to denote that $s = C[l\sigma]_p$ and $t = C[r\sigma]_p$ with $l \to r \in \mathcal{R}$ for some context C. Often in the literature, a rewrite step $\langle s, t, \to_{\mathcal{R}} \rangle$ and the statement $\langle s, t \rangle \in \to_{\mathcal{R}}$ (holds or not) are used interchangingly under the same notation $s \to_{\mathcal{R}} t$; however, we will need to distinguish them to overcome some subtleties of the UNR property. The following notations are introduced also for this purpose.

A non-empty (alternating) sequence $t_0\ R_0\ t_1\ R_1\ \cdots\ R_{n-1}\ t_n$ ($n \geq 0$) of terms and relations $R_i \in \{\to_{\mathcal{R}}, \leftarrow_{\mathcal{R}}\}$ is called a $\mathcal{R}$-*conversion over a signature* $\mathcal{F}$ if $t_i \in \mathrm{T}(\mathcal{F}, \mathcal{V})$ for all $0 \leq i \leq n$ and $t_i\ R_i\ t_{i+1}$ holds for each $0 \leq i < n$. The set of $\mathcal{R}$-conversions over $\mathcal{F}$ is denoted by $\mathrm{Conv}(\mathcal{F}, \mathcal{R})$. When $\mathcal{R}$ and/or $\mathcal{F}$ are clear from the context, we refer an $\mathcal{R}$-conversion over $\mathcal{F}$ as an $\mathcal{R}$-conversion, or a conversion over $\mathcal{F}$, or even just a conversion. We write $\rho = t_0\ R_0\ t_1\ R_1\ \cdots\ R_{n-1}\ t_n$ as $\rho = (t_i\ R_i\ t_{i+1})_{0 \leq i < n}$. An $\mathcal{R}$-conversion $(t_i\ R_i\ t_{i+1})_{0 \leq i < n}$ is said to be a *(finite)* $\mathcal{R}$-*reduction* if $R_i = \to_{\mathcal{R}}$ for all $0 \leq i < n$.[2]

Lemma 1 (closure of conversion). $\mathrm{Conv}(\mathcal{F}, \mathcal{R})$ *is closed under contexts and substitutions, namely, if* $\rho = (t_i\ R_i\ t_{i+1})_{0 \leq i < n} \in \mathrm{Conv}(\mathcal{F}, \mathcal{R})$ *then* $C[\rho\sigma] = (C[t_i\sigma]\ R_i\ C[t_{i+1}\sigma])_{0 \leq i < n} \in \mathrm{Conv}(\mathcal{F}, \mathcal{R})$ *for any context* C *and substitution* σ *over* $\mathcal{F}$.

Lemma 2 (embedding of conversion). *Let* $\mathcal{F}, \mathcal{F}'$ *be signatures such that* $\mathcal{F} \subseteq \mathcal{F}'$. *Then*, $\mathrm{Conv}(\mathcal{F}, \mathcal{R}) \subseteq \mathrm{Conv}(\mathcal{F}', \mathcal{R})$.

A *form* of conversion $\rho = (t_i\ R_i\ t_{i+1})_{0 \leq i < n}$ is a predicate φ built over $t_0, \ldots, t_n, \xrightarrow{*}_{\mathcal{R}}, \xleftarrow{*}_{\mathcal{R}}, \xleftrightarrow{*}_{\mathcal{R}}, \ldots$ such that ρ satisfies φ. We write $\rho : \varphi$ if ρ has the form φ. For example, we write $\rho : t_0 \xrightarrow{*}_{\mathcal{R}} t_n$ if $\rho = t_0 \to_{\mathcal{R}} \cdots \to_{\mathcal{R}} t_n$. Similarly, we write $\rho : t_0 \xleftarrow{*}_{\mathcal{R}} t_n$, $\rho : t_0 \xleftrightarrow{*}_{\mathcal{R}} t_n$, $\rho : t_0 \xleftarrow{*}_{\mathcal{R}} \circ \xrightarrow{*}_{\mathcal{R}} t_n$, etc. A conversion is sometimes referred to without an explicit name; that is, instead of saying a conversion $\rho : s \xleftrightarrow{*}_{\mathcal{R}} t$, we say a conversion $s \xleftrightarrow{*}_{\mathcal{R}} t$.

A *term rewriting system* (*TRS*, for short) is a pair $\langle \mathcal{F}, \mathcal{R} \rangle$ of a signature $\mathcal{F}$ and a set $\mathcal{R}$ of rewrite rules such that $\mathcal{R} \subseteq \mathrm{T}(\mathcal{F}, \mathcal{V})^2$. We refer the following four properties to *normal form properties*:

1. A TRS $\langle \mathcal{F}, \mathcal{R} \rangle$ is *confluent* ($\mathrm{CR}(\langle \mathcal{F}, \mathcal{R} \rangle)$) if for any conversion $s \xleftarrow{*}_{\mathcal{R}} \circ \xrightarrow{*}_{\mathcal{R}} t$ over $\mathcal{F}$ there exists a conversion $s \xrightarrow{*}_{\mathcal{R}} \circ \xleftarrow{*}_{\mathcal{R}} t$ over $\mathcal{F}$.
2. A TRS $\langle \mathcal{F}, \mathcal{R} \rangle$ *has unique normal forms w.r.t. reduction* ($\mathrm{UNR}(\langle \mathcal{F}, \mathcal{R} \rangle)$) if $s = t$ holds for any conversion $s \xleftarrow{*}_{\mathcal{R}} \circ \xrightarrow{*}_{\mathcal{R}} t$ over $\mathcal{F}$ with $s, t \in \mathrm{NF}(\mathcal{R})$.
3. A TRS $\langle \mathcal{F}, \mathcal{R} \rangle$ *has unique normal forms w.r.t. conversion* ($\mathrm{UNC}(\langle \mathcal{F}, \mathcal{R} \rangle)$) if $s = t$ holds for any conversion $s \xleftrightarrow{*}_{\mathcal{R}} t$ over $\mathcal{F}$ with $s, t \in \mathrm{NF}(\mathcal{R})$.
4. A TRS $\langle \mathcal{F}, \mathcal{R} \rangle$ *has the normal form property* ($\mathrm{NFP}(\langle \mathcal{F}, \mathcal{R} \rangle)$) if for any conversion $s \xleftrightarrow{*}_{\mathcal{R}} t$ over $\mathcal{F}$ with $t \in \mathrm{NF}(\mathcal{R})$, there exists a conversion $s \xrightarrow{*}_{\mathcal{R}} t$ over $\mathcal{F}$.

---
[2] For $\mathcal{R}$-reduction, we usually allow infinite $\mathcal{R}$-reductions $t_0 \to_{\mathcal{R}} t_1 \to_{\mathcal{R}} \cdots$.

Non-Φ ($\Phi \in \{\text{CR}, \text{NFP}, \text{UNC}, \text{UNR}\}$) stands for the negation of Φ; for example, "a TRS $\langle \mathcal{F}, \mathcal{R} \rangle$ is non-UNR" means "a TRS $\langle \mathcal{F}, \mathcal{R} \rangle$ is not UNR".

A TRS $\langle \mathcal{F}, \mathcal{R} \rangle$ is said to be *left-linear/linear/non-erasing* if so are all $l \to r \in \mathcal{R}$. A TRS $\langle \mathcal{F}, \mathcal{R} \rangle$ is *strongly closed* if so is the rules part $\mathcal{R}$; it is known that linear strongly closed TRSs are confluent [8]. A TRS $\langle \mathcal{F}, \mathcal{R} \rangle$ is *terminating* if the associated relation $\to_\mathcal{R}$ is well-founded.

Proposition 1 ([6,13]). *The following sequence of implications holds for any TRS $\langle \mathcal{F}, \mathcal{R} \rangle$: $\text{CR}(\langle \mathcal{F}, \mathcal{R} \rangle) \implies \text{NFP}(\langle \mathcal{F}, \mathcal{R} \rangle) \implies \text{UNC}(\langle \mathcal{F}, \mathcal{R} \rangle) \implies \text{UNR}(\langle \mathcal{F}, \mathcal{R} \rangle)$. Inverse implications do not hold in general, but hold if $\langle \mathcal{F}, \mathcal{R} \rangle$ is terminating.*

We now explain some important properties that distinguish UNR from more well-known properties CR and UNC.

Proposition 2 ([12,13]). *UNR is modular for the direct sums of left-linear TRSs, that is, $\text{UNR}(\langle \mathcal{F}, \mathcal{R} \rangle)$ and $\text{UNR}(\langle \mathcal{F}', \mathcal{R}' \rangle)$ imply $\text{UNR}(\langle \mathcal{F} \cup \mathcal{F}', \mathcal{R} \cup \mathcal{R}' \rangle)$ for left-linear TRSs $\langle \mathcal{F}, \mathcal{R} \rangle$, $\langle \mathcal{F}', \mathcal{R}' \rangle$ such that $\mathcal{F} \cap \mathcal{F}' = \emptyset$. On the other hand, UNR is not modular for the direct sums of TRSs, in general.*

On the contrary, CR and UNC are modular for the direct sums of TRSs [13,17].

Corollary 1. *Let $\langle \mathcal{F}, \mathcal{R} \rangle$ be a TRS and suppose $\mathcal{F} \subseteq \mathcal{F}'$. If $\langle \mathcal{F}, \mathcal{R} \rangle$ is either left-linear or terminating, then $\text{UNR}(\langle \mathcal{F}, \mathcal{R} \rangle)$ implies $\text{UNR}(\langle \mathcal{F}', \mathcal{R} \rangle)$.*

Can the restrictions in this corollary be dropped? In fact, this question is negatively answered. A property Φ of a TRS is said be *closed under signature extensions* if $\Phi(\langle \mathcal{F}, \mathcal{R} \rangle)$ implies $\Phi(\langle \mathcal{F}', \mathcal{R} \rangle)$ for any $\mathcal{F}' \supseteq \mathcal{F}$.

Proposition 3 ([10]). *UNR is not closed under signature extensions*[3].

A property Φ of TRSs *implies UNR* if $\Phi(\langle \mathcal{F}, \mathcal{R} \rangle)$ implies $\text{UNR}(\langle \mathcal{F}, \mathcal{R} \rangle)$ for any TRS $\langle \mathcal{F}, \mathcal{R} \rangle$. We note that such properties include CR, NFP and UNC.

Proposition 4. *Let $\langle \mathcal{F}, \mathcal{R} \rangle$ be a TRS and suppose $\mathcal{F} \subseteq \mathcal{F}'$. Let Φ be a property that implies UNR and is closed under signature extensions. Then, $\Phi(\langle \mathcal{F}, \mathcal{R} \rangle)$ implies $\text{UNR}(\langle \mathcal{F}', \mathcal{R} \rangle)$.*

We next consider opposite direction of signature extensions. Since a counterexample of UNR over the signature $\mathcal{F}$ is again a counterexample of UNR over the signature $\mathcal{F}' \supseteq \mathcal{F}$, we have the following.

Proposition 5. *non-UNR is closed under signature extensions. Thus, for any TRSs $\langle \mathcal{F}, \mathcal{R} \rangle$ and $\langle \mathcal{F}', \mathcal{R} \rangle$ such that $\mathcal{F} \subseteq \mathcal{F}'$, $\text{UNR}(\langle \mathcal{F}', \mathcal{R} \rangle)$ implies $\text{UNR}(\langle \mathcal{F}, \mathcal{R} \rangle)$.*

To end the section, let us explain celebrated counterexamples from [10,13].

---
[3] For CR and UNC, modularity for the direct sums of TRSs [13,17] implies they are closed under signature extensions. NFP is not closed under signature extensions [10].

Example 1. Let $\mathcal{F}_0 = \{\mathsf{A}^0, \mathsf{B}^0, \mathsf{C}^0, \mathsf{D}^0, \mathsf{E}^0\}$ and $\mathcal{R}_0 = \{\mathsf{A} \to \mathsf{B}, \mathsf{A} \to \mathsf{C}, \mathsf{D} \to \mathsf{C}, \mathsf{D} \to \mathsf{E}, \mathsf{C} \to \mathsf{C}\}$. Here, the notation f^i stands for that the arity of the function symbol f is i. It is easy to check by case analysis that $\langle \mathcal{F}_0, \mathcal{R}_0 \rangle$ is UNR, as terms over $\mathcal{F}_0$ are variables, or constants $\mathsf{A}, \ldots, \mathsf{E}$. Let $\mathcal{F}_1 = \{\mathsf{F}^2, \mathsf{G}^0\}$ and $\mathcal{R}_1 = \{\mathsf{F}(x,x) \to \mathsf{G}\}$. Since $\langle \mathcal{F}_1, \mathcal{R}_1 \rangle$ is confluent (as it is terminating and has no critical pairs), $\langle \mathcal{F}_1, \mathcal{R}_1 \rangle$ is UNR. On the other hand, $\langle \mathcal{F}_0 \cup \mathcal{F}_1, \mathcal{R}_0 \cup \mathcal{R}_1 \rangle$ is not UNR, as witnessed by the following conversion over $\mathcal{F}_0 \cup \mathcal{F}_1$

$$\mathsf{F}(\mathsf{B},\mathsf{E}) \xleftarrow{*}_{\mathcal{R}_0 \cup \mathcal{R}_1} \mathsf{F}(\mathsf{A},\mathsf{D}) \xrightarrow{*}_{\mathcal{R}_0 \cup \mathcal{R}_1} \mathsf{F}(\mathsf{C},\mathsf{C}) \to_{\mathcal{R}_0 \cup \mathcal{R}_1} \mathsf{G}$$

between $\mathsf{F}(\mathsf{B},\mathsf{E}), \mathsf{G} \in \mathrm{NF}(\mathcal{R}_0 \cup \mathcal{R}_1)$. Since $\mathcal{F}_0 \cap \mathcal{F}_1 = \emptyset$, this is a counterexample of modularity for the direct sum [13]. Next, consider $\mathcal{F}_2 = \mathcal{F}_0 \cup \mathcal{F}_1$ and $\mathcal{R}_2 = \mathcal{R}_0 \cup \mathcal{R}_1 \cup \{\mathsf{F}(Z,x) \to \mathsf{G} \mid Z \in \{\mathsf{A},\mathsf{B},\mathsf{C},\mathsf{D},\mathsf{E}\}\} \cup \{\mathsf{F}(x,Z) \to \mathsf{G} \mid Z \in \{\mathsf{A},\mathsf{B},\mathsf{C},\mathsf{D},\mathsf{E}\}\}$. Consider a conversion $\rho : t_1 \xleftarrow{*}_{\mathcal{R}_2} s \xrightarrow{*}_{\mathcal{R}_2} t_2$ over the signature $\mathcal{F}_2$ with $t_1, t_2 \in \mathrm{NF}(\mathcal{R}_2)$ and $t_1 \neq t_2$. Now, it is easy to check that the term s cannot be a constant or a variable. Thus, we know that s must be a term in a (tree) form of internal nodes labeled by F and leaves labeled by a variable or a constant. Then one knows that, because of the rules in $\mathcal{R}_2 \setminus (\mathcal{R}_0 \cup \mathcal{R}_1)$, t_1, t_2 can not contain constants $\mathsf{A}, \ldots, \mathsf{E}$. Also, the difference of constants $\mathsf{A}, \ldots, \mathsf{E}$ does not affect the applicability of rules in $\mathcal{R}_2 \setminus \mathcal{R}_0$. Hence, rules in $\mathcal{R}_0$ are redundant in this conversion, that is, we have a conversion $\rho' : t_1 \xleftarrow{*}_{\mathcal{R}_2 \setminus \mathcal{R}_0} s \xrightarrow{*}_{\mathcal{R}_2 \setminus \mathcal{R}_0} t_2$. Then, since $\mathcal{R}_2 \setminus \mathcal{R}_0$ is terminating and all critical pairs are joinable, it is confluent. Thus, $t_1 = t_2$, which is a contradiction. Hence, we conclude $\langle \mathcal{F}_2, \mathcal{R}_2 \rangle$ is UNR. On the other hand, $\langle \mathcal{F}_2 \cup \{\mathsf{H}^1\}, \mathcal{R}_2 \rangle$ is not UNR as witnessed by the following conversion:

$$\mathsf{F}(\mathsf{H}(\mathsf{B}),\mathsf{H}(\mathsf{E})) \xleftarrow{*}_{\mathcal{R}_2} \mathsf{F}(\mathsf{H}(\mathsf{A}),\mathsf{H}(\mathsf{D})) \xrightarrow{*}_{\mathcal{R}_2} \mathsf{F}(\mathsf{H}(\mathsf{C}),\mathsf{H}(\mathsf{C})) \to_{\mathcal{R}_2} \mathsf{G}.$$

This is a counterexample of closure under signature extensions of UNR [10].

3 Transformation by Rule Addition and Elimination

In this section, we study some transformations of TRSs that preserve UNR or non-UNR. In particular, we explore what kinds of rewrite rules can be added or eliminated without losing UNR. This kind of transformations of TRSs via addition and elimination of rewrite rules that preserves CR and UNC has been studied in [3,14], respectively. Such transformations are employed in some tools for (dis)proving CR or UNC.

First, we explore what kind of auxiliary rules can be added without losing the guarantee of UNR of the original TRS.

Theorem 1. *Let $\langle \mathcal{F}, \mathcal{R} \rangle$, $\langle \mathcal{F}', \mathcal{R}' \rangle$ be TRSs. Suppose that $l' \in \mathrm{RED}(\mathcal{R})$ for any $l' \to r' \in \mathcal{R}'$. Then, $\mathrm{UNR}(\langle \mathcal{F} \cup \mathcal{F}', \mathcal{R} \cup \mathcal{R}' \rangle)$ implies $\mathrm{UNR}(\langle \mathcal{F}, \mathcal{R} \rangle)$.*

Next, we explore the other direction, that is, what kind of rules can be eliminated without losing the guarantee of UNR of the original TRS.

Theorem 2. Let $\langle \mathcal{F}, \mathcal{R} \rangle$, $\langle \mathcal{F}, \mathcal{R}' \rangle$ be TRSs. Suppose $l' \xrightarrow{*}_{\mathcal{R}} r'$ over $\mathcal{F}$ for each $l' \to r' \in \mathcal{R}'$. Then, $\mathrm{UNR}(\langle \mathcal{F}, \mathcal{R} \rangle)$ implies $\mathrm{UNR}(\langle \mathcal{F}, \mathcal{R} \cup \mathcal{R}' \rangle)$.

Note that, in the theorem above, one can not extend $\langle \mathcal{F}, \mathcal{R}' \rangle$ to general TRSs $\langle \mathcal{F}', \mathcal{R}' \rangle$, which would contradicts Proposition 3 by taking $\mathcal{R}' = \emptyset$. One possible way to replace $\langle \mathcal{F}, \mathcal{R}' \rangle$ with $\langle \mathcal{F}', \mathcal{R}' \rangle$ is to strengthen the condition $\mathrm{UNR}(\langle \mathcal{F}, \mathcal{R} \rangle)$ with other properties which are closed under signature extensions:

Corollary 2. Let $\langle \mathcal{F}, \mathcal{R} \rangle$, $\langle \mathcal{F}', \mathcal{R}' \rangle$ be TRSs. Suppose $l' \xrightarrow{*}_{\mathcal{R}} r'$ over $\mathcal{F}'$ for each $l' \to r' \in \mathcal{R}'$. Let Φ be a property that is closed under signature extensions and implies UNR. Then, $\Phi(\langle \mathcal{F}, \mathcal{R} \rangle)$ implies $\mathrm{UNR}(\langle \mathcal{F} \cup \mathcal{F}', \mathcal{R} \cup \mathcal{R}' \rangle)$.

For concrete examples of such a property Φ, one can employ CR and UNC by Proposition 1 and closedness of these properties under signature extensions.

We now give a corollary that can be obtained by using both Theorems 1 and 2. For this, let us use the following notion of linearization of terms. Consider a term $l = C[x_1, \ldots, x_n]$ where all variable occurrences in l are displayed (i.e. $\mathcal{V}(C) = \emptyset$). Here note that it may be the case $x_i = x_j$ ($i \neq j$), as we don't assume that $x_1, \ldots, x_n$ are mutually distinct. A *linearization* l' of l is a linear term $l' = C[y_1, \ldots, y_n]$ such that $\{x_1, \ldots, x_n\} \subseteq \{y_1, \ldots, y_n\}$ and $y_i \in \{x_1, \ldots, x_n\}$ implies $y_i = x_i$, for all $1 \leq i \leq n$. Remark here that $y_1, \ldots, y_n$ are mutually distinct because of the linearity of l'. For example, $\mathsf{f}(x, x', x'', y, y', z)$ as well as $\mathsf{f}(x', x, x'', y, y, z)$ are linearizations of $\mathsf{f}(x, x, x, y, y, z)$, but neither $\mathsf{f}(x, x', x', y, y', z)$ nor $\mathsf{f}(x, x', x'', y, y', z')$ are. Finally, we note that $l' \trianglelefteq l$ for any linearization l' of l.

Since non-left-linear rules in $\mathcal{R}$ often impede applying criteria for normal form properties, the following transformation is sometimes useful.

Corollary 3. Let $l \to r \in \mathcal{R}$. Suppose l' is a linearization of l such that $l' \notin \mathrm{NF}(\mathcal{R})$, and let $\mathcal{R}'$ be the TRS obtained from $\mathcal{R}$ by replacing $l \to r$ by $l' \to r$. Then $\mathrm{UNR}(\langle \mathcal{F}, \mathcal{R}' \rangle)$ implies $\mathrm{UNR}(\langle \mathcal{F}, \mathcal{R} \rangle)$.

Remark 1. As one can easily see in the proof, the similar preservation theorems hold also for UNC. That is, (1) if $l' \in \mathrm{RED}(\mathcal{R})$ for any $l' \to r' \in \mathcal{R}'$, then $\mathrm{UNC}(\langle \mathcal{F} \cup \mathcal{F}', \mathcal{R} \cup \mathcal{R}' \rangle)$ implies $\mathrm{UNC}(\langle \mathcal{F}, \mathcal{R} \rangle)$, and (2) if $l' \xrightarrow{*}_{\mathcal{R}} r'$ over $\mathcal{F}$ for all $l' \to r' \in \mathcal{R}'$, then $\mathrm{UNC}(\langle \mathcal{F}, \mathcal{R} \rangle)$ implies $\mathrm{UNC}(\langle \mathcal{F}, \mathcal{R} \cup \mathcal{R}' \rangle)$. Suppose $\mathcal{R}'$ is obtained from $\mathcal{R}$ by the transformations. Then, if $\mathcal{R}$ is a TRS with UNR but non-UNC, then again $\mathcal{R}'$ is such a TRS. Thus, the transformations are not useful to close the gap between UNR and UNC properties. In other words, unless we have a technique to show UNR of non-UNC TRSs, these transformations do not improve the situation of attempting to prove UNR of such a TRS $\mathcal{R}$.

Remark 2. We now remark the possible effects of Theorems 1 and 2 on proving (automatically) UNR, via proving of UNC and CR based on Proposition 1. As pointed out in the previous Remark 1, as far as our primary technique for proving UNR is to infer it from UNC, these transformations are not useful (that is, if the UNC proving method includes these transformations). In contrast, the similar preservation theorems do not hold for CR. Thus, when one infers UNR by proving CR, these transformations may improve the power of UNR proving.

4 Abstract Criterion for UNR-Preservation

As explained in Remark 1, the transformations in the previous section do not improve the situation so far that there are no general methods for proving UNR of non-UNC TRSs. In order to investigate general methods for proving UNR of non-UNC TRSs, we here consider an abstract setting for UNR-preserving transformations.

An apparent approach to obtain such a criterion is to go back to the definition of the UNR property. For this, we introduce a couple of notions. We denote by WN($\mathcal{R}$) the set of terms t having a reduction to a normal form starting from t.

Definition 1 (UNR-preserving transformation). *Let $\langle \mathcal{F}, \mathcal{R} \rangle$, $\langle \mathcal{F}', \mathcal{R}' \rangle$ be TRSs, and let $\varphi : \mathrm{T}(\mathcal{F}, \mathcal{V}) \to \mathrm{T}(\mathcal{F}', \mathcal{V})$ be a partial mapping. Then φ is said to be a UNR-preserving transformation from $\langle \mathcal{F}, \mathcal{R} \rangle$ to $\langle \mathcal{F}', \mathcal{R}' \rangle$ if (i) WN($\mathcal{R}$) $\subseteq$ dom(φ), where dom(φ) denotes the domain of φ, (ii) φ preserves reductions to normal forms i.e. for any reduction $s \xrightarrow{*}_{\mathcal{R}} t$ with $t \in$ NF($\mathcal{R}$) over $\mathcal{F}$ we have a reduction $\varphi(s) \xrightarrow{*}_{\mathcal{R}'} \varphi(t)$ with $\varphi(t) \in$ NF($\mathcal{R}'$) over $\mathcal{F}'$, and (iii) φ is injective on normal forms i.e. for any $s, t \in$ NF($\mathcal{R}$), $\varphi(s) = \varphi(t)$ implies $s = t$.*

Now we can state a general criterion for UNR-preserving transformations.

Theorem 3. *Let $\langle \mathcal{F}, \mathcal{R} \rangle$, $\langle \mathcal{F}', \mathcal{R}' \rangle$ be TRSs. Suppose that there exists a UNR-preserving transformation $\varphi : \mathrm{T}(\mathcal{F}, \mathcal{V}) \to \mathrm{T}(\mathcal{F}', \mathcal{V})$ from $\langle \mathcal{F}, \mathcal{R} \rangle$ to $\langle \mathcal{F}', \mathcal{R}' \rangle$. Then, UNR($\langle \mathcal{F}', \mathcal{R}' \rangle$) implies UNR($\langle \mathcal{F}, \mathcal{R} \rangle$).*

Example 2. Let $\langle \mathcal{F}, \mathcal{R} \rangle$ be a TRS with $\mathcal{F} = \{\mathsf{f}^1, \mathsf{h}^2, \mathsf{a}^0, \mathsf{b}^0, \mathsf{c}^0\}$ and $\mathcal{R} = \{\mathsf{h}(\mathsf{f}(x), \mathsf{b}) \to \mathsf{c},\ \mathsf{b} \to \mathsf{a},\ \mathsf{c} \to \mathsf{f}(\mathsf{c})\}$. From a term $\mathsf{h}(\mathsf{f}(x), \mathsf{b})$, one can consider the following two kinds of reductions: $\rho_1 : \mathsf{h}(\mathsf{f}(x), \mathsf{b}) \to_{\mathcal{R}} \mathsf{h}(\mathsf{f}(x), \mathsf{a})$ and $\rho_2 : \mathsf{h}(\mathsf{f}(x), \mathsf{b}) \to_{\mathcal{R}} \mathsf{c} \to_{\mathcal{R}} \mathsf{f}(\mathsf{c}) \to_{\mathcal{R}} \cdots \to_{\mathcal{R}} \mathsf{f}(\cdots(\mathsf{f}(\mathsf{c}))\cdots)$. Clearly, reductions of the latter type do not reach to a normal form. In fact, it is observed that any term containing a constant c does not reduce to a normal form. In the view of the abstract criterion above, we can take $\mathcal{F}' = \{\mathsf{f}^1, \mathsf{h}^2, \mathsf{a}^0, \mathsf{b}^0\}$, $\mathcal{R}' = \{\mathsf{b} \to \mathsf{a}\}$, and φ as the identity mapping on $\mathrm{T}(\mathcal{F}', \mathcal{V})$. Thus, UNR($\langle \mathcal{F}, \mathcal{R} \rangle$) is inferred from UNR($\langle \mathcal{F}', \mathcal{R}' \rangle$).

One might be tempting to conjecture that if we have a rule like $\mathsf{c} \to \mathsf{f}(\mathsf{c})$ then rules containing c can be removed. This is, however, not true in general:

Example 3. Consider a TRS $\langle \mathcal{F}, \mathcal{R} \rangle$ with $\mathcal{F} = \{\mathsf{f}^1, \mathsf{h}^2, \mathsf{a}^0, \mathsf{b}^0, \mathsf{c}^0\}$, $\mathcal{R} = \{\mathsf{h}(x, \mathsf{b}) \to \mathsf{c},\ \mathsf{b} \to \mathsf{a},\ \mathsf{f}(x) \to \mathsf{a},\ \mathsf{c} \to \mathsf{f}(\mathsf{c})\}$. By removing the rules containing c one obtains $\mathcal{R}' = \{\mathsf{b} \to \mathsf{a},\ \mathsf{f}(x) \to \mathsf{a}\}$, for which $\langle \mathcal{F}, \mathcal{R}' \rangle$ is UNR. However, $\langle \mathcal{F}, \mathcal{R} \rangle$ is not UNR as witnessed by reductions $\rho_1 : \mathsf{h}(x, \mathsf{b}) \to_{\mathcal{R}} \mathsf{h}(x, \mathsf{a})$ and $\rho_2 : \mathsf{h}(x, \mathsf{b}) \xrightarrow{*}_{\mathcal{R}} \mathsf{a}$.

Remark 3. A similar but different situation is considered in the context of *uniform normalization* [11]. We denote by SN($\mathcal{R}$) the set of terms t having no infinite reduction starting from t. A TRS $\langle \mathcal{F}, \mathcal{R} \rangle$ is said to be *uniformly normalizing* if WN($\mathcal{R}$) $\subseteq$ SN($\mathcal{R}$). It is easy to see that a TRS is uniformly normalizing iff there does not exist a rewrite step $s \to_{\mathcal{R}} t$ such that $t \in$ SN($\mathcal{R}$) but $s \notin$ SN($\mathcal{R}$). In this context, a term l is said to be *perpetual* if, for any rewrite step $s \to_{l \to r} t$, $t \in \mathrm{SN}(\mathcal{R})$ implies $s \in \mathrm{SN}(\mathcal{R})$. In particular, if we have a rule $\mathsf{c} \to \mathsf{f}(\mathsf{c}) \in \mathcal{R}$ then c is perpetual. In contrast, as we saw above, $\mathsf{c} \to \mathsf{f}(\mathsf{c}) \in \mathcal{R}$ alone is not sufficient to conclude that we can eliminate the rule. To be precise, here we want to select a rewrite rule $l \to r$ such that $t \notin \mathrm{WN}(\mathcal{R})$ for any rewrite step $s \to_{l \to r} t$.

5 Collapsing Mapping

The transformation criterion in the previous section is too abstract to use in proving UNR automatically. In this and the next sections, we present more dedicated TRS transformations that are useful for proving UNR of non-UNC TRSs. The transformation presented in this section is inspired by the existence of a special constant in Example 2. For generality, we first explore the case c is not only a constant but an arbitrary linear term.

Definition 2 (collapsing mapping). *Let $\langle \mathcal{F}, \mathcal{R} \rangle$ be a TRS. Suppose that there exists a linear term c such that (i) $c \notin \mathrm{NF}(\mathcal{R})$, (ii) for all $l \to r \in \mathcal{R}$, $c \trianglelefteq l$ implies $c \trianglelefteq r$, and (iii) for all $l \to r \in \mathcal{R}$ such that $c \ntrianglelefteq r$ the following are satisfied: (a) $l \to r$ is non-erasing, (b) for any substitution σ, if $l\sigma$ overlaps on c at the position p, then $r\sigma$ overlaps on c at p, and (c) if c overlaps on l at the position p then $l|_p \trianglelefteq r$. Then, we define a collapsing mapping φ_c for $\mathcal{R}$ as follows.*

$$\varphi_c(\mathcal{R}) = \{l \to r \in \mathcal{R} \mid c \ntrianglelefteq r\} \cup \{l \to l \mid c \ntrianglelefteq r, c \ntrianglelefteq l, l \to r \in \mathcal{R}\} \cup \{c \to c\}$$

The next lemmas will be used for showing the preservation of UNR under the transformation φ_c.

Lemma 3. *Let $\langle \mathcal{F}, \mathcal{R} \rangle$ be a TRS. Suppose a linear term c satisfies the conditions in Definition 2. (1) If $s \to_\mathcal{R} t$ and $c \trianglelefteq s$, then $c \trianglelefteq t$. (2) If $s \xrightarrow{*}_\mathcal{R} t$ and $c \ntrianglelefteq t$, then $c \ntrianglelefteq s$.*

Lemma 4. *Let $\langle \mathcal{F}, \mathcal{R} \rangle$ be a TRS. Suppose a linear term c satisfies the conditions in Definition 2. (1) $\mathrm{NF}(\mathcal{R}) \subseteq \mathrm{NF}(\varphi_c(\mathcal{R}))$. (2) If $t \in \mathrm{NF}(\varphi_c(\mathcal{R}))$ and $c \ntrianglelefteq t$ then $t \in \mathrm{NF}(\mathcal{R})$.*

Theorem 4 (UNR by collapsing mapping). *Let $\langle \mathcal{F}, \mathcal{R} \rangle$ be a TRS. For any collapsing mapping φ_c for $\mathcal{R}$, we have $\mathrm{UNR}(\langle \mathcal{F}, \varphi_c(\mathcal{R}) \rangle)$ iff $\mathrm{UNR}(\langle \mathcal{F}, \mathcal{R} \rangle)$.*

Corollary 4 (collapsing maping for constants). *Let $\langle \mathcal{F}, \mathcal{R} \rangle$ be a TRS. Suppose that there exists a constant c such that (i) $c \notin \mathrm{NF}(\mathcal{R})$, (ii) for all $l \to r \in \mathcal{R}$, $c \notin \mathcal{F}(l) \setminus \mathcal{F}(r)$, and (iii) for all $l \to r \in \mathcal{R}$, if $c \notin \mathcal{F}(r)$ then $l \to r$ is non-erasing. Then, $\mathrm{UNR}(\langle \mathcal{F} \setminus \{c\}, \varphi_c(\mathcal{R}) \setminus \{c \to c\} \rangle)$ iff $\mathrm{UNR}(\langle \mathcal{F}, \mathcal{R} \rangle)$.*

Example 4. We consider the TRS $\mathcal{R}$ given in Example 2, and the translation φ_c for taking a constant c for c. We here also note that the TRS $\mathcal{R}$ does not have the UNC property. For, we have $\mathsf{h}(\mathsf{f}(x), \mathsf{a}) \leftarrow \mathsf{h}(\mathsf{f}(x), \mathsf{b}) \to \mathsf{c} \leftarrow \mathsf{h}(\mathsf{f}(y), \mathsf{b}) \to \mathsf{h}(\mathsf{f}(y), \mathsf{a})$ and $\mathsf{h}(\mathsf{f}(x), \mathsf{a}), \mathsf{h}(\mathsf{f}(y), \mathsf{a}) \in \mathsf{NF}(\mathcal{R})$. Thus, any technique for proving UNC or CR is not applicable for proving UNR of this TRS. We obtain $\mathcal{R}' = \varphi_c(\mathcal{R}) \setminus \{c \to c\} = \{h(f(x), b) \to h(f(x), b), b \to a\}$. It is easy to check all conditions of Corollary 4 are satisfied. Also, $\mathcal{R}'$ is linear strongly closed, and hence CR. Thus, $\mathcal{R}'$ is UNR by Proposition 1; hence, so is $\mathcal{R}$ by Corollary 4.

As witnessed in Example 4, the transformation can be applied successfully for showing UNR of a non-UNC TRS. One also observes in these examples that neither the implication $\mathrm{UNC}(\varphi_c(\mathcal{R})) \implies \mathrm{UNC}(\mathcal{R})$ nor $\mathrm{CR}(\varphi_c(\mathcal{R})) \implies \mathrm{CR}(\mathcal{R})$ hold. This is in contrast to the transformations given in Sect. 3 (see Remark 1).

6 SRR Elimination

The idea of collapsing mapping is to eliminate a pattern c which does not contribute to the reductions to normal forms. However, the applicability of the criterion depends on the existence of such a special pattern c. In this section, we explore the idea of eliminating rewrite rules that do not contribute to the reduction to normal forms.

Definition 3. *Let $\mathcal{R}$ be a set of rewrite rules. A term t is* strongly $\mathcal{R}$-reducible *if no instance of t has a normal form, i.e. for any σ, $\exists t'. \, t\sigma \to^* t' \in \mathrm{NF}(\mathcal{R})$ does not hold. We say $l \to r \in \mathcal{R}$ is* strongly $\mathcal{R}$-reducible *if r is strongly $\mathcal{R}$-reducible.*

Remark that strongly $\mathcal{R}$-reducibility is undecidable in general. Thus, we need to employ some approximation in implementation.

Example 5. Let $\mathcal{R} = \{(1) \, a \to b, \, (2) \, f(g(b), b) \to f(g(a), a), \, (3) \, f(g(x), a) \to f(g(a), a)\}$. Since $\{t \mid f(g(a), a) \xrightarrow{*}_\mathcal{R} t\} = \{f(g(u), v) \mid u, v \in \{a, b\}\}$, no instantiation of $f(g(a), a)$ possibly reduces to a normal form, i.e. $f(g(a), a)$ is a strongly $\mathcal{R}$-reducible term. Hence, rules (2), (3) are strongly $\mathcal{R}$-reducible. We here also note that $\mathcal{R}$ is not UNC, as witnessed by $f(g(x), b) \leftarrow_\mathcal{R} f(g(x), a) \to_\mathcal{R} f(g(a), a) \leftarrow_\mathcal{R} f(g(y), a) \to_\mathcal{R} f(g(y), b)$ and $f(g(x), b), f(g(y), b) \in \mathrm{NF}(\mathcal{R})$. Thus, any technique for proving UNC or CR is not applicable for proving UNR of this TRS. Also, none of the linear patterns $a, b, f(x, a)$, etc. satisfies the conditions of Definition 2; thus the collapsing mapping transformation is not applicable. However, our technique below can show that this TRS is UNR.

To present our technique, let us recall the notion of descendants. Let $\mathcal{R}$ be a set of rewrite rules. Suppose that p is a position in a term s, and $\tau : s \to_{q, l \to r} t$ is a rewrite step of $\mathcal{R}$. Then, the set p/τ of the descendants of the position p by the rewrite step τ is defined as follows:

$$p/\tau = \begin{cases} \{p\} & \text{if } p \parallel q \text{ or } p \leq q \\ \{q.o'.(p \setminus (q.o)) \mid o' \in \mathrm{Pos}(r), r|_{o'} = l|_o\} & \text{if } \exists o. \, l|_o = x \in \mathcal{V} \wedge q.o \leq p \\ \emptyset & \text{otherwise} \end{cases}$$

For $\Phi \subseteq \mathrm{Pos}(s)$, we define $\Phi/\tau = \bigcup_{p \in \Phi} p/\tau$. Then for a rewrite sequence $\rho : s = t_0 \to_\mathcal{R} t_1 \to_\mathcal{R} \cdots \to_\mathcal{R} t_k = t$, we define Φ/ρ inductively as follows: $\Phi/\rho = \Phi$ if $k =$

0, otherwise $\Phi/\rho = (\Phi/\rho_{hd})/\rho_{tl}$ where $\rho_{hd} : t_0 \to_{\mathcal{R}} t_1$ and $\rho_{tl} : t_1 \to_{\mathcal{R}} \cdots \to_{\mathcal{R}} t_k$ are the parts of ρ as specified. Finally, we put $p/\rho = \{p\}/\rho$ for $p \in \text{Pos}(s)$ and $\rho : s \xrightarrow{*} t$. The idea of descendants is tracing each subterm occurrence of the initial term along the rewrite sequence—each subterm may change by rewriting its subterm; it also may be copied along the rewrite sequence (by the rewrite step using duplicating rules at the position above that subterm). For example, for $\mathcal{R} = \{\mathsf{a} \to \mathsf{b}, \mathsf{f}(x,y) \to \mathsf{g}(x,x)\}$ and $\rho : \mathsf{f}(\mathsf{a},\mathsf{b}) \to_\varepsilon \mathsf{g}(\mathsf{a},\mathsf{a}) \to_1 \mathsf{g}(\mathsf{b},\mathsf{a})$, we have $\varepsilon/\rho = \{\varepsilon\}$, $1/\rho = \{1,2\}$, and $2/\rho = \emptyset$.

The following lemma concerning descendants is not difficult to show.

Lemma 5. *Let $\rho : s \xrightarrow{*}_{\mathcal{R}} t$ and $p \in \text{Pos}(s)$.*

1. $s|_p \xrightarrow{*}_{\mathcal{R}} t|_q$ for any $q \in p/\rho$.
2. If $\mathcal{R}$ is non-erasing and $p/\rho = \emptyset$, then there exists a prefix part $\rho' : s \xrightarrow{*}_{\mathcal{R}} u$ of ρ and $q \in p/\rho'$ such that $u|_q$ properly overlaps with some $l \in \text{LHS}(\mathcal{R})$.

Theorem 5 (UNR by elimination of strongly $\mathcal{R}$-reducible rules (SRR elimination)). *Let $\langle \mathcal{F}, \mathcal{R} \rangle$ be a TRS. Suppose that (i) all rules in $\mathcal{R}' \subseteq \mathcal{R}$ are strongly $\mathcal{R}$-reducible, (ii) $\mathcal{R} \setminus \mathcal{R}'$ is non-erasing, and (iii) for any $l' \to r' \in \mathcal{R}'$ and $l \to r \in \mathcal{R} \setminus \mathcal{R}'$, if $r'\sigma \xrightarrow{*}_{\mathcal{R}} t'$ then t' does not properly overlap with l. Then, $\text{UNR}(\langle \mathcal{F}, \mathcal{R} \setminus \mathcal{R}' \rangle)$ implies $\text{UNR}(\langle \mathcal{F}, \mathcal{R} \rangle)$.*

Corollary 5. *Let $\langle \mathcal{F}, \mathcal{R} \rangle$ be a TRS. Suppose that conditions (i)–(iii) of Theorem 5 hold. Let $\mathcal{R}'' = (\mathcal{R} \setminus \mathcal{R}') \cup \{l \to l \mid l \to r \in \mathcal{R}'\}$. Then, $\text{UNR}(\langle \mathcal{F}, \mathcal{R}'' \rangle)$ iff $\text{UNR}(\langle \mathcal{F}, \mathcal{R} \rangle)$.*

Example 6 (cont'd from Example 5). Let $\mathcal{R}$ be the one given in Example 5. Let $\mathcal{R}' = \{(2), (3)\} \subseteq \mathcal{R}$. Because $\mathcal{R} \setminus \mathcal{R}' = \{\mathsf{a} \to \mathsf{b}\}$ is orthogonal, it is CR, and hence UNR. It is easy to see that the conditions (i)—(iii) of Theorem 5 are satisfied (note that (iii) follows because $\text{LHS}(\mathcal{R} \setminus \mathcal{R}')$ contains only constants). Thus, by Theorem 5, $\langle \{\mathsf{a}, \mathsf{b}, \mathsf{g}, \mathsf{f}\}, \mathcal{R} \rangle$ is UNR.

As witnessed in Examples 5 and 6, the transformation can be applied successfully for showing UNR of a non-UNC TRS, and the similar implications of Theorem 5 do not hold for UNC and CR, similarly to the case of soundness of collapsing mapping φ_c in Sect. 5. Again, this is in contrast to the transformations given in Sect. 3 (see Remark 1).

7 Implementation and Experiments

We have implemented the transformations and criteria that have been presented in previous sections on our tool ACP [4]. ACP originally intends to (dis)prove confluence of TRSs. It started dealing with the UNC property of TRSs from 2018 [3]. We now have added UNR proving mechanisms to ACP based on the techniques presented in the paper. In this section, we report on our implementation and experiments, aiming to estimate effectiveness of the presented criteria.

Table 1. Summary of experiments

	Str	Dev	Str&Dev	Col.Map.	SRR elim.	all
CR	62	52	73			
UNR (w.app)	89 (89)	88 (89)	112 (113)	44	28	159
non-UNR (w.app)	58 (63)	64 (71)	65 (72)	0	0	72

Input: a TRS $\langle \mathcal{F}, \mathcal{R} \rangle$
Output: YES or NO or Failure

  Step 1 Let $R := \mathcal{R}$ and $S := \emptyset$.
  Step 2. If $R \cup S$ is linear strongly closed then return YES.
  Step 3. Compute $\mathrm{CP}(R \cup S)$. For each $\langle s, t \rangle \in \mathrm{CP}(R \cup S)$ which is not strongly closed:
    (a) Case $s, t \in \mathrm{NF}(\mathcal{R})$. Return NO if $\langle s, t \rangle \in \mathrm{CP}(R)$.
    (b) Case $s, t \notin \mathrm{NF}(\mathcal{R})$.
      (b.1) Compute some $X \subseteq \{s' \mid s \xrightarrow{*}_\mathcal{R} s'\}$. If $t \in X$ and $\langle s, t \rangle \in \mathrm{CP}(R)$ then add $s \to t$ to R else add $s \to t$ to S.
      (b.2) Compute some $X \subseteq \{t' \mid t \xrightarrow{*}_\mathcal{R} t'\}$. If $s \in X$ and $\langle s, t \rangle \in \mathrm{CP}(R)$ then add $t \to s$ to R else add $t \to s$ to S.
    (c) Case $s \notin \mathrm{NF}(\mathcal{R})$ and $t \in \mathrm{NF}(\mathcal{R})$.
      (c.1) Search for a normal form of s that is different from t. If there is one, then return NO if $\langle s, t \rangle \in \mathrm{CP}(R)$.
      (c.2) Same as (b.1).
    (d) Case $s \in \mathrm{NF}(\mathcal{R})$ and $t \notin \mathrm{NF}(\mathcal{R})$. Similar to the Case (c).
  Step 4. If no new rules are added, then return Failure. Otherwise, go to Step 2.

Fig. 1. UNR proving procedure with linear strongly closed criterion

For the experiments, we use the latest confluence database Cops[4] which contains 577 TRSs[5]. A summary of our experiments is shown in Table 1, which will be explained below. Details of experiments are found at http://www.nue.ie.niigata-u.ac.jp/experiments/24LOPSTR/.

7.1 UNR-Preserving Transformation

In the view of Remark 2, we consider an UNR (dis)proving procedure that aims for finding a UNR-preserving transformation to some confluent TRS. Below we present UNR (dis)proving procedures based on UNR-preserving transformations, whose heuristics is designed so as to obtain a TRS that satisfies some specific confluence criteria from [8, 19], which are based on closedness of critical pairs.

In Fig. 1, we present one based on that linear strongly closed TRSs are confluent [8]. Our procedure iteratively computes a pair of sets of rewrite rules $\langle R, S \rangle$ satisfying $\mathcal{R} \subseteq R \subseteq \xrightarrow{*}_\mathcal{R}$, $S \subseteq \xleftrightarrow{*}_\mathcal{R}$ and $l \notin \mathrm{NF}(\mathcal{R})$ for all $l \to r \in R \cup S$.

---
[4] http://cops.uibk.ac.at/.
[5] The database contains 1655 problems (as of writing), but the remaining problems are not TRSs such as CTRSs, MSTRSs, infeasibility problems, etc.

aiming for a set $S \cup R$ of rewrite rules that is linear strongly closed by adding rewrite rules to $\mathcal{R}$, so that we can infer UNR of the input TRS $\langle \mathcal{F}, \mathcal{R} \rangle$ using Theorem 1. Since the strongly closedness requires (not only $u \xrightarrow{*}_\mathcal{R} \circ \xleftarrow{*}_\mathcal{R} v$ but) $u \xrightarrow{*}_\mathcal{R} \circ \xleftarrow{=}_\mathcal{R} v$ and $u \xrightarrow{=}_\mathcal{R} \circ \xleftarrow{*}_\mathcal{R} v$ for each $\langle u, v \rangle \in \mathrm{CP}(\mathcal{R})$, adding rewrite rules may positively affect in the view of replacing $s \xleftrightarrow{*}_\mathcal{R} t$ with $s \to_R t$ but negatively by the increase of $\mathrm{CP}(\mathcal{R})$ to $\mathrm{CP}(R)$. In the procedure, S and R rules are separately considered, in order to also detect a counterexample (if possible) in the course of the procedure—during the procedure, any counterexample for UNR of $\langle \mathcal{F}, R \rangle$ is a counterexample for UNR of the input TRS $\langle \mathcal{F}, \mathcal{R} \rangle$.

The results obtained by this procedure are shown in the column titled "Str" in Table 1. As shown on the first row, 62 TRSs are already checked to be confluent by the linear strongly closed criterion. Now, our procedure can detect UNR of 89 TRSs, and non-UNR of 58 TRSs (as shown on the second and third rows). In particular, there are 27 more TRSs that can be proved to be UNR by the UNR-preserving transformation. In the course of the addition of rules, linear approximations can be made in the light of Corollary 3; this does not increase the number of UNR but the number of non-UNR by 5.

We have also implemented a similar procedure based on that left-linear development closed TRSs are confluent [19]. The results are shown in the column titled "Dev" in Table 1. Finally, we tested the combination of these procedures so as to estimate the effect of overlap. The results are shown in the column titled "Str&Dev" in Table 1. One can observe that on one hand TRSs successfully proved to be UNR by these two criteria are disjoint as in the CR case, and TRSs successfully proved to be non-UNR largely overlap on the other hand.

7.2 Collapsing Mapping

Our current implementation of collapsing mapping is based on Corollary 4; fully exploring the power of Theorem 4 remains in our future work. Thus, the technique depends on the existence of a special constant. A specific constant c that satisfies the conditions of Corollary 4 is found in 45 TRSs.

To check UNR of the transformed TRS $\varphi_c(\mathcal{R})$, we use the several methods for proving UNR (note that the transformation only does not suffices to show UNR): (1) Combination of two transformation methods in Subsect. 7.1(with linear approximations), (2) UNR by $\varphi_c(\mathcal{R}) \subseteq \Delta$ where $\Delta = \{\langle t, t \rangle \mid t \in \mathrm{T}(\mathcal{F}_{univ}, \mathcal{V})\}$, and (3) Knuth-Bendix criterion for $\varphi_c(\mathcal{R}) \setminus \Delta$ (but non-UNR only when $\varphi_c(\mathcal{R}) \cap \Delta = \emptyset$). Then, among these 45 TRSs, our implementation could detect UNR of 44 TRSs (the column titled "Col.Map." in Table 1).

7.3 SRR Elimination

We now explain our transformation procedure that is used with Theorem 5. First, note that there may be a flexibility in choosing $\mathcal{R}'$ that satisfies conditions (i) and (ii) of Theorem 5, and that the condition (iii) of Theorem 5 is a criterion between rules in $\mathcal{R}'$ and rules in $\mathcal{R} \setminus \mathcal{R}'$. Our procedure is presented in Fig. 2. In

Input: a TRS $\langle \mathcal{F}, \mathcal{R} \rangle$
Output: a pair of sets of rules $\langle \mathcal{R} \setminus \mathcal{R}', \mathcal{R}' \rangle$ or Failure

Step 1. Check that all erasing rules are strongly $\mathcal{R}$-reducible. If it is not the case return Failure.
Step 2. We let $\mathcal{R}_{sr} = \{l \to r \in \mathcal{R} \mid l \to r \text{ is strongly } \mathcal{R}\text{-reducible}\}$ and $\mathcal{R}_{ne} = \mathcal{R} \setminus \mathcal{R}_{sr}$. Note that $\mathcal{R}_{ne}$ is non-erasing by the check of Step 1.
Step 3. Check for each $l' \to r' \in \mathcal{R}_{sr}$, whether there exists σ, t' such that $r'\sigma \xrightarrow{*}_{\mathcal{R}} t'$ and t' properly overlap with $l \in \text{LHS}(\mathcal{R}_{ne})$. Let S be the set of such rules $l' \to r'$.
  Step 3-1. if $S = \emptyset$, then return $\langle \mathcal{R}_{ne}, \mathcal{R}_{sr} \rangle$.
  Step 3-2. if S contains an erasing rule, then return Failure.
  Step 3-3. Otherwise, move all $l' \to r' \in S$ from $\mathcal{R}_{sr}$ to $\mathcal{R}_{ne}$. Return to Step 2.

Fig. 2. A procedure for SRR elimination

the procedure, checking whether $l \to r$ is strongly $\mathcal{R}$-reducible is undecidable, we only consider the case r is a ground term in our implementation.

The strategy employed in the procedure is as follows. First, compute maximal set $\mathcal{R}' \subseteq \mathcal{R}$ that satisfies conditions (i) and (ii) of the theorem (Steps 1 and 2). Then, iteratively check the condition (iii) of the theorem and remove $l' \to r' \in \mathcal{R}'$ if $r'\sigma \xrightarrow{*} t'$ and t' properly overlaps with some $l \to r \in \mathcal{R} \setminus \mathcal{R}'$.

In our experiments, we check UNR of the transformed TRS $(\mathcal{R} \setminus \mathcal{R}') \cup \{l \to l \mid l \to r \in \mathcal{R}\}$ based on Corollary 5, using the same procedure for the results of collapsing mapping transformation explained in Subsect. 7.2. There are 30 TRSs for which the transformation can be performed, and our implementation could detect UNR of 28 TRSs (the column titled "SRR elm." in Table 1).

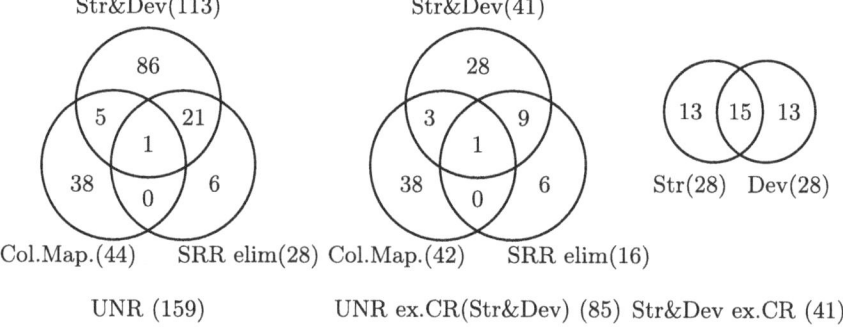

Fig. 3. Summary of overlaps between different techniques

Combing all techniques presented in the paper, we can prove UNR of 159 TRSs and non-UNR of 72 TRSs (the column titled "all" in Table 1). Figure 3 summarises overlaps between different techniques in these 159 TRSs proved to be UNR. The leftmost one shows the overlaps between different UNR proving techniques Str&Dev(w.app), Col.Map., and SRR elim. (shown in the row titled "UNR" in Table 1) in these 159 TRSs. The middle one shows the results of removing the effect of the number of 73 TRSs proved to be CR by Str&Dev. The rightmost one shows the overlaps between Str and Dev. We can see each of techniques contributes differently.

Finally, among these 159 TRSs, there are 105 TRSs which are proved to be confluent by the state-of-art confluence tools, according to CoCo 2023 full run competition result. Also, there are 212 TRSs which are proved to be non-confluent according to CoCo 2023 full run competition result—among which 50 UNR and 64 non-UNR have been proved by our methods, and the rest are yet to be explored whether UNR or non-UNR. To end the section, we present a TRS in Cops which is non-confluent but successfully proved to be UNR.

Example 7. (Cops #243). $\mathcal{R} = \{\mathsf{a} \to \mathsf{h}(\mathsf{g}(\mathsf{b})),\ \mathsf{a} \to \mathsf{h}(\mathsf{c}),\ \mathsf{b} \to \mathsf{g}(\mathsf{b}),\ \mathsf{h}(\mathsf{g}(x)) \to \mathsf{g}(\mathsf{h}(x)),\ \mathsf{g}(x) \to \mathsf{h}(x)\}$. This TRS $\mathcal{R}$ is known to be non-confluent. We can successfully prove UNR of $\mathcal{R}$, using the collapsing mapping technique.

8 Conclusion

We have presented automated verification methods for uniqueness of normal form w.r.t. reduction (UNR) of TRSs. We have given sufficient criteria of transformations that preserve UNR under addition and elimination of rules. These transformations are not useful when combined with UNC proving, but possibly useful when combined with CR proving. Since these transformations do not improve the situation that there are no (general) methods for proving UNR of TRSs which are not UNC, more dedicated transformations that are useful for showing UNR of non-UNC TRSs have been investigated. For this, we first have given an abstract criterion for UNC-preserving transformation. Then, we have given two further transformations of TRSs, namely the collapsing mapping transformations and the SRR elimination transformations. With these transformations, one can prove UNR of TRSs that are not UNC. We have reported on an implementation of UNR proving procedures based the presented methods, and have reported on the experiments. In particular, we have given UNR proving procedures based on the UNR-preserving transformations which are guided by some confluence criteria, and a procedure for searching successful applications of collapsing mapping and SRR elimination. Our experiments have shown that all the presented methods can be successfully applied for showing UNR (and non-UNR) of non-small numbers of TRSs from the benchmark.

Acknowledgment. The author thanks Kouya Okamoto for preliminary discussions on the topic. Thanks are also due to anonymous reviewers for comments. This work was partially supported by JSPS KAKENHI Grant Numbers 21K11750, 24K14817.

References

1. Aoto, T., Toyama, Y.: Top-down labelling and modularity of term rewriting systems. Research Report IS-RR-96-0023F, School of Information Science, JAIST (1996)
2. Aoto, T., Toyama, Y.: On composable properties of term rewriting systems. In: Hanus, M., Heering, J., Meinke, K. (eds.) ALP/HOA -1997. LNCS, vol. 1298, pp. 114–128. Springer, Heidelberg (1997). https://doi.org/10.1007/BFb0027006
3. Aoto, T., Toyama, Y.: Automated proofs of unique normal forms w.r.t. conversion for term rewriting systems. In: Herzig, A., Popescu, A. (eds.) FroCoS 2019. LNCS (LNAI), vol. 11715, pp. 330–347. Springer, Cham (2019). https://doi.org/10.1007/978-3-030-29007-8_19
4. Aoto, T., Yoshida, J., Toyama, Y.: Proving confluence of term rewriting systems automatically. In: Treinen, R. (ed.) RTA 2009. LNCS, vol. 5595, pp. 93–102. Springer, Heidelberg (2009). https://doi.org/10.1007/978-3-642-02348-4_7
5. Baader, F., Nipkow, T.: Term Rewriting and All That. Cambridge University Press, Cambridge (1998)
6. Felgenhauer, B.: Deciding confluence and normal form properties of ground term rewrite systems efficiently. Logical Methods Comput. Sci. **14**(4:7), 1–35 (2018). https://doi.org/10.23638/LMCS-14(4:7)2018
7. Godoy, G., Jacquemard, F.: Unique normalization for shallow TRS. In: Treinen, R. (ed.) RTA 2009. LNCS, vol. 5595, pp. 63–77. Springer, Heidelberg (2009). https://doi.org/10.1007/978-3-642-02348-4_5
8. Huet, G.: Confluent reductions: abstract properties and applications to term rewriting systems. J. ACM **27**(4), 797–821 (1980). https://doi.org/10.1145/322217.322230
9. Kahrs, S., Smith, C.: Non-ω-overlapping TRSs are UN. In: Proceedings of 1st FSCD. LIPIcs, vol. 52, pp. 22:1–17. Schloss Dagstuhl (2016). https://doi.org/10.4230/LIPIcs.FSCD.2016.22
10. Kennaway, R., Klop, J.W., Sleep, R., de Vries, F.J.: Comparing curried and uncurried rewriting. J. Symb. Comput. **21**, 15–39 (1996). https://doi.org/10.1006/jsco.1996.0002
11. Khasidashvili, Z., Ogawa, M., van Oostrom, V.: Perpetuality and uniform normalization in orthogonal rewrite systems. Inf. Comput. **164**(1), 118–151 (2001). https://doi.org/10.1006/inco.2000.2888
12. Marchiori, M.: On the modularity of normal forms in rewriting. J. Symb. Comput. **22**, 143–154 (1996). https://doi.org/10.1006/jsco.1996.0045
13. Middeldorp, A.: Modular aspects of properties of term rewriting systems related to normal forms. In: Dershowitz, N. (ed.) RTA 1989. LNCS, vol. 355, pp. 263–277. Springer, Heidelberg (1989). https://doi.org/10.1007/3-540-51081-8_113
14. Nagele, J., Felgenhauer, B., Middeldorp, A.: Improving automatic confluence analysis of rewrite systems by redundant rules. In: Proceedings of 26th RTA. LIPIcs, vol. 36, pp. 257–268. Schloss Dagstuhl (2015). https://doi.org/10.4230/LIPIcs.RTA.2015.257
15. Radcliffe, N., Moreas, L., Verma, R.: Uniqueness of normal forms for shallow term rewrite systems. ACM Trans. Comput. Logic **18**(2), 17:1–17:20 (2017). https://doi.org/10.1145/3060144
16. Sato, Y., Aoto, T.: Undecidability of some properties related to the uniqueness of normal forms for flat term rewriting systems. IPSJ Trans. Program. **14**(2), 15–24 (2021). (in Japanese)

17. Toyama, Y.: On the Church-Rosser property for the direct sum of term rewriting systems. J. ACM **34**(1), 128–143 (1987). https://doi.org/10.1145/7531.7534
18. Toyama, Y., Oyamaguchi, M.: Conditional linearization of non-duplicating term rewriting systems. IEICE Trans. Inf. Syst. **E84-D**(4), 439–447 (2001)
19. van Oostrom, V.: Developing developments. Theoret. Comput. Sci. **175**(1), 159–181 (1997). https://doi.org/10.1016/S0304-3975(96)00173-9
20. de Vrijer, R.: Conditional linearization. Indag. Math. **10**(1), 145–159 (1999). https://doi.org/10.1016/S0019-3577(99)80012-3
21. Yamaguchi, M., Aoto, T.: A fast decision procedure for uniqueness of normal forms w.r.t. conversion of shallow term rewriting systems. In: Proceedings of 5th FSCD, LIPIcs, vol. 167, pp. 11:1–23. Schloss Dagstuhl (2020). https://doi.org/10.4230/LIPIcs.FSCD.2020.11

Rewriting Induction for Higher-Order Constrained Term Rewriting Systems

Kasper Hagens[(✉)] and Cynthia Kop

Radboud University Nijmegen, Nijmegen, The Netherlands
kasper.hagens@ru.nl, c.kop@cs.ru.nl

Abstract. Logically Constrained Term Rewriting Systems (LCTRSs) provide a framework very suitable for modeling both imperative and functional languages. One may convert programs in traditional languages into LCTRSs, and then use methods from term rewriting to analyze properties such as termination or program equivalence.

In particular in functional programming, higher-order constructs arise naturally. These have been studied using *higher-order* term rewriting. The recent definition of LCSTRSs combines higher-order rewriting with logical constraints, which creates the framework to closely model functional programs, but very few methods for their analysis have thus far been defined. Here, we study program equivalence for LCSTRSs, combining the definition of *rewriting induction* for first-order constrained rewriting with insights from unconstrained higher-order equivalence analysis.

Keywords: Term Rewriting · LCSTRSs · higher-order LCTRSs · Rewriting Induction · Inductive Theorems

1 Introduction

Consider the following two Haskell definitions of sumfun :: (Int -> Int) -> Int -> Int which computes the function $(f, x) \mapsto \sum_{i=0}^{x} f(i)$ for all $x \geq 0$.
(SFa) sumfun f x
         | x ≤ 0   = f x
         | otherwise = (f x) + (sumfun f (x - 1))
(SFb) sumfun f x = foldl (+) 0 (map f [x,x-1..0])

By human reasoning we know these implementations produce the same output for all inputs with $x \geq 0$. The general problem of deciding whether two arbitrary programs produce the same output, for all possible inputs that satisfy some condition, is known as *program equivalence*. This is a challenging problem, which naturally arises in software development. For example, code may be refactored for optimization purposes, to improve code maintainability, or in preparation for later updates [14]. To guarantee preservation of reliability and functionality, such transformations are expected to retain equivalence.

There is a variety of methods to prove equivalence of programs automatically, e.g. abstract interpretation [6,22], Hoare-style proof rules [10], constrained Horn clauses [1,8], and Rewriting Induction (RI) [7,9,20]. This paper builds on the latter approach. RI [23] is a proof system to prove/disprove convertibility of two terms. The idea is to translate two versions of the same program (or: program fragment) into a single term rewriting system, and use RI to prove equivalence of the terms corresponding to, for instance, the two different sumfun functions.

In particular, this line of work considers translations to extensions of traditional term rewriting systems with support for integers and booleans, as well as logical constraints to naturally model control flow – for example, rules like $\text{sum}(x) \to x + \text{sum}(x-1)$ $[x > 0]$. We will focus on *Logically Constrained Term Rewriting Systems (LCTRSs)* [16], a unifying formalism that supports arbitrary theories (e.g., bitvectors, floating point numbers or integer arrays). Programs in (fragments of) imperative languages may be translated into LCTRSs automatically (see, e.g., [9,21]) and be analyzed using rewriting methods.

However, when translating *functional* programs we soon encounter the problem of higher-order constructs: functions like foldl and map, which take functions as arguments, have no counterpart in first-order term rewriting. They are, however, naturally modeled using *higher-order* term rewriting. A recent definition of *Logically Constrained Simply typed Term Rewriting Systems* (LCSTRSs) [12] combines higher-order rewriting with native support for theories and constraints. The question arises whether we can also define RI in this setting – and if so, if this is usable for program analysis.

Bringing together higher-order rewriting with (constrained) RI poses new challenges. The papers [3,4] illustrate that, already for unconstrained higher-order rewriting, it is not easy to define what equivalence of terms even means. For first-order rewriting, this notion is straightforward, but higher-order rewriting admits multiple possible definitions – each of which comes with limitations, or loses important properties which makes them harder to analyze. Thus, a core task lies in finding definitions that allow us to not only adapt RI to the higher-order setting, but also have it usable in practice.

Paper Overview and Contributions. After some preliminaries (Sect. 2), we build on the unconstrained literature to propose a basic definition of *higher-order inductive theorems* for higher-order LCTRSs (Sect. 3). We then extend RI for constrained TRSs to this new setting (Sect. 4). Unfortunately, the basic definition lacks the property of *extensibility*; to solve this, we introduce a notion of *global inductive theorems* (Sect. 5) and show how to make it compatible with RI (Sect. 6). We conclude with some thoughts on future work (Sect. 7).

Scientific Context. Please note that the purpose of this paper is *not* to investigate equivalence in Haskell in particular: we focus on *constrained higher-order term rewriting systems*. Translating Haskell and other (functional and imperative) languages into term rewriting is a topic of active research and beyond the scope of this paper. Here, we hope to provide a foundation for a form of higher-order analysis that in the future can be used as part of a larger toolbox to analyze programs, for example in Haskell, Scala or OCaml. We have chosen the LCSTRS formalism since this is the first higher-order extension of LCTRSs, and comes with existing (fully automated) support for termination analysis.

Although LCSTRSs do not support lambda-expressions (an important structure in functional programs), these can typically be encoded, and it seems likely that the theory will extend naturally when these are included in the future.

2 Preliminaries

2.1 Logically Constrained Simply Typed Rewriting Systems

We will recap LCSTRSs [12], a higher-order extension of LCTRSs. This considers *applicative* higher-order rewriting (without λ) and *first-order* constraints.

Types and Terms. Assume given a set of sorts (base types) $\mathcal{S}$; the set $\mathcal{T}$ of types is defined by the grammar $\mathcal{T} ::= \mathcal{S} \mid \mathcal{T} \to \mathcal{T}$. Here, $\to$ is right-associative, so all types may be written as $type_1 \to \ldots \to type_m \to sort$ with $m \geq 0$.

We assume given a signature Σ of *function symbols* and a disjoint set $\mathcal{V}$ of variables, and a function *typeof* from $\Sigma \cup \mathcal{V}$ to $\mathcal{T}$; we require that there are infinitely many variables of all types. The set of terms $T(\Sigma, \mathcal{V})$ over Σ and $\mathcal{V}$ are the expressions in $\mathbb{T}$ – defined by the grammar $\mathbb{T} ::= \Sigma \mid \mathcal{V} \mid \mathbb{T}\,\mathbb{T}$ – that are *well-typed*: if $s :: \sigma \to \tau$ and $t :: \sigma$ then $s\,t :: \tau$, and $a :: typeof(a)$ for $a \in \Sigma \cup \mathcal{V}$. For a term t, let $\mathtt{Var}(t)$ be the set of variables in t. A term t is *ground* if $\mathtt{Var}(t) = \emptyset$. It is *linear* if no variable occurs more than once in t.

We also assume given a subset $\mathcal{S}_{theory} \subseteq \mathcal{S}$ of *theory sorts* (e.g., int and bool), and define the set of *theory types* by the grammar $\mathcal{T}_{theory} ::= \mathcal{S}_{theory} \mid \mathcal{S}_{theory} \to \mathcal{T}_{theory}$. Each sort ι is associated with a non-empty set $\mathcal{I}_\iota$ (e.g., $\mathcal{I}_{int} = \mathbb{Z}$, the set of all integers), and we let $\mathcal{I}_{\iota \to \sigma}$ be the set of functions from $\mathcal{I}_\iota$ to $\mathcal{I}_\sigma$.

We assume that Σ is the disjoint union $\Sigma_{theory} \uplus \Sigma_{terms}$ of two sets, where $typeof(\mathsf{f}) \in \mathcal{T}_{theory}$ for all $\mathsf{f} \in \Sigma_{theory}$. Each $\mathsf{f} \in \Sigma_{theory}$ comes with an interpretation $[\![\mathsf{f}]\!] \in \mathcal{I}_{typeof(\mathsf{f})}$. For example, with a theory symbol $* :: \mathsf{int} \to \mathsf{int} \to \mathsf{int}$ its interpretation may be multiplication on $\mathbb{Z}$. Symbols in Σ_{terms} do not have an interpretation since their behavior will be defined through the rewriting system. Values are theory symbols of base type, i.e. $\mathcal{V}al = \{v \in \Sigma_{theory} \mid typeof(v) \in \mathcal{S}_{theory}\}$. There should be exactly one value for each element of $\mathcal{I}_\iota$ ($\iota \in \mathcal{S}_{theory}$).

The set of *theory terms* is $T(\Sigma_{theory}, \mathcal{V})$. For *ground* theory terms, we define $[\![s\,t]\!] = [\![s]\!]([\![t]\!])$, thus mapping each term of type σ to an element of $\mathcal{I}_\sigma$.

We fix a theory sort bool with $\mathcal{I}_{\mathsf{bool}} = \{\top, \bot\}$. A *constraint* is a theory term s of type bool, such that $typeof(x) \in \mathcal{S}_{theory}$ for all $x \in \mathtt{Var}(s)$.

Example 1. In all examples in this paper, we will use $\mathcal{S}_{theory} = \{\mathsf{int}, \mathsf{bool}\}$ and $\Sigma_{theory} = \{+, -, *, <, \leq, >, \geq, =, \wedge, \vee, \neg, \mathtt{true}, \mathtt{false}\} \cup \{\mathsf{n} \mid n \in \mathbb{Z}\}$, with $+, -, * :: \mathsf{int} \to \mathsf{int} \to \mathsf{int}$, $<, \leq, >, \geq, = \; :: \mathsf{int} \to \mathsf{int} \to \mathsf{bool}$, $\wedge, \vee :: \mathsf{bool} \to \mathsf{bool} \to \mathsf{bool}$, $\neg :: \mathsf{bool} \to \mathsf{bool}$, $\mathtt{true}, \mathtt{false} :: \mathsf{bool}$ and $\mathsf{n} :: \mathsf{int}$. We let $\mathcal{I}_{\mathsf{int}} = \mathbb{Z}$, $\mathcal{I}_{\mathsf{bool}} = \{\top, \bot\}$ and interpret all symbols as expected. We use infix notation for the binary symbols, or use [f] for prefix or partially applied notation (e.g., $[+]\,x\,y$ and $x + y$ are the same). The values are $\mathtt{true}, \mathtt{false}$ and all n. Theory terms are for instance $x + 3$, $\mathtt{true}$ and $-7 * 0$. The latter two are ground. We have $[\![-7 * 0]\!] = 0$. The theory term $x > 0$ is a constraint, but the theory term

$(x\ y) > 0$ with $x :: \mathsf{int} \to \mathsf{int}$ is not, nor is $[>]\ 0 :: \mathsf{int} \to \mathsf{bool}$ (constraints are first-order terms).

Remark 1. Most programming languages have pre-defined (non-recursive) data structures and operators, e.g. the integers with a multiplication operator $*$. This makes it possible to for instance define the factorial function without first defining multiplication. This is exactly what an LCSTRS seeks to replicate: we can think of Σ_{theory} as the set of such pre-defined operators, including constants.

Substitutions, Contexts and Positions. A substitution is a type-preserving mapping $\gamma : \mathcal{V} \to T(\Sigma, \mathcal{V})$. The domain of a substitution is defined as $dom(\gamma) = \{x \in \mathcal{V} \mid \gamma(x) \neq x\}$, and the image of a substitution as $im(\gamma) = \{\gamma(x) \mid x \in dom(\gamma)\}$. A substitution on finite domain $\{x_1, \ldots, x_n\}$ is often denoted $[x_1 := s_1, \ldots, x_n := s_n]$. A substitution γ is extended to a function $s \mapsto s\gamma$ on terms by placewise substituting variables in the term by their image: **(i)** $t\gamma = t$ if $t \in \Sigma$, **(ii)** $t\gamma = \gamma(t)$ if $t \in \mathcal{V}$, and **(iii)** $(t_0\ t_1)\gamma = (t_0\gamma)\ (t_1\gamma)$. If $M \subseteq T(\Sigma, \mathcal{V})$ then $\gamma(M)$ denotes the set $\{t\gamma \mid t \in M\}$. A *unifier* of terms s, t is a substitution γ such that $s\gamma = t\gamma$; a *most general unifier* or *mgu* is a unifier γ such that all other unifiers are instances of γ. For unifiable terms, an mgu always exists.

Let $\square_1, \ldots, \square_n$ be fresh, typed constants ($n \geq 1$). A context $C[\square_1, \ldots, \square_n]$ (or just: C) is a term in $T(\Sigma \cup \{\square_1, \ldots, \square_n\}, \mathcal{V})$ in which each $\square_i$ occurs exactly once. (They may occur at the head of an application.) The term obtained from C by replacing each $\square_i$ by a term t_i of the same type is denoted by $C[t_1, \ldots, t_n]$.

For a term $t = a\ t_1 \cdots t_n$ with $a \in \Sigma \cup \mathcal{V}$ and $n \geq 0$ (all terms can be denoted this way), the set of positions $Pos(t) \subseteq \mathbb{N}^*$ is defined by: $Pos(t) = \{\epsilon\} \cup \bigcup_{i=1}^n \{i \cdot p \mid p \in Pos(t_i)\}$. We define the subterm $t|_p$ of t at position $p \in Pos(t)$ as follows: **(i)** $t|_\epsilon = t$, **(ii)** $(a\ t_1 \cdots t_n)|_{i \cdot p} = t_i|_p$. If t, s are terms and $p \in Pos(t)$ then we define $t[s]_p$ as the term obtained from t by replacing $t|_p$ by s.

Rules and Reduction. A rule is an expression $\ell \to r\ [\varphi]$. Here ℓ and r are terms of the same type, ℓ has a form $\mathsf{f}\ \ell_1 \cdots \ell_k$ with $k \geq 0$ and $\mathsf{f} \in \Sigma$, φ is a constraint, and for $x \in \mathsf{Var}(r) \setminus \mathsf{Var}(\ell)$, $typeof(x) \in \mathcal{S}_{theory}$. If $\varphi = \mathsf{true}$, we may denote the rule as just $\ell \to r$. Define $LVar(\ell \to r\ [\varphi]) = \mathsf{Var}(\varphi) \cup (\mathsf{Var}(r) \setminus \mathsf{Var}(\ell))$. A substitution γ *respects* $\ell \to r\ [\varphi]$ if $\gamma(LVar(\ell \to r\ [\varphi])) \subseteq \mathcal{V}al$ and $[\![\varphi\gamma]\!] = \top$.

We assume given a set of logically constrained rewrite rules $\mathcal{R}$ such that for $\ell \to r\ [\varphi] \in \mathcal{R}$, the left-hand side ℓ is not a theory term. In addition, let $\mathcal{R}_{calc}$ be the set containing, for every $\mathsf{f} \in \Sigma_{theory} \setminus \mathcal{V}al$ with $typeof(\mathsf{f}) = \iota_1 \to \ldots \to \iota_m \to \kappa$ ($\kappa \in \mathcal{S}_{theory}$) a rule $\mathsf{f}\ x_1 \cdots x_m \to y\ [y = \mathsf{f}\ x_1 \cdots x_m]$. We call these *calculation rules*. The reduction relation $\to_\mathcal{R}$ is defined by:

$$C[\ell\gamma] \to_\mathcal{R} C[r\gamma] \text{ if } \ell \to r\ [\varphi] \in \mathcal{R} \cup \mathcal{R}_{calc} \text{ and } \gamma \text{ respects } \ell \to r\ [\varphi]$$

We say that s *has normal form* t if $s \to_\mathcal{R}^* t$ and t cannot be reduced.

For a fixed set of rules $\mathcal{R}$, let $\mathcal{D} = \{\mathsf{f} \in \Sigma \mid \text{there is a rule } \mathsf{f}\ \ell_1 \cdots \ell_k \to r\ [\varphi] \in \mathcal{R}\}$; we call the elements of $\mathcal{D}$ *defined symbols*, and the elements of $\mathcal{C} = \mathcal{V}al \cup (\Sigma_{terms} \setminus \mathcal{D})$ *constructors*. The elements of $\Sigma_{theory} \setminus \mathcal{V}al$ are called *calculation symbols*. A term in $T(\mathcal{C}, \mathcal{V})$ is called a *constructor term*. A *ground constructor substitution* is a substitution γ such that $im(\gamma) \subseteq T(\mathcal{C}, \emptyset)$.

For a rule $\ell \to r\ [\varphi]$ define $head(\ell \to r\ [\varphi]) = \mathsf{f}$ if ℓ is of the shape $\mathsf{f}\ \ell_1 \cdots \ell_k$. For $\mathsf{f} \in \Sigma$ define $\mathcal{R}_\mathsf{f} = \{\ell \to r\ [\varphi] \in \mathcal{R} \mid head(\ell) = \mathsf{f}\}$ (so $\mathcal{R}_\mathsf{f} = \emptyset$ if $\mathsf{f} \notin \mathcal{D}$).

A *Logically Constrained Simply-typed Term Rewriting System (LCSTRS)* is a pair $(T(\Sigma, \mathcal{V}), \to_\mathcal{R})$ generated by $(\mathcal{S}, \mathcal{S}_{theory}, \Sigma_{terms}, \Sigma_{theory}, \mathcal{V}, typeof, \mathcal{I}, \llbracket \cdot \rrbracket, \mathcal{R})$. To refer to an LCSTRS we often supply just Σ and $\mathcal{R}$ and leave the rest implicit.

Example 2. Haskell function (SFa) can be modeled by $\mathcal{S} = \mathcal{S}_{theory} = \{\mathsf{int}, \mathsf{bool}\}$, $\Sigma_{terms} = \{\mathsf{sumfun} :: (\mathsf{int} \to \mathsf{int}) \to \mathsf{int} \to \mathsf{int}\}$, Σ_{theory} from Example 1, and
$$\mathcal{R} = \begin{cases} (R1).\ \mathsf{sumfun}\ f\ x \to f\ x & [x \leq 0] \\ (R2).\ \mathsf{sumfun}\ f\ x \to [+]\ (f\ x)\ (\mathsf{sumfun}\ f\ (x-1)) & [x > 0] \end{cases}$$
Then $\mathcal{D} = \{\mathsf{sumfun}\}$, and $\mathcal{C} = \mathcal{V}al = \{\mathsf{true}, \mathsf{false}\} \cup \{\mathsf{n} \mid n \in \mathbb{Z}\}$. We have $LVar((R1)) = \{x\}$ and $[x := 0]$ respects $(R1)$. We have $\mathsf{sumfun}\ f\ 0 \to_\mathcal{R} f\ 0$. An example of a rewrite sequence, computing a normal form: $\mathsf{sumfun}\ ([*]\ 2)\ 1 \to_{(R2)}$ $[+]\ ((([*]\ 2)\ 1)\ (\mathsf{sumfun}\ ([*]\ 2)\ (1-1)) \to_{\mathcal{R}_{calc}} [+]\ 2\ (\mathsf{sumfun}\ ([*]\ 2)\ (1-1)) \to_{\mathcal{R}_{calc}}$ $[+]\ 2\ (\mathsf{sumfun}\ ([*]\ 2)\ 0) \to_{(R1)} [+]\ 2\ (([*]\ 2)\ 0) \to_{\mathcal{R}_{calc}} [+]\ 2\ 0 \to_{\mathcal{R}_{calc}} 2$.

2.2 Rewriting Induction

Equations. An equation is a triple $s \approx t\ [\varphi]$ with s, t terms of the same type and φ a constraint. A substitution γ respects φ if $\gamma(\mathsf{Var}(\varphi)) \subseteq \mathcal{V}al$ and $\llbracket \varphi\gamma \rrbracket = \top$. A substitution γ respects $s \approx t\ [\varphi]$ if γ respects φ and $\mathsf{Var}(s) \cup \mathsf{Var}(t) \subseteq dom(\gamma)$.

Example 3. The function (SFb) from the introduction is modeled as follows
  (R3). $\mathsf{fold}\ g\ v\ \mathsf{nil} \to v$ \qquad\qquad (R6). $\mathsf{map}\ f\ \mathsf{nil} \to \mathsf{nil}$
  (R4). $\mathsf{fold}\ g\ v\ (h:t) \to \mathsf{fold}\ g\ (g\ v\ h)\ t$ \qquad (R7). $\mathsf{map}\ f\ (h:t) \to (f\ h) : \mathsf{map}\ f\ t$
  (R5). $\mathsf{init}\ n \to \mathsf{nil}\quad [n < 0]$ \qquad\qquad (R8). $\mathsf{init}\ n \to n : \mathsf{init}\ (n-1)\ [n \geq 0]$
Now the equivalence mentioned in the introduction is expressed as below.
$$\mathsf{sumfun}\ f\ n \approx \mathsf{fold}\ [+]\ 0\ (\mathsf{map}\ f\ (\mathsf{init}\ n))\quad [n \geq 0]$$
Substitution $[n := 0]$ does not respect the equation, but $[n := 0,\ f := [*]\ 2]$ does.

Inductive Theorems. Equivalence is defined via inductive theorems. For first-order rewriting, an equation $s \approx t\ [\varphi]$ is an inductive theorem if $s\gamma \leftrightarrow^*_\mathcal{R} t\gamma$ for every ground substitution γ that respects this equation. Here, $\leftrightarrow_\mathcal{R}$ is the union $\to_\mathcal{R} \cup \leftarrow_\mathcal{R}$, and $\leftrightarrow^*_\mathcal{R}$ is its transitive, reflexive closure.

Rewriting Induction. Rewriting Induction (RI) is a proof system for showing that equations are inductive theorems. It proceeds by transforming proof states: pairs $(\mathcal{E}, \mathcal{H})$ where $\mathcal{E}$ is a set of equations and $\mathcal{H}$ a set of rewrite rules. For such a proof state, we can think of $\mathcal{E}$ as the set containing all current proof goals, and of $\mathcal{H}$ as the set of induction hypotheses (oriented equations) that have been assumed in the process leading to this proof state. At the start $\mathcal{E}$ consists of all equations that we want to prove to be inductive theorems, and $\mathcal{H} = \emptyset$. The goal of RI is to find a deduction sequence $(\mathcal{E}, \emptyset) \vdash^* (\emptyset, \mathcal{H})$, for some set $\mathcal{H}$.

There are subtle variations in the rules defining $\vdash$ between different first-order variants of RI (e.g. in [7,9,23]). We will not state the rules here, but provide a higher-order extension in Sect. 4. They typically satisfy the following property:

Let $\mathcal{R}$ be a terminating, quasi-reductive LCTRS and $\mathcal{E}$ a set of equations. If $(\mathcal{E}, \emptyset) \vdash^* (\emptyset, \mathcal{H})$ for some $\mathcal{H}$, then every equation in $\mathcal{E}$ is an inductive theorem.

The proof of this result relies on well-founded induction over the relation $\to_{\mathcal{R} \cup \mathcal{H}}$, and therefore the method is limited to *terminating* LCTRSs (i.e. there are no infinite reductions $s_0 \to_\mathcal{R} s_1 \to_\mathcal{R} s_2 \to_\mathcal{R} \ldots$). The result also relies on *quasi-reductivity*; that is, that every ground normal form is a constructor term. Quasi-reductivity ensures that evaluation on ground terms cannot get "stuck"; roughly, that pattern matching is exhaustive. Termination and quasi-reductivity together ensure that every ground term reduces to a constructor term.

3 Higher-Order Inductive Theorems

While the first-order definition of inductive theorems is straightforward, it is not immediately obvious how it should be extended to a higher-order setting. In particular, the question of *extensionality* comes into play. We will present some ideas from the literature, and then posit our definition, before also extending the notions of constructor term and quasi-reductivity, which will be needed for RI.

3.1 Inductive Theorems and Extensionality

A first definition of higher-order inductive theorems (without constraints) appears in [4]. Aside from letting $s \approx t$ be an inductive theorem if $s\gamma \leftrightarrow^*_\mathcal{R} t\gamma$ for all ground substitutions γ, the authors consider two functions equivalent if they are *extensionally* equivalent: their value on all inputs is equivalent. That is, for terms s, t of type $\sigma_1 \to \ldots \to \sigma_m \to \iota$, they consider $s \approx t$ an inductive theorem if $(s\ x_1 \cdots x_m)\gamma \leftrightarrow^*_\mathcal{R} (t\ x_1 \cdots x_m)\gamma$. Hence, e.g. map $([+]\ 0) \approx$ map $([*]\ 1)$ is an inductive theorem since map $([+]\ 0)\ l \leftrightarrow^*_\mathcal{R}$ map $([*]\ 1)\ l$ for all ground terms l. This is intuitive since functions are often viewed in an extensional way.

Unfortunately, it comes at a price, since this definition violates *monotonicity*: the property that if $s \approx t$ is an inductive theorem and C a context, then $C[s] \approx C[t]$ is also an inductive theorem. This is illustrated by the following example:

Example 4. Let $\Sigma_{terms} = \{\text{add} :: \text{nat} \to \text{nat} \to \text{nat}, \text{s} :: \text{nat} \to \text{nat}, 0 :: \text{nat}, \text{fnil} :: \text{funclist}, \text{fcons} :: (\text{nat} \to \text{nat}) \to \text{funclist} \to \text{funclist}\}$, $\Sigma_{theory} = \emptyset$ and $\mathcal{R} = \{\text{add } x\ 0 \to x, \text{add } x\ (\text{s } y) \to \text{s } (\text{add } x\ y)\}$. Every ground term of type nat has a normal form of the shape $\text{s}^n\ 0$, for some $n \in \mathbb{N}$, so we can restrict to ground substitutions of the shape $\gamma_n = [x := \text{s}^n\ 0]$. Note that (add (s 0) $x)\gamma_n \to^*_\mathcal{R} \text{s}^{n+1}\ 0 \leftarrow^*_\mathcal{R} (\text{s } x)\gamma_n$. Hence add (s 0) $\approx$ s is an extensional inductive theorem. However, for $C[\Box] = \text{fcons } \Box$ fnil we do *not* have $C[\text{add (s 0)}] \leftrightarrow^*_\mathcal{R} C[\text{s}]$.

The authors of [4] tackle the problem by imposing limitations on the systems they consider. Aside from other consequences (which are similar to the ones we will consider in Sect. 5), their restrictions essentially block constructors from taking higher-order arguments – thus for instance disallowing lists of functions as used in Example 4. Since such constructs naturally occur in functional programs, we consider this restriction too severe, and have elected not to go into the extensional direction. Instead, we keep the first-order definition, which also makes sense in the higher-order setting and does satisfy monotonicity:

Definition 1 (Higher-order inductive theorems). *An equation $s \approx t \ [\varphi]$ is a higher-order inductive theorem of an LCSTRS with rules $\mathcal{R}$ if $s\gamma \leftrightarrow^*_\mathcal{R} t\gamma$ for every ground substitution γ that respects this equation.*

Discussion. The choice whether to consider extensionality ties in to a larger discussion on the semantics of equations and (constrained higher-order) rewriting. Traditionally (in unconstrained rewriting), rules are seen as *oriented equations*. Ground terms may be interpreted in a *model*, and an equation $s \approx t$ holds in a model if for all ground instances $s\gamma \approx t\gamma$ of the equation, the interpretation of $s\gamma$ and $t\gamma$ is the same. We say that $\mathcal{R} \vDash s \approx t$ if $s \approx t$ holds in every model for which the rules of $\mathcal{R}$ all hold. In such a semantics, a term of higher type would typically be mapped to a function, so it is natural to use an extensional perspective where two terms are equivalent if their result on all input is equivalent.

The authors of [2] define such a semantics for first-order LCTRSs, and prove that convertibility of ground terms corresponds to their semantic notion; i.e., an equation $s \approx t \ [\varphi]$ is "CE-valid" (which corresponds to our notion of inductive theorem) if and only if $\mathcal{R} \vDash s \approx t \ [\varphi]$. However, in *higher-order* rewriting, this does not typically hold – even if we include abstraction and the η rule scheme.

A very relevant paper with regards to higher-order equivalence is [3]. This paper defines "extensional theorems" (for unconstrained higher-order rewriting) as equivalence in a model, and shows that this semantic equivalence corresponds to syntactic equivalence of ground instances of equations in an inference system. This definition solves the monotonicity problem of [4] as monotonicity is built in, but it loses the direct correspondence to the convertibility relation $\leftrightarrow^*_\mathcal{R}$.

We have elected not to follow this example because our primary application domain is not equational reasoning – where a semantic definition is the natural choice – but functional programming, where the syntactic notion of convertibility seems preferable. Definition 1 has the benefit of minimality: any equivalence relation that includes $\rightarrow_\mathcal{R}$ must include $\leftrightarrow^*_\mathcal{R}$. Hence, any higher-order inductive theorem is also an extensional theorem, and the method of rewriting induction defined in this paper can be used to derive extensional theorems as well.

It is worth noting that if a system is *ground confluent* – i.e., if $s \rightarrow^*_\mathcal{R} t$ and $s \rightarrow^*_\mathcal{R} u$ for a ground term s, then there is a term w such that $t \rightarrow^*_\mathcal{R} w$ and $u \rightarrow^*_\mathcal{R} w$ – two ground terms are convertible if and only if they reduce to the same term (using $\rightarrow^*_\mathcal{R}$). This property is typically satisfied in LCSTRSs obtained from (deterministic) programs. Thus, in a terminating and ground confluent system, two terms are convertible if and only if they compute the same result.

3.2 Higher-Order Quasi-Reductivity

Quasi-reductivity is an essential component of rewriting induction, since it allows us to reduce any ground term to a constructor term. Yet in higher-order rewriting this property is hard to obtain, since it is usually possible for ground terms of function type to not be constructor terms; e.g., the term $[+]\ 1$, or the term init from Example 3. And Example 4, which seems relatively innocuous, even admits base-type ground non-constructor normal forms (e.g., fcons (add (s 0)) fnil).

Thus, we will update our definition to include partially applied function symbols, both at the root and below constructors. For the notion of "partially applied" to make sense, we impose a mild restriction on the LCSTRSs we consider:

(Rule Arity) every $f \in \mathcal{D}$ with $f :: \sigma_1 \to \ldots \to \sigma_m \to \iota$, $\iota \in \mathcal{S}$ has a *rule arity* $k = ar(f) \leq m$, meaning that every rule in $\mathcal{R}_f$ is of the shape $f\ l_1\ \ldots\ l_k \to r\ [\varphi]$.

That is, we do not for instance have both a rule add (s x) $y \to$ s (add x y) where add takes two arguments, and a rule add $0 \to$ id where it takes only one. This is not a significant restriction because we can simply alter the second rule to add $0\ x \to$ id x without changing its meaning. With this restriction, a symbol is partially applied if it has fewer than $ar(f)$ arguments. This allows us to define:

Definition 2 (Semi-constructor terms). *Let $\mathcal{L}$ be some LCSTRS $(\Sigma, \mathcal{R})$ (leaving $\mathcal{V}, typeof$, etc. implicit). The semi-constructor terms over $\mathcal{L}$, notation $\mathcal{SCT}_\mathcal{L}$, are defined by (i) $\mathcal{V} \subseteq \mathcal{SCT}_\mathcal{L}$, (ii) if $f \in \Sigma$ with $f :: \sigma_1 \to \ldots \to \sigma_m \to \iota$, $\iota \in \mathcal{S}$ and $s_1 :: \sigma_1, \ldots, s_n :: \sigma_n \in \mathcal{SCT}_\mathcal{L}$ with $n \leq m$, then $f\ s_1 \cdots s_n \in \mathcal{SCT}_\mathcal{L}$ if: (ii.a) $f \in \mathcal{C}$, or (ii.b) $f \in \mathcal{D}$ and $n < ar(f)$, or (ii.c) $f \in \Sigma_{theory} \setminus \mathcal{D}$ and $n < m$. The set $\mathcal{SCT}_\mathcal{L}^\emptyset$ refers to* ground *semi-constructor terms (built without (i)).*

Note that all constructor terms are semi-constructor terms, by (i) and (ii.a). In our previous examples, $[+]\ 1$ and init and fcons (add (s 0)) fnil, are all (ground) semi-constructor terms as well. Non-examples of semi-constructor terms are for instance $[+]\ 0\ y$ and fcons (add (add 0 0)). Using (ii.b) we easily obtain:

Lemma 1. *Every semi-constructor term is in normal form.*

Now, ground semi-constructor terms are the higher-order counterpart of ground constructor terms, as intuitively, in an LCSTRS without missing cases in the pattern matching, the only ground normal forms are in $\mathcal{SCT}_\mathcal{L}^\emptyset$.

Definition 3 (Quasi-reductivity). *An LCSTRS $\mathcal{L} = (\Sigma, typeof, \mathcal{R})$ is quasi-reductive if for every $t \in T(\Sigma, \emptyset)$ we have $t \in \mathcal{SCT}_\mathcal{L}^\emptyset$ or t reduces with $\to_\mathcal{R}$.*

Note that for quasi-reductive LCSTRSs, we can limit the substitutions in Definition 1 to substitutions such that $im(\gamma) \subseteq \mathcal{SCT}_\mathcal{L}^\emptyset$, without loss of generality. We call such a substitution a *ground semi-constructor substitution*.

4 Higher-Order Rewriting Induction

We will now give the derivation rules for higher-order RI. These are obtained from first-order RI [7,9,16,20,23], and adapted to the higher-order setting. We particularly build on RI, as defined in [9,16]. As a running example, we will use the LCSTRS and equation from Example 3. Thus, we start with the proof state:

$$(\mathcal{E}_0, \emptyset) := (\{ \text{ (A) sumfun } f \ n \approx \text{ fold } [+] \ 0 \ (\text{map } f \ (\text{init } n)) \ [n \geq 0] \} \ , \ \emptyset)$$

and show that we can derive some proof state $(\emptyset, \mathcal{H})$. Theorem 1, presented in Sect. 4.4, will guarantee the correctness of this procedure.

4.1 Simplifying Equations

The first, core rule of rewriting induction is to rewrite an equation using a rule in $\mathcal{R} \cup \mathcal{R}_{calc} \cup \mathcal{H}$. We view an equation with variables as a way to represent all *ground semi-constructor instances* of that equation. Hence, we can use "constrained term reduction" [16], which takes the constraint of the equation into account.

Simplification. Let $\ell \to r \ [\varphi] \in \mathcal{R} \cup \mathcal{R}_{calc} \cup \mathcal{H}$ with C a context, δ a substitution such that $\delta(LVar(\ell \to r \ [\varphi])) \subseteq \mathcal{V}al \cup \text{Var}(\psi)$, and $C[\ell\delta] \approx t \ [\psi]$ an equation. If the implication $\psi \Longrightarrow \varphi\delta$ is valid, then

$$(\mathcal{E} \uplus \{C[\ell\delta] \approx t \ [\psi]\}, \mathcal{H}) \vdash (\mathcal{E} \cup \{C[r\delta] \approx t \ [\psi]\}, \mathcal{H})$$

There is an analogous Simplification rule to apply a rewrite rule to the right-hand side of an equation.

Example 5. Starting with our running example, simplifying on the right-hand side of (A) using rule (R8) yields $(\mathcal{E}_0, \emptyset) \vdash (\mathcal{E}_1, \emptyset)$ with $\mathcal{E}_1 = \{(B) \text{ sumfun } f \ n \approx \text{ fold } [+] \ 0 \ (\text{map } f \ (n : (\text{init } (n-1)))) \ [n \geq 0] \}$. Using subsequent Simplification steps with (R7) and (R4), we have $(\mathcal{E}_1, \emptyset) \vdash^* (\mathcal{E}_3, \emptyset)$ with $\mathcal{E}_3 = \{(C) \text{ sumfun } f \ n \approx \text{ fold } [+] \ (0 + (f \ n)) \ (\text{map } f \ (\text{init } (n-1))) \ [n \geq 0] \}$.

4.2 Expanding Equations (Doing a Case Analysis)

After a few simplifications, we typically end up in a state where nothing can be done without knowing how the variables are instantiated. Our second rule allows us to do a case analysis and create an induction hypothesis at the same time.

(Expansion). Let $s \approx t \ [\varphi]$ be an equation and $p \in Pos(s)$ a position such that $s|_p = f \ s_1 \cdots s_n$ with $f \in \mathcal{D}$, $n \geq k = ar(f)$ and every argument s_i is a semi-constructor term. Suppose that $\mathcal{R} \cup \mathcal{H} \cup \{s \to t \ [\varphi]\}$ is terminating. Then

$$(\mathcal{E} \uplus \{s \approx t \ [\varphi]\}, \mathcal{H}) \vdash (\mathcal{E} \cup Expd(s \approx t \ [\varphi], p), \mathcal{H} \cup \{s \to t \ [\varphi]\})$$

where $Expd(s \approx t \ [\varphi], p)$ is the set:

$$\{s[r \ s_{k+1} \cdots s_n]_p \gamma \approx t\gamma \ [(\varphi\gamma) \wedge (\psi\gamma)] \mid \ell \to r \ [\psi] \in \mathcal{R}, \text{mgu}(f \ s_1 \cdots s_k, \ell) = \gamma\}$$

There is an analogous rule for performing Expansion on the right-hand side of an equation. In that case, $t \to s \ [\varphi]$ is added to $\mathcal{H}$.

Example 6. In our running example, we expand the left-hand side of (C) at position ϵ. This gives $(\mathcal{E}_3, \emptyset) \vdash (\mathcal{E}_4, \mathcal{H}_1)$, where:

$$\mathcal{E}_4 = \left\{ \begin{array}{l} \text{(D) } f\ n \approx \text{ fold } [+]\ (0 + (f\ n))\ (\text{map } f\ (\text{init } (n-1)))\ [n \geq 0 \wedge n \leq 0] \\ \text{(E) } (f\ n)\ +\ (\text{sumfun } f\ (n-1))\ \approx \\ \qquad \text{fold } [+]\ (0 + (f\ n))\ (\text{map } f\ (\text{init } (n-1)))\ [n \geq 0 \wedge n > 0] \end{array} \right\}$$

$\mathcal{H}_1 = \{$ (C') sumfun $f\ n\ \to\ $ fold $[+]\ (0 + (f\ n))\ (\text{map } f\ (\text{init } (n-1)))\ [n \geq 0]\ \}$

Viewing an equation as a way to represent a set of ground equations, this definition essentially allows us to split this set into multiple subsets, by considering the possible instances at position p of s. Since we have assumed quasi-reductivity, some rule is applicable at position p of $s\gamma$ for any ground semi-constructor substitution γ. The set $Expd(s \approx t\ [\varphi], p)$ contains a representative result equation for any of the rules that might have been chosen.

Note that, after applying Expansion, the equation $s \approx t\ [\varphi]$ becomes an induction hypothesis, because we add it to $\mathcal{H}$ as an oriented equation. Therefore, we can think of Expansion as starting an induction proof on the subterm $s|_p$.

4.3 Altering (and Generalizing) Equations

In Example 6, intuitively we should be able to simplify init $(n-1)$ in the first equation of $\mathcal{E}_4$ to nil, as the constraint implies $n - 1 < 0$. However, the Simplification rule does not allow this: the variable in the init rule must be instantiated by a value or variable. Nor can we apply Simplification with a calculation rule.

Of course we could adapt the definition of Simplification, but this is actually part of a larger pattern: since our derivation rules rely on the shape of an equation, it is often useful to alter an equation to a semantically equivalent one. That is, since an equation represents the set of its ground semi-constructor instances, we should be able to replace it by an equation that represents the same set.

We define two very similar derivation rules: one that lets us replace an equation by another equation that represents the *same* set, and a second that lets us replace an equation by another that may represent a *larger* set.

(Generalize). Suppose that for every ground semi-constructor substitution (gsc) substitution γ that respects $s \approx t\ [\varphi]$ there exists a substitution δ that respects $u \approx v\ [\psi]$ such that $s\gamma = u\delta$ and $t\gamma = v\delta$. Then

$$(\mathcal{E} \uplus \{s \approx t\ [\varphi]\}, \mathcal{H}) \vdash (\mathcal{E} \cup \{u \approx v\ [\psi]\}, \mathcal{H})$$

If, moreover, for every gsc substitution δ that respects $u \approx v\ [\psi]$ there exists a substitution γ that respects $s \approx t\ [\varphi]$ such that $s\gamma = u\delta$ and $t\gamma = v\delta$, then we refer to the deduction step as **(Alter)** instead.

Example 7. Since $n \geq 0 \wedge n \leq 0$ is logically equivalent to $-1 = n - 1$, we can use Alter to replace (D) by $f\ n \approx $ fold $[+]\ (0 + (f\ n))\ (\text{map } f\ (\text{init } (n-1)))\ [-1 = n - 1]$. Then we can use Simplification using the calculation rule $x - y \to z\ [z = x - y]$ and substitution $\gamma = [x := n, y := 1, z := -1]$ to obtain

$f\ n \approx \text{fold } [+]\ (0 + (f\ n))\ (\text{map } f\ (\text{init } (-1)))\ [-1 = n-1]$.
Similarly, since $n \geq 0 \wedge n > 0 \iff \exists m[n > 0 \wedge m = n-1]$, we can change the constraint of (E) to $[n > 0 \wedge m = n-1]$, and use two calculation steps to obtain:
$(f\ n) + (\text{sumfun } f\ m) \approx \text{fold } [+]\ (0 + (f\ n))\ (\text{map } f\ (\text{init } m))\ [n > 0 \wedge m = n-1]$
We use Simplification a few more times to eventually end up at $(\{(F), (G)\}, \mathcal{H}_1)$:

(F) $f\ n \approx 0 + (f\ n)\ [-1 = n-1]$
(G) $(f\ n)\ + (\text{sumfun } f\ m) \approx \text{fold } [+]$
$\quad ((0 + (f\ n)) + (f\ m))\ (\text{map } f\ (\text{init } (m-1)))\ [n > 0 \wedge m = n-1]$

Neither Simplification nor Expansion can be applied to (F), and it is not obviously removable (see Sect. 4.4) since f could be instantiated by anything in $\mathcal{SCT}_\mathcal{L}^\emptyset$. However, using Generalize, we can replace (F) by: $x \approx (0+x)$. Then, using Alter and Simplification with a calculation rule, we obtain: (H) $x \approx x\ [x = 0+x]$.

Discussion. Despite their similarity, the Alter and Generalize rules are used in very different ways. Alter is more innocent: replacing an equation by an equivalent one cannot harm the proof process – in contrast with Generalize, which can easily replace an equation that is an inductive theorem by one that is not.

In practice, Alter is typically used in combination with other derivation rules, e.g., to set up a Simplification step with a calculation as done in Example 7. We most often use Alter to replace $s \approx t\ [\varphi]$ by $u \approx v\ [\psi]$ in the following scenarios:

I. *(replacing a constraint by an equi-satisfiable one)* $s = u$ and $t = v$ and $(\exists \boldsymbol{x}.\varphi) \iff (\exists \boldsymbol{y}.\psi)$ is valid, where $\{\boldsymbol{x}\} = \text{Var}(\varphi) \setminus (\text{Var}(s) \cup \text{Var}(t))$ and $\{\boldsymbol{y}\} = \text{Var}(\psi) \setminus (\text{Var}(u) \cup \text{Var}(v))$. (This is particularly done before a Simplification step, to put the constraint in the right shape.)
II. *(replacing variables/values by equivalent ones)* $s = C[x_1, \ldots, x_n]$ and $u = C[y_1, \ldots, y_n]$ and $t = v$ and $\varphi = \psi$, if all x_i, y_i are values or variables in $\text{Var}(\varphi)$, and $\varphi \implies \bigwedge_{i=1}^n x_i = y_i$ is valid. For example, replacing $f\ n \approx f\ 0\ [n = 0]$ by $f\ 0 \approx f\ 0\ [n = 0]$. (This is particularly useful when $t = C[y_1, \ldots, y_n]$.)
III. *(adding safe variables to the constraint)* $s = u$ and $t = v$ and $\psi = \varphi \wedge (x_1 = x_1) \wedge \cdots \wedge (x_n = x_n)$, where $x_1, \ldots, x_n$ are variables in s, u that do not occur in φ, but whose type is a theory sort $\iota \in \mathcal{S}_{theory}$ such that no constructors of a type $\sigma_1 \to \ldots \to \sigma_m \to \iota$ exist other than values (and therefore, any ground semi-constructor instance of this variable must be a value).

On the other hand, Generalize is primarily used as a form of *lemma generation*: as we will see in Sect. 4.4, it is sometimes needed to generalize an equation to obtain a stronger induction hypothesis. Finding suitable lemmas is a core challenge in inductive theorem proving. Generalization is also useful to abstract away from variable applications, as done for $f\ n$ in Example 7.

4.4 Finishing up

Thus far, we have only modified equations; to remove them from $\mathcal{E}$, we can use:

(Deletion). Let $s \approx t \ [\varphi]$ be such that either $s = t$ or φ is unsatisfiable, then

$$(\mathcal{E} \uplus \{s \approx t \ [\varphi]\}, \mathcal{H}) \vdash (\mathcal{E}, \mathcal{H})$$

Example 8. We apply Deletion to (H) and obtain $(\mathcal{E}_6, \mathcal{H}_1)$ with $\mathcal{E}_6 = \{(\text{G})\}$.

We have now defined all the inference rules necessary to complete our running example. The process is *mostly* straightforward; we detail only the harder steps. First, in equation (G), we use Simplification with the induction rule (C') to get:
$$(f\ n)\ +\ (\text{fold}\ [+]\ (0+(f\ m))\ (\text{map}\ f\ (\text{init}\ (m-1)))) \approx$$
$$\text{fold}\ [+]\ ((0+(f\ n))+(f\ m))\ (\text{map}\ f\ (\text{init}\ (m-1)))\ [n > 0 \wedge m = n - 1]$$
After Alter and a calculation step, we arrive at:
$$(f\ n)\ +\ (\text{fold}\ [+]\ (0+(f\ m))\ (\text{map}\ f\ (\text{init}\ k))) \approx$$
$$\text{fold}\ [+]\ ((0+(f\ n))+(f\ m))\ (\text{map}\ f\ (\text{init}\ k))\ [n > 0 \wedge m = n - 1 \wedge k = m - 1]$$
Now, we could continue to do Expansions, but doing so would result in a loop: none of the induction rules this process generates ends up being applicable. To avoid this issue, we instead use Generalize to obtain:
$$x + (\text{fold}\ [+]\ (0+y)\ l) \approx \text{fold nil+}\ ((0+x)+y)\ l$$
In the next Expansion step, the termination requirement forces us to expand on the left rather than the right, creating an induction rule:
$$x + (\text{fold nil+}\ y\ l) \rightarrow \text{fold nil+}\ z\ l\ [z = x + y]$$
This rule has a calculation symbol $(+)$ as the root symbol on the left, which is non-standard, but allowed in LCSTRSs, and termination can be proved. The rest of the proof is entirely straightforward.

Although we did not need it for our running example, we also adapt the Constructor rule of [9] because it changes in the higher-order setting.

(Semi-constructor). Let $s = s_1 \ldots s_n$ and $t = t_1 \ldots t_n$ be terms with $n > 0$. If c is either a variable, a constructor, a defined symbol with $ar(\mathsf{c}) > n$ or a non-defined calculation symbol of type $\sigma_1 \rightarrow \ldots \rightarrow \sigma_m \rightarrow \iota$ with $m > n$ then

$$(\mathcal{E} \uplus \{\mathsf{c}\ s \approx \mathsf{c}\ t\ [\varphi]\}, \mathcal{H}) \vdash (\mathcal{E} \cup \{s_i \approx t_i\ [\varphi] \mid 1 \leq i \leq n\}, \mathcal{H})$$

Example 9. In an extension of the LCSTRS of Example 2 with extra symbols, we could deduce $(\{\text{fold g (h 0 } x) \approx \text{fold h (g 0 } x)\}, \emptyset) \vdash (\{\text{g} \approx \text{h, h 0 } x \approx \text{g 0 } x\}, \emptyset)$.

Theorem 1. *Let $\mathcal{L}$ be a terminating, quasi-reductive LCSTRS and let $\mathcal{E}$ be a set of equations. If, by higher-order rewriting induction, $(\mathcal{E}, \emptyset) \vdash^* (\emptyset, \mathcal{H})$, for some set $\mathcal{H}$, then every equation in $\mathcal{E}$ is a higher-order inductive theorem of $\mathcal{L}$.*

The proof of Theorem 1 (in [13, Appendix A]) follows the same outline as in first-order RI [9,15]. It proceeds by showing that certain properties are invariant through every proof step, and uses an induction on $\rightarrow_{\mathcal{R} \cup \mathcal{H}}$ to prove $\leftrightarrow_{\mathcal{E}} \subseteq \leftrightarrow_{\mathcal{R}}$.

4.5 Comparison to the First-Order Literature

Surprisingly few changes were needed to adapt the first-order definitions in [9] to the higher-order setting. The most important changes are the new definitions of quasi-reductivity and semi-constructor terms. The proof was also adapted to take these changes into account, but its overall structure remains the same.

The most significant change compared to [9] could already have been made in the first-order setting: the introduction of Alter, and updating Generalize to quantify over ground *semi-constructor* substitutions, rather than *all* substitutions. In [9], scenario I was combined with Simplification, and a separate rule (Eq-Deletion) was used to handle II, but III was not supported – thus leaving it impossible to prove for instance init $(n + 1) \approx$ init $(1 + n)$ if n was not in the constraint. This limitation is particularly relevant in the higher-order setting: due to proof states with higher-order variables (such as (F)), the Generalize rule is needed much more often than in first-order RI, and to progress the proof further we need to be able to move the resulting variables into the constraint.

5 Global Induction Theorems

A very desirable property we do not yet have is *extensibility*. This means that if an equation is an inductive theorem in $\mathcal{R}$, it remains an inductive theorem in any reasonable (i.e., adding defined symbols, not constructors) LCSTRS extending $\mathcal{R}$. In terms of functional programming, it should be possible to import functions from external modules without breaking any equivalence. Extensibility allows for more local reasoning: to prove properties about a small part of a larger system, it is very desirable to only have to consider the rules that are directly related. This property is also used in some existing methods for program transformations (see, e.g., [5]). The authors of [3] give the following example to illustrate the issue:

Example 10. Let $\Sigma_{terms} = \{$zero :: (nat $\rightarrow$ nat) $\rightarrow$ nat, add :: nat $\rightarrow$ nat $\rightarrow$ nat, s :: nat $\rightarrow$ nat, 0 :: nat$\}$, $\Sigma_{theory} = \emptyset$ and let $\mathcal{R}$ consist of

$$\text{add 0 } y \rightarrow y \quad \text{add (s } x) \ y \rightarrow \text{s (add } x \ y) \quad \text{zero s} \rightarrow 0$$

Then add x $y \approx$ add y x is an inductive theorem, since (add x $y)\gamma \leftrightarrow^*_\mathcal{R}$ (add y $x)\gamma$ holds for any ground substitution γ. However, if we introduce a new defined symbol f :: nat $\rightarrow$ nat and rule f $x \rightarrow 0$ then add x $y \approx$ add y x is not an inductive theorem since add 0 (zero f) $\leftrightarrow^*_{\mathcal{R}'}$ add (zero f) 0 does not hold.

A key problem in Example 10 is that the extension breaks quasi-reductivity: by importing f we create a missing pattern in $\mathcal{R}_{\text{zero}}$. This is caused by *pattern matching on a function*: the last rule matches on the expression s which is a non-variable term of type nat $\rightarrow$ nat. If we now import any new symbol of this type, no matter how innocent, it creates a new pattern; and thus quasi-reductivity is lost. From the perspective of functional programming, rules like this seem very unnatural; it is not typically *allowed* for a pattern to have a higher-order subterm that is not a variable. Thus, we argue that the original system is inherently problematic. To prevent such pathological examples, we extend the definition of quasi-reductivity to exclude this program structure.

Definition 4 (CHV term). *Let $\mathcal{L}$ be an LCSTRS. A Constructor term with (only) Higher-order Variables (CHV term) over $\mathcal{L}$ is a constructor term s over $\mathcal{L}$ such that $\text{Var}(s)$ contains only variables of higher type.*

Definition 5 (Strong quasi-reductivity). *An LCSTRS $\mathcal{L} = (\Sigma, \mathcal{R})$ with defined symbols $\mathcal{D}$ is strong quasi-reductive if any term of the form $\mathsf{f}\ s_1 \cdots s_n$ with $\mathsf{f} \in \mathcal{D}$, $n \geq \text{ar}(\mathsf{f})$ and each s_i a CHV term over $\mathcal{L}$ reduces with $\rightarrow_{\mathcal{R}}$.*

Any strong quasi-reductive LCSTRS is also quasi-reductive (see [13, Appendix B]). An LCSTRS is certainly strong quasi-reductive if it has exhaustive pattern matching, left-linear rules and all strict higher-order subterms of left-hand sides are variables. Strong quasi-reductivity is close to (and implies) the *quasi-reducibility* notion in [3]. On the other hand, strong quasi-reductivity is weaker than the notion of higher-order sufficient completeness (HSC) in [4].

Unfortunately, limiting interest to strong quasi-reductive systems does not suffice to obtain extensibility:

Example 11. Let $\Sigma_{terms} = \{\mathsf{a} :: \mathsf{A},\ \mathsf{b} :: \mathsf{A},\ \mathsf{c} :: \mathsf{C},\ \mathsf{f} :: (\mathsf{C} \rightarrow \mathsf{A}) \rightarrow \mathsf{A},\ \mathsf{g} :: \mathsf{C} \rightarrow \mathsf{A}\}$, $\Sigma_{theory} = \emptyset$ and consider the LCSTRS with rules $\mathsf{f}\ F \rightarrow F\ \mathsf{c}$ and $\mathsf{g}\ x \rightarrow \mathsf{b}$. Then $\mathsf{f}\ F \approx \mathsf{b}$ is an inductive theorem, since the only ground term that can instantiate F is g, and indeed $\mathsf{f}\ \mathsf{g} \rightarrow_{\mathcal{R}} \mathsf{g}\ \mathsf{c} \rightarrow_{\mathcal{R}} \mathsf{b}$. However, this is not an inductive theorem if we extend the signature with a defined symbol $\mathsf{h} :: \mathsf{C} \rightarrow \mathsf{A}$ and rule $\mathsf{h}\ x \rightarrow \mathsf{a}$.

Here, $\mathsf{f}\ F \approx \mathsf{b}$ is a (naive) inductive theorem because of a *global* reasoning over the original signature: there is only one possible instance of F. This is of course no longer true in the extension. To avoid examples like this, we follow the approach of [3] and directly define a kind of inductive theorems that are preserved under extensions – provided they satisfy reasonable restrictions:

Definition 6 (Natural extensions). *An LCSTRS $\mathcal{L}'$ (generated by $\mathcal{S}', \Sigma', \mathcal{R}'$, etc.) is a natural extension of $\mathcal{L}$ (generated by $\mathcal{S}, \Sigma, \mathcal{R}$, etc.) if:*

- $\mathcal{S}' \supseteq \mathcal{S}$ *and* $\mathcal{S}'_{theory} \supseteq \mathcal{S}_{theory}$ *and* $\Sigma'_{theory} \supseteq \Sigma_{theory}$ *and* $\Sigma'_{terms} \supseteq \Sigma_{terms}$ *and* $\mathcal{V}' \supseteq \mathcal{V}$ *and* $\mathcal{R}' \supseteq \mathcal{R}$
- $\mathcal{I}'_\iota = \mathcal{I}_\iota$ *for all $\iota \in \mathcal{S}_{theory}$, and* $[\![\mathsf{f}]\!]' = [\![\mathsf{f}]\!]$ *for all* $\mathsf{f} \in \Sigma_{theory}$
- *for all* $a \in \Sigma \cup \mathcal{V}$: $\text{typeof}'(a) = \text{typeof}(a)$
- *for all* $\mathsf{f} \in \Sigma$: $\mathcal{R}'_\mathsf{f} = \mathcal{R}_\mathsf{f}$ *(so $\mathcal{R}' \setminus \mathcal{R}$ does not define any of the constructors in $\mathcal{L}$, nor add cases to a defined symbol or calculation symbol)*
- *for all* $\mathsf{f} :: \sigma_1 \rightarrow \ldots \rightarrow \sigma_m \rightarrow \iota \in \Sigma'$, *all i: there is a ground term of type σ_i*
- *for all constructor symbols* $\mathsf{c} :: \sigma_1 \rightarrow \ldots \rightarrow \sigma_m \rightarrow \iota \in \mathcal{C}' \setminus \mathcal{C}$ *we have* $\iota \notin \mathcal{S}$

Hence, a natural extension can add more rules, but cannot interfere with the meaning of the original LCSTRS, nor add new patterns to its sorts. Note that, by the last restriction, any ground constructor term of a sort ι that occurs in the original signature can only use constructors in this signature.

Definition 7 (Global inductive theorems). *An equation $s \approx t\ [\varphi]$ over a terminating, strong quasi-reductive LCSTRS $\mathcal{L}$ is a global inductive theorem of $\mathcal{L}$ if for every terminating, quasi-reductive natural extension $\mathcal{L}'$ with rules $\mathcal{R}'$ and every ground substitution γ over $\mathcal{L}'$ that respects this equation: $s\gamma \leftrightarrow^*_{\mathcal{R}'} t\gamma$.*

6 Global Rewriting Induction

We now aim to extend higher-order RI in such a way that it proves equations to be *global* inductive theorems. Largely, this is straightforward (as we can mostly ignore rules whose defined symbols do not occur inside the equation), but a major problem arises with the Expansion rule: we now have to prove termination of $\mathcal{R}' \cup \mathcal{H}$ for *any* natural extension $\mathcal{R}'$ of $\mathcal{R}$. This is in general not possible.

To handle this issue, we use a specific, more manageable kind of extension:

Definition 8 (Oracle extension). *An* Oracle extension *of an LCSTRS* $\mathcal{L} = (\Sigma, \mathcal{R})$ *is a natural extension* $\mathcal{Q} = (\Sigma^{\mathcal{Q}}, \mathcal{R}^{\mathcal{Q}})$ *such that all rules in* $\mathcal{R}^{\mathcal{Q}} \setminus \mathcal{R}$ *have a form* $f \; v_1 \cdots v_m \to w$ *where* $f :: \sigma_1 \to \ldots \to \sigma_m \to \iota$ $(\iota \in \mathcal{S}')$, *all* v_i *ground, and each* w *is a ground semi-constructor term over* $\mathcal{Q}$ *that contains no defined symbols of* $\mathcal{R}$. *Moreover,* $\mathcal{Q}$ *is quasi-reductive and terminating.*

Thus, an Oracle extension adds functions that, given ground arguments, compute a semi-constructor result in exactly one step. Moreover, their right-hand sides do not use defined symbols in $\mathcal{R}$, thus removing any dependency. We call the rules of $\mathcal{R}^{\mathcal{Q}} \setminus \mathcal{R}$ *oracle rules*. There are typically infinitely many.

The idea is that a natural extension $\mathcal{L}'$ of $\mathcal{L}$ may be translated into an Oracle extension essentially by taking, for every defined symbol f of $\mathcal{R}' \setminus \mathcal{R}$ and ground terms $v_1, \ldots, v_m$, the normal form w of $f \; v_1 \cdots v_m$, and including $f \; v_1 \cdots v_m \to w$ as a rule. To ensure that the right-hand sides of the oracle rules do not use the defined symbols of $\mathcal{R}$, we also include copies versions of these defined symbols, and corresponding rules. The full construction is in [13, Appendix C.1].

Lemma 2. *An equation* $s \approx t \; [\varphi]$ *over a terminating, strong quasi-reductive LCSTRS is a global inductive theorem of* $\mathcal{L}$ *if for every Oracle extension* $\mathcal{Q}$ *and ground substitution* γ *over* $\mathcal{Q}$ *that respects this equation:* $s\gamma \leftrightarrow^*_{\mathcal{R}^{\mathcal{Q}}} t\gamma$.

We now update higher-order RI in such a way that we can prove global inductive theorems of terminating, strong quasi-reductive LCSTRSs. Since the only function symbols occurring in equations and rules are those in the original signature, the Simplification rule is unchanged. The Deletion and Semi-constructor rules are also the same. For the Alter and Generalize rule, we now quantify over all ground semi-constructor substitutions in the extended signature, but scenarios I–III all still apply. Hence, the only rule that changes is Expansion.

Global Expansion. Let $s \approx t \; [\varphi]$ be an equation and $p \in Pos(s)$ a position such that $s|_p = f \; s_1 \cdots s_n$ with $f \in \mathcal{D}$, $n \geq k = ar(f)$ and for all $1 \leq i \leq k$, $q \in Pos(s_i)$: if $s_i|_q$ has base type and is not a variable, then $s_i|_q$ has a form $c \; t_1 \cdots t_m$ with c a constructor symbol. If $\mathcal{R}^{\mathcal{Q}} \cup \mathcal{H} \cup \{s \to t \; [\varphi]\}$ is terminating for every Oracle extension $\mathcal{Q}$ of $\mathcal{L}$ then

$$(\mathcal{E} \uplus \{s \approx t \; [\varphi]\}, \mathcal{H}) \vdash (\mathcal{E} \cup Expd(s \approx t \; [\varphi], p), \mathcal{H} \cup \{s \to t \; [\varphi]\})$$

Compared to the original Expansion rule, the requirement on the shape of the s_i is weaker than before; this is possible due to the strong quasi-reductivity requirement. While the termination requirement *is* harder to check, this could be done either through dynamic dependency pairs [18,19] (since the oracle rules do not generate any dependency pairs), or, if certain (reasonable) restrictions on the original system are satisfied, using static dependency pairs. An automated variation of the latter approach is available for LCSTRSs [11]. We use Oracle extensions, rather than arbitrary (terminating, strong quasi-reductive) natural extensions, because the extra rules do not depend on the defined symbols of $\mathcal{L}$, which is what makes it feasible to prove termination results.

6.1 Soundness Result

We let "global rewriting induction" be the proof process obtained from the Simplification, Deletion, Semi-constructor and updated Generalization and Alter rules, along with Global Expansion. We then obtain the main result:

Theorem 2. *Let $\mathcal{L}$ be a terminating, strong quasi-reductive LCSTRS and let $\mathcal{E}$ be a set of equations. If, by global rewriting induction, $(\mathcal{E}, \emptyset) \vdash^* (\emptyset, \mathcal{H})$, for some set $\mathcal{H}$, then every equation in $\mathcal{E}$ is a global inductive theorem of $\mathcal{L}$.*

The proof of Theorem 2 ([13, Appendix C]) follows a very similar outline as the soundness proof of Theorem 1: for an arbitrary Oracle extension $\mathcal{Q}$ of $\mathcal{L}$ we use induction on $\rightarrow_{\mathcal{R}_\mathcal{Q} \cup \mathcal{H}}$ to prove that $\leftrightarrow_{\mathcal{E}} \subseteq \leftrightarrow^*_{\mathcal{R}_\mathcal{Q}}$. Using this and Lemma 2 we find that the same holds for any quasi-reductive and terminating natural extension. Compared to the proof in [13, Appendix A] the main changes are:

- In every step, we consider the Oracle extension rather than $\mathcal{L}$ directly.
- In the proof that the Global Expansion rule maintains the invariants, we use the definition of strong quasi-reductivity to show that a ground base-type term of the shape $f\ s_1 \cdots s_n$ (with $f \in \mathcal{D}$) can be reduced at the root if $s_i|_q$ has a constructor as head symbol whenever $s_i|_q$ has base type.

7 Discussion and Future Work

In this paper we proposed two variations of higher-order rewriting induction for constrained term rewriting systems. This includes two adaptations of inductive theorems, based on quasi-reductivity for higher-order LCSTRSs, and in the latter case, also on extensibility. We do not claim that the proof system is finished, but it provides a solid foundation for further work.

An obvious extension is to use rewriting induction not just to prove that equations are (global) inductive theorems, but also to prove that they are not. The mechanism for this exists [9] (in first-order RI) and we do not foresee major issues. It uses an additional flag in proof states to keep track of when we are allowed to derive non-equivalence (since the Generalize rule sometimes creates unsolvable equations). This extension does require *ground confluence* (as defined in Sect. 3.1). A new challenge is whether we can also prove that something is not a global inductive theorem, even if it *is* an inductive theorem.

A second idea is to admit extensionality; that is, to allow a rule (or: induction rule) $s\ x_1 \cdots x_n \approx t\ x_1 \cdots x_n\ [\varphi]$ to be used to reduce a term $C[s\gamma]$. If we restrict constructors to have base-type arguments, we postulate that such a deduction rule is also sound in our setting (perhaps under additional restrictions like ground confluence). It could also be used in an alternative rewriting induction approach designed for proving *extensional inductive theorems* following [3].

A very useful extension could be to weaken the termination requirement. In many cases, an obvious lemma cannot be used because the resulting induction rule would not be terminating. We postulate that, under reasonable restrictions, termination of such a rule is unnecessary if the rule is always followed by Deletion.

Related, our current definitions only support finite data. Using *coinduction* rather than *induction* may allow us to consider systems with streams, and replace the termination requirement by one of productivity.

We intend to implement rewriting induction in our tool Cora, which already supports (Oracle) termination. Fully automatic proof search could build on the ideas in [17], but will require more work on automatic strategies.

References

1. De Angelis, E., Fioravanti, F., Pettorossi, A., Proietti, M.: Relational verification through horn clause transformation. In: Rival, X. (ed.) SAS 2016. LNCS, vol. 9837, pp. 147–169. Springer, Heidelberg (2016). https://doi.org/10.1007/978-3-662-53413-7_8
2. Aoto, T., Nishida, N., Schöpf, J.: Equational theories and validity for logically constrained term rewriting. In: Proceedings of the FSCD 2024. LIPIcs, vol. 299, pp. 31:1–31:21 (2024). https://doi.org/10.4230/LIPICS.FSCD.2024.31
3. Aoto, T., Yamada, T., Chiba, Y.: Natural inductive theorems for higher-order rewriting. In: Proceedings of the RTA 2011. LIPIcs, vol. 10, pp. 107–121 (2011). https://doi.org/10.4230/LIPIcs.RTA.2011.107
4. Aoto, T., Yamada, T., Toyama, Y.: Inductive theorems for higher-order rewriting. In: van Oostrom, V. (ed.) RTA 2004. LNCS, vol. 3091, pp. 269–284. Springer, Heidelberg (2004). https://doi.org/10.1007/978-3-540-25979-4_19
5. Chiba, Y., Aoto, T., Toyama, Y.: Program transformation templates for tupling based on term rewriting. IEICE Trans. Inf. Syst. **E93-D**(5), 963–973 (2010). https://doi.org/10.1587/transinf.E93.D.963
6. Delmas, D., Miné, A.: Analysis of software patches using numerical abstract interpretation. In: Chang, B.-Y.E. (ed.) SAS 2019. LNCS, vol. 11822, pp. 225–246. Springer, Cham (2019). https://doi.org/10.1007/978-3-030-32304-2_12
7. Falke, S., Kapur, D.: Rewriting induction + linear arithmetic = decision procedure. In: Gramlich, B., Miller, D., Sattler, U. (eds.) IJCAR 2012. LNCS (LNAI), vol. 7364, pp. 241–255. Springer, Heidelberg (2012). https://doi.org/10.1007/978-3-642-31365-3_20
8. Felsing, D., Grebing, S., Klebanov, V., Rümmer, P., Ulbrich, M.: Automating regression verification. In: Proceedings of the ASE 2014, pp. 349–360. ACM (2014). https://doi.org/10.1145/2642937.2642987
9. Fuhs, C., Kop, C., Nishida, N.: Verifying procedural programs via constrained rewriting induction. ACM Trans. Comput. Logic (TOCL) **18**(2), 14:1–14:50 (2017). https://doi.org/10.1145/3060143

10. Godlin, B., Strichman, O.: Inference rules for proving the equivalence of recursive procedures. Acta Informatica **45**(6), 403–439 (2008). https://doi.org/10.1007/s00236-008-0075-2
11. Guo, L., Hagens, K., Kop, C., Vale, D.: Higher-order constrained dependency pairs for (universal) computability. In: Proceedings of the MFCS 2024 (2024, to Appear). https://doi.org/10.48550/arXiv.2406.19379
12. Guo, L., Kop, C.: Higher-order LCTRSs and their termination. In: Weirich, S. (ed.) ESOP 2024. LNCS, vol. 14577, pp. 331–357. Springer, Cham (2024). https://doi.org/10.1007/978-3-031-57267-8_13
13. Hagens, K., Kop, C.: Rewriting induction for higher-order constrained term rewriting systems (2024). https://www.cs.ru.nl/~cynthiakop/lopstr24.pdf; pre-editing copy of this paper including appendix
14. Huth, M., Ryan, M.D.: Modelling and Reasoning About Systems. Cambridge University Press (2004). https://doi.org/10.1017/CBO9780511810275
15. Koike, H., Toyama, Y.: Inductionless induction and rewriting induction. Comput. Softw. **17**(6), 509–520 (2000). https://doi.org/10.11309/jssst.17.509. (in Japanese)
16. Kop, C., Nishida, N.: Term rewriting with logical constraints. In: Fontaine, P., Ringeissen, C., Schmidt, R.A. (eds.) FroCoS 2013. LNCS (LNAI), vol. 8152, pp. 343–358. Springer, Heidelberg (2013). https://doi.org/10.1007/978-3-642-40885-4_24
17. Kop, C., Nishida, N.: Constrained term rewriting tooL. In: Davis, M., Fehnker, A., McIver, A., Voronkov, A. (eds.) LPAR 2015. LNCS, vol. 9450, pp. 549–557. Springer, Heidelberg (2015). https://doi.org/10.1007/978-3-662-48899-7_38
18. Kop, C., Raamsdonk, F.v.: Dynamic dependency pairs for algebraic functional systems. LMCS **8**(2), 1–51 (2012). https://doi.org/10.2168/LMCS-8(2:10)2012
19. Kusakari, K.: On proving termination of term rewriting systems with higher-order variables. IPSJ Trans. Program. **42**(7), 35–45 (2001). http://id.nii.ac.jp/1001/00016864/
20. Nakabayashi, N., Nishida, N., Kusakari, K., Sakabe, T., Sakai, M.: Lemma generation method in rewriting induction for constrained term rewriting systems. Comput. Softw. **28**(1), 173–189 (2010). https://www.trs.css.i.nagoya-u.ac.jp/crisys/nakabayashi10.pdf
21. Nishida, N., Kojima, M., Kato, T.: On transforming imperative programs into logically constrained term rewrite systems via injective functions from configurations to terms. In: Proceedings of the WPTE 2022 (2022). https://wvvw.easychair.org/publications/preprint_download/DbM2
22. Partush, N., Yahav, E.: Abstract semantic differencing via speculative correlation. In: Proceedings of the OOPSLA 2014, pp. 811–828. ACM (2014). https://doi.org/10.1145/2660193.2660245
23. Reddy, U.S.: Term rewriting induction. In: Stickel, M.E. (ed.) CADE 1990. LNCS, vol. 449, pp. 162–177. Springer, Heidelberg (1990). https://doi.org/10.1007/3-540-52885-7_86

Introducing Quantification into a Hierarchical Graph Rewriting Language

Haruto Mishina[✉] and Kazunori Ueda[ⁱᵈ]

Waseda University, Tokyo 169-8555, Japan
{mishina,ueda}@ueda.info.waseda.ac.jp

Abstract. LMNtal is a programming and modeling language based on hierarchical graph rewriting that uses logical variables to represent connectivity and membranes to represent hierarchy. On the theoretical side, it allows logical interpretation based on intuitionistic linear logic; on the practical side, its full-fledged implementation supports a graph-based parallel model checker and has been used to model diverse applications including various computational models. This paper discuss how we extend LMNtal to QLMNtal (LMNtal with Quantification) to further enhance the usefulness of hierarchical graph rewriting for high-level modeling by introducing quantifiers into rewriting as well as matching. Those quantifiers allows us to express universal quantification, cardinality and non-existence in an integrated manner. Unlike other attempts to introduce quantifiers into graph rewriting, QLMNtal has term-based syntax, whose semantics is smoothly integrated into the small-step semantics of the base language LMNtal. The proposed constructs allow combined and nested use of quantifiers within individual rewrite rules.

Keywords: quantification · graph rewriting language · language design

1 Introduction

1.1 Background and Objectives

Graph rewriting languages express computation as successive and possibly concurrent updating of graph structures. Graphs are highly general means for modeling diverse structures in the real world. They also generalize algebraic data types such as trees and lists in programming, and designing high-level programming languages for handling graphs safely and clearly is an important challenge.

A specific challenge towards an expressive graph rewriting language, both in theory and in practice, is the ability to handle the "quantities"—broadly construed—of graph elements. They include existential and universal quantification, non-existence, and cardinality. Existing tools for graph rewriting have proposed various methods including the combination of rewrite rules and the introduction of control structures (see Sect. 7 for related work), but how to provide those features in the standard setting of programming languages, i.e., inductively defined syntax and structural operational semantics, has been an open question.

LMNtal [18] is a programming and modeling language that handles hierarchical undirected graphs. Its distinguished features include:

- It is based on inductively defined syntax and small-step semantics, with structural congruence (as with process calculi) to characterize its data structures which are often referred to as *port graphs* (as opposed to standard graphs in graph theory). Links (which are unlabeled edges) interconnect two ports of atoms (which are labeled nodes).
- It features *membranes* as a means of hierarchization (of nodes), representation of first-class multisets, and localization of rewriting.
- The de facto standard runtime system, SLIM [11], provides an LTL model checker as well as (don't-care non-deterministic) concurrent rewriting.

Historically, LMNtal was born as an attempt to unify constraint-based concurrency [19] and Constraint Handling Rules [7], the two notable extensions to (concurrent) logic programming. This attempt then resulted in the unified handling of symbols representing programs and those representing data, which are interconnected by logical variables that could be interpreted as graph edges. LMNtal programs allow interpretation based on intuitionistic linear logic [18].

Unlike many other tools for graph rewriting that provide control structures (like sequencing), computation of LMNtal is basically controlled by subgraph matching which can be regarded essentially as existential quantification and which acts as a *synchronization mechanism*. The implementation of LMNtal supports constructs for handling a limited form of negative information including the non-equality checking of node labels, but constructs for handling various sorts of quantities has been an open issue, which is the subject of this paper.

1.2 Contributions

The objective of this work is to introduce constructs for specifying quantities into a graph rewriting language and to formulate the syntax and semantics of the extended language. The contributions of our approach are summarized below.

Introducing Cardinality, Non-existence and Universal Quantification. Pattern matching in LMNtal is to find a subgraph which is structurally congruent to the LHS (left-hand side) of a rewrite rule and can be viewed as existential quantification. We first introduce constructs for specifying the quantities of graph elements. *Cardinality quantification* specifies the minimum and the maximum numbers of specified subgraph occurrences in the target graph and rewrites them in a single step. *Non-existence quantification* ensures the non-existence of a specified subgraph in the target graph. *Universal quantification* finds all (non-overlapping) subgraphs and rewrite them in a single step.

Relating Different Quantifications. One of the technical challenges is to introduce the above quantification systematically in such a way that different constructs are related to each other. We identify cardinality and non-existence as two basic quantification and universal quantification as a derived construct. Also, we show that *labelling* of quantification plays a key role in controlling the (in)dependence of quantification.

Abstract State Space. Although various forms of quantification could be programmed using multiple rewrite rules and/or built-in control structures, they complicate the state space of model checking. Expressive quantification within individual rules will keep the state space at the right level of abstraction.

Combination and Nesting of Quantification. Thanks to the approach based on structural operational semantics, the semantics of combined and nested use of quantification can be given in a systematic manner. We show that the proposed framework can express typical use cases of nested quantification (Sect. 6).

The rest of this paper is organized as follows. Section 2 describes LMNtal, the base language of this study. Section 3 introduces QLMNtal's functionalities using simple examples. The syntax of QLMNtal is given Sect. 4 and the operational semantics in Sect. 5. Section 6 describes two examples showing the expressive power of the constructs. Section 7 discusses related work and our future work.

2 LMNtal

This section outlines (a fragment of) LMNtal [18] as the basis of QLMNtal. Due to its background as a concurrent language, its main construct is called a *process*, whose structure formed by atoms, links and membranes will evolve by another construct called *(rewrite) rules*. The original definition [18] handles rewrite rules as part of a process so that they can be placed in membranes to express local rewriting inside them, but here we handle only global rewrite rules and separate them from processes as in [21].

2.1 Syntax of LMNtal

An LMNtal program discussed in this paper is the pair of a *process* and a set of *(rewrite) rules* defined as in Fig. 1, where p stands for a name starting with a lowercase letter or a number, and X_i stands for a name starting with an uppercase letter. Processes and rules are subject to the Link Condition:

Definition 1 (Link Condition). *Each link name can occur at most twice in a process, and each link name in a rule must occur exactly twice in the rule.* □

An m-ary *atom* $p(X_1, \ldots, X_m)$ represents a node (labelled by the name p) of a port graph with m totally ordered ports to which the links $X_1, \ldots, X_m$ are connected.

A *link* represents an undirected edge. A link occurring twice in a process is called a *local link*, while a link occurring once in a process is called a *free link*.

Parallel composition glues two processes to build a larger process. Note that, if each of P_1 and P_2 has a free link with the same name, it becomes a local link in (P_1, P_2). A reader may notice that the Link Condition may not always allow us

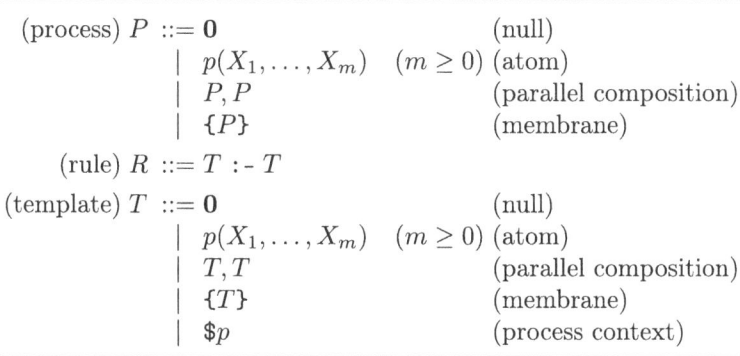

Fig. 1. Syntax of LMNtal

to form (P_1, P_2) from two graphs P_1 and P_2 each satisfying the Link Condition. How LMNtal handles this point will be discussed in Sect. 2.2.

A special binary atom, called a *connector* =(X, Y), also written as $X = Y$, fuses (or glues) two links X and Y.

A *membrane* acts as a means to group atoms and other membranes and forms a hierarchical structure. A link may cross membranes to connect two atoms belonging to two different places of the hierarchical membrane structure.

A *(rewrite) rule* consists of two templates, called the *head* (the left-hand side) and the *body* (the right-hand side) of the rule, respectively.

A *process context* p in the head of a rule acts as the *wildcard* of a membrane (which contains the p at its top level; i.e., not inside another membrane contained by the aforementioned membrane) in pattern matching. A process context p with the name p may appear (i) at most once in the head of a rule as the sole process context of some membrane (i.e., a membrane can contain at most one process context at its top level), and (ii) if it does, it must appear exactly once in the body of the rule.

Note that a process context in (full) LMNtal [18] may optionally specify the list of free links of the process context, but we omit the description of this feature because it is not used in the examples of the present paper. We just note that a process context of the form p may match a process with any number of free links.

A term representing a process is subject to Structural Congruence defined in Sect. 2.2, which then stands for an undirected multigraph, i.e., a graph that allows multi-edges and self-loops.

Figure 2 shows a simple process and its visualization as a port graph, in which the local link name Z is insignificant and not shown. Unlike LMNtal atoms, nodes of standard graphs in graph theory come with unordered edges and the graphs are often directed, as in Fig. 3. Those standard graphs can be represented in LMNtal as in Fig. 4, where each node is represented by a membrane containing an atom representing the node label, and each edge is represented by a link connecting a unary s(ource) atom and a unary t(arget) atom.

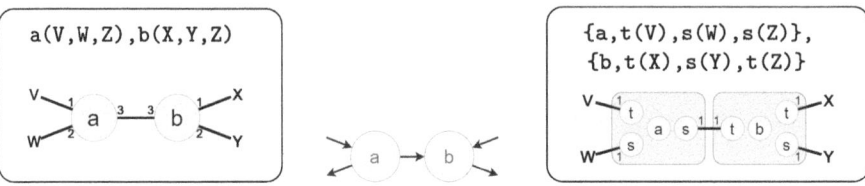

Fig. 2. An LMNtal process and its visualization   **Fig. 3.** A standard directed graph   **Fig. 4.** An LMNtal representation of Fig. 3

(E1)	$0, P$	$\equiv$	P
(E2)	P, Q	$\equiv$	Q, P
(E3)	$P, (Q, R)$	$\equiv$	$(P, Q), R$
(E4)	P	$\equiv$	$P[Y/X]$ if X is a local link of P
(E5)	$P \equiv P'$	$\Rightarrow$	$P, Q \equiv P', Q$
(E6)	$P \equiv P'$	$\Rightarrow$	$\{P\} \equiv \{P'\}$
(E7)	$X = X$	$\equiv$	$\mathbf{0}$
(E8)	$X = Y$	$\equiv$	$Y = X$
(E9)	$X = Y, P$	$\equiv$	$P[Y/X]$ if P is an atom and X is a free link of P
(E10)	$\{X = Y, P\}$	$\equiv$	$X = Y, \{P\}$ if exactly one of X and Y is a free link of P

Fig. 5. Structural congruence on LMNtal processes

2.2 Semantics of LMNtal

The semantics of LMNtal consists of *structural congruence* and a *reduction relation*.

Structural Congruence. The syntax defined above does not yet characterize LMNtal graphs because the port graph of Fig. 2 allows other syntactic representations, e.g., b(X,Y,U),a(V,W,U).

Figure 5 defines an equivalence relation, called *structural congruence*, to absorb the syntactic variations. $P[Y/X]$ stands for the renaming of link X in process P to Y. The rules apply only when each process satisfies the Link Condition. The rules are as in [18] and readers familiar with LMNtal may skip the details.

(E1)–(E3) characterize atoms as multisets. (E5) and (E6) are structural rules to make $\equiv$ a congruence. (E4) is α-conversion of local link names, where Y must be a fresh link because of the Link Condition. (E7)–(E10) are rules for connectors. They play key roles in many LMNtal programs, but in this paper we just refer the readers to [18] because connectors do not appear in this paper.

(R1) $\dfrac{P \xrightarrow{R} P'}{P,Q \xrightarrow{R} P',Q}$ (R3) $\dfrac{Q \equiv P \quad P \xrightarrow{R} P' \quad P' \equiv Q'}{Q \xrightarrow{R} Q'}$

(R4) $\{X = Y, P\} \xrightarrow{R} X = Y, \{P\}$ if X and Y are free links of $\{X = Y, P\}$

(R5) $X = Y, \{P\} \xrightarrow{R} \{X = Y, P\}$ if X and Y are free links of P

(R6) $T\theta \xrightarrow{T:-U} U\theta$

Fig. 6. Reduction relation on LMNtal processes

For convenience, the following term notation is introduced.

Definition 2 (Term Notation). *We allow*

$p(X_1,\ldots,X_{k-1},L,X_{k+1},\ldots,X_m), q(Y_1,\ldots,Y_{n-1},L)$ $(1 \leq k \leq m, 1 \leq n)$,

to be written as $p(X_1,\ldots,X_{k-1},q(Y_1,\ldots,Y_{n-1}),X_{k+1},\ldots,X_m)$. □

For instance, `p(X,Y),a(X),b(Y)` can be written also as `p(a,Y),b(Y)` and then as `p(a,b)`, which is much more concise and looks like a standard term.

Reduction Relation. The reduction relation by a rule R, denoted $\xrightarrow{R}$, is the minimum binary relation satisfying the rules in Fig. 6.

(R1) and (R3) are structural rules.[1] (R4) and (R5) handle interaction between connectors and membranes, and we leave the details to [18]. The key rule, (R6), handles rewriting by a rule. The substitution θ handles how the process contexts in T are instantiated; it is a finite set $\{P_1/\$p_1, P_2/\$p_2, \ldots\}$ of substitution elements of the form $P_i/\$p_i$ that assigns P_i to each process context $\$p_i$ occurring in T. Finally, given a set S of rewrite rules, we write $P \xrightarrow{S} P'$ if $\exists R \in S, P \xrightarrow{R} P'$. We also write $P \xrightarrow{T:-U} \!\!\!\!\!/\,\,$ if $\neg \exists P'(P \xrightarrow{T:-U} P')$. Since the body U of the rule is irrelevant to this non-reducibility property, we may also write it as $P \xrightarrow{T:-} \!\!\!\!\!/\,\,$, meaning that P is not reducible by a rule with the head T.

3 Introductory Examples of QLMNtal

Before formally defining the syntax and the semantics of QLMNtal, we introduce its features by means of simple examples. Further examples will be given in Sect. 6, and detailed explanation of how the examples work is given in a full version of this paper at arXiv.org.

---

[1] (R2) for handling autonomous process evolution inside a membrane is irrelevant in the present formulation (which handles global rewrite rules only) and is omitted. In the present setting, rewriting of the contents of a membrane must be specified as global rewrite rules that explicitly mention the membrane.

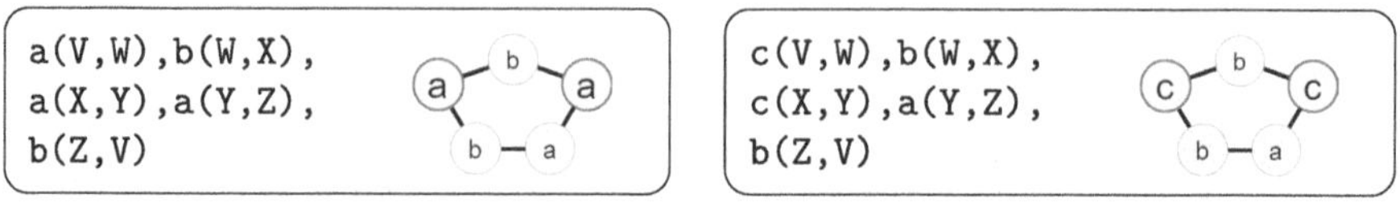

```
a(V,W),b(W,X),
a(X,Y),a(Y,Z),
b(Z,V)
```

(a) Process before rewriting

```
c(V,W),b(W,X),
c(X,Y),a(Y,Z),
b(Z,V)
```

(b) Process after rewriting

Fig. 7. Rewriting with the rule <1,3>a(X,Y) :- <1,3>c(X,Y)

3.1 Cardinality Quantification

In QLMNtal, cardinality quantification tells how many copies of the specified process may appear within the process to be rewritten, and the matching is performed in a non-greedy manner in the sense that a cardinality-quantified process in the head can match (or 'grab') *any* number (within the specified range) of processes. Then the corresponding cardinality-quantified process in the body, if any, represents the same number of processes as the number of processes grabbed by the cardinality-quantified process in the head. For example, the QLMNtal rule

<1,3>a(X,Y) :- <1,3>c(X,Y)

says "rewrite one to three copies of (binary) a to (binary) c".

Two notes are appropriate here:

1. The semantics in Sect. 5 appropriately renames links of the three copies.
2. The two quantifiers are interdependent; as will be explained in Sect. 5, they are given the same 'empty' label indicating the dependence between quantifiers.

Figure 7 shows an example of rewriting two a's to two c's. Since the rewriting is nondeterministic, it is also possible that one or three a's are rewritten.

A cardinality quantifier <1,3> specifies the range of the number of matching processes. When the range is a single number, one may write <2> instead of <2,2>. When one allows any cardinality, the quantifier can be written as <?>.

3.2 Non-existence Quantification

In QLMNtal, non-existence quantification is the basic construct for expressing so-called negative application conditions (NACs). For example, the rule

<^>{a,a,$p} :- ok

says "if no membrane contains two or more a's, generate an ok," where <^> stands for a non-existence quantifier. Since non-existence quantification is used to express matching conditions, it is meaningless to use non-existence quantification explicitly in the body, and the Link Condition in Sect. 2.1 does not apply to links occuring under non-existence quantification. Figure 8 shows an example of generating an ok because a membrane containing two or more a's does not exist.

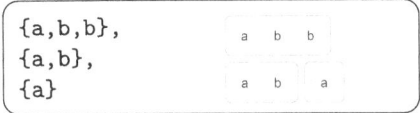

(a) Process before rewriting   (b) Process after rewriting

Fig. 8. Rewriting with the rule <^>{a,a,$p} :- ok

3.3 Universal Quantification

Universal quantification in QLMNtal implies greedy matching, that is, a universally quantified process in the head grabs the largest possible number of processes, and the corresponding universally-quantified process in the body represents the same number of processes as the number of processes grabbed in the head. For example, the rule

<*>a(X,Y) :- <*>c(X,Y)

says "rewrite all binary a's to binary c's." This rule rewrites

a(V,W),b(W,X),a(X,Y),a(Y,Z),b(Z,V)   (Fig. 7(a))

to c(V,W),b(W,X),c(X,Y),c(Y,Z),b(Z,V) .

It is important to note that the quantifier '<*>' is provided for notational convenience in the sense that the same rule can be written also as

<?>a(X,Y),<^>a(X0,Y0) :- <?>c(X,Y),<^>c(X0,Y0) ,

where the non-existence quantification tells that the <?>a(X,Y) should grab the largest possible number of the binary a's. That is, if <?>a(X,Y) does not grab all the a's, the pattern matching fails due to <^>a(X0,Y0).

QLMNtal provides another (derived) universal quantifier, '<+>', corresponding to $\forall^{>0}$ in logic, which is useful when one wishes to avoid vacuous rule application when there are zero instances of the specified process.

3.4 Combined Use of Quantifiers

QLMNtal allows quantifiers to be used in combination. For example, the rule

<*>{a,b,$p1},<^>{b,$p2} :- <*>{a,b,$p1},ok

says "generate an ok if all membranes containing b contain a as well." This rule rewrites {a,b,b},{a,b},{a} (Fig. 8(a)) to {a,b,b},{a,b},{a},ok.

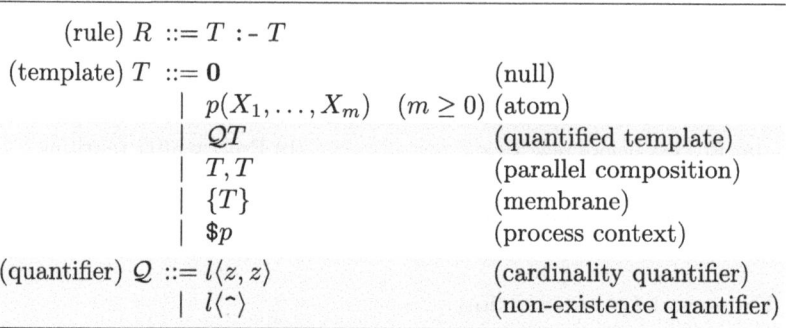

Fig. 9. Syntax of QLMNtal

4 Syntax of QLMNtal

We first introduce the following syntactic convention.

Definition 3 (vector notation). *For some syntactic entity E, $\vec{E}$ stands for a sequence $E_1, \ldots, E_n$ for some $n(\geq 0)$. When we wish to mention the indices explicitly, $E_1, \ldots, E_n$ will also be denoted as $\vec{E_i}^i$.* □

The syntax of QLMNtal is shown in Fig. 9, which adds to the syntax of LMNtal (Fig. 1) notations for the cardinality quantifier and the non-existence quantifier for process templates. Note that the syntax of processes remains exactly the same and is omitted. In the syntax of the quantifier $\mathcal{Q}$,

- l stands for a *quantifier label* whose concrete syntax is either (i) an empty name or (ii) a name starting with a capital letter;
- z stands for an integer or ∞, where we assume $\infty - 1 = \infty$ and $\forall n(n < \infty)$;
- $\mathcal{Q}$ binds tighter than ',' for parallel composition; for example, <3,5>a,a,a stands for (<3,5>a),a,a and not <3,5>(a,a,a).

$\mathcal{Q}T$ is a construct encompassing the *cardinality quantification template* $l\langle z_1, z_2\rangle T$ and the *non-existence quantification template* $l\langle \char`\~\rangle T$. In a quantified template $\mathcal{Q}T$, $\mathcal{Q}$ is called the *(labeled) quantifier* and T is called the *quantified part*. Quantified templates can be nested as in $\mathcal{Q}_0 \mathcal{Q}_1 \mathcal{Q}_2 T$. The quantifier of a quantified template that is not inside another quantifier is called the *outermost quantifier*. In the sequel, $\mathcal{Q}_0$ (a quantifier with the suffix 0) denotes an outermost quantifier.

A quantified template must observe the following syntactic conditions.

Definition 4 (Syntactic Conditions for Quantified Templates). *Links or process contexts appearing in a quantified template may appear only in (the same or different) quantified templates with the same quantifier.* □

An important feature of QLMNtal is the *labeling of quantifiers*. We often wish to *jointly* quantify entities distributed over isolated places of the membrane structure and entities on both sides (head and body) of a rule. However, it is difficult to achieve this in textual syntax. Accordingly, we introduce *labels* to associate physically distributed but logically related quantifiers. For instance, in `L<^>a(X),{L<^>b(X),$p}`, a(X) and b(X) are quantified collectively.

Cardinality Quantification Template. The cardinality quantification template $l\langle z_1, z_2\rangle T$ stands for either of $z_1, \ldots, z_2$ copies of T by the extended structural congruence rule to be given in Sect. 5.1, where the links of the T's are appropriately renamed by (E4). The l is a (possibly empty) label name used to distinguish between independent cardinality quantifiers.

Note that a cardinality quantification template does not itself enforce a greedy match. For example, given a process `a,a,a,a`, the template `<1,3>a` can match either one, two, or three of the four a's. Table 1 summarizes the notations of cardinality quantification templates.

Table 1. Notations of Cardinality Quantification Templates

Notation	Shorthand	Meaning	Example
$l\langle 0, \infty\rangle T$	$l\langle ?\rangle T$	any number of T's	`<?>a(X)`
$l\langle z, \infty\rangle T$	$l\langle z,\rangle T$	not less than z T's	`<3,>a(X)`
$l\langle z_1, z_2\rangle T$	$l\langle z_1, z_2\rangle T$	not less than z_1 and not more than z_2 T's	`<3,5>a(X)`
$l\langle z, z\rangle T$	$l\langle z\rangle T$	exactly z T's	`<3>a(X)`

The outermost cardinality quantifier, denoted as $l\langle z_1, z_2\rangle_0$, plays an important role in the structural congruence and the reduction relation.

Non-existence Quantification Template. $l\langle \hat{}\,\rangle T$ is a template that indicates the non-existence of a process, and the pattern matching fails if there is a process that matches the template T. The l is a (possibly empty) label name used to distinguish between unrelated non-existence quantifiers. (Non-existence and cardinality quantifiers are never related to each other.) The outermost non-existence quantifier, denoted as $l\langle \hat{}\,\rangle_0$, plays an important role in the reduction relation.

We note that $l\langle \hat{}\,\rangle T$ appearing in the body may be omitted as already explained in Sect. 3.2. For example, `<^>a(X) :- <^>b(X)` can be written also as `<^>a(X) :- .` This is because the non-existence quantification in the body of a rule does not play any role (other than to satisfy the occurrence conditions of free links and process contexts) in the semantics defined in Sect. 5. This abbreviation may omit free links and process contexts in the body, but we do not consider it as violating the syntactic conditions of rules.

4.1 Representation of Universal Quantification

As explained in Sect. 3.3, universal quantification of QLMNtal can be represented by combining cardinality quantification and non-existence quantification.
To enable concise description, we provide shorthands '$\langle * \rangle$' and '$\langle + \rangle$' given in Table 2. In order to satisfy the Link Condition and the occurrence conditions of process contexts (Sect. 2.1), T' stands for a process template obtained from T by renaming links and process contexts using fresh names. Likewise, each label name l is renamed to its 'primed' version l' to distinguish it from the label name of the cardinality quantification template and the non-existence quantification template. This way, if the same label name l is used to write $l\langle * \rangle T_1, l\langle \char`\^ \rangle T_2$ as in the example of Sect. 3.4, the label names do not conflict because they are translated to $(l'\langle 0, \infty \rangle T_1, l'\langle \char`\^ \rangle T_1, l\langle \char`\^ \rangle T_2)$.

Table 2. Notations for expressing Universal Quantification

Notation	Shorthand	Meaning	Example
$l'\langle 0, \infty \rangle T, l'\langle \char`\^ \rangle T'$	$l\langle * \rangle T$	maximum number of T's	<*>a(X)
$l'\langle 1, \infty \rangle T, l'\langle \char`\^ \rangle T'$	$l\langle + \rangle T$	maximum number (≥ 1) of T's	<+>a(X)

5 Semantics of QLMNtal

5.1 Structural Congruence of QLMNtal

One of the key ideas towards using quantified rewrite rules is to give a law for "unrolling" cardinality quantification. This is achieved by an additional equivalence relation (EQ) shown below:

$$(\text{EQ}) \quad T \mathbin{:\text{-}} U \equiv (T \mathbin{:\text{-}} U)\left[\overline{(l\langle z_1-1, z_2-1\rangle_0 T_i, T_i') / l\langle z_1, z_2\rangle_0 \overrightarrow{T_i}^i}\right]$$

where $\overrightarrow{l\langle z_1, z_2\rangle_0 T_i}^i$ stands for all the templates quantified with $l\langle z_1, z_2\rangle_0$, $\overrightarrow{T_i'}^i$ stands for templates in which all the links, process contexts, and labels appearing in $\overrightarrow{T_i}^i$ are replaced with fresh names not appearing in $T \mathbin{:\text{-}} U$

(EQ) states that all T_i's quantified by $l\langle z_1, z_2\rangle_0$ (i.e., the outermost occurrences of the quantifier $l\langle z_1, z_2\rangle$) in a rule can be expanded to $\bigl(l\langle (z_1-1), (z_2-1)\rangle_0 T_i, T_i'\bigr)$. That is, the outermost cardinality quantification template $l\langle z_1, z_2\rangle_0 T_i$ can be unrolled by replicating T_i and decrementing z_1 and z_2. In order to maintain consistency, all the outermost occurrences of the cardinality quantifier with the same l, z_1, z_2 must be replicated at the same time. Furthermore, all the link names, process context names and label names inside the replicated T_i must be renamed with fresh names.

For example, the following rules are structurally congruent, where the underline indicates the process template replicated by (EQ).

$$\begin{array}{l}\texttt{<0,}\infty\texttt{>\{a(X),<3,3>b,\$c\} :- <0,}\infty\texttt{>\{a(X),\$c\}}\\ \stackrel{(EQ)}{\equiv} \texttt{<-1,}\infty\texttt{>\{a(X),<3,3>b,\$c\}, \underline{\{a(X1),M<3,3>b,\$c1\}}}\\ \texttt{:- <-1,}\infty\texttt{>\{a(X),\$c\},\underline{\{a(X1),\$c1\}}}\\ \stackrel{(EQ)}{\equiv} \texttt{<-1,}\infty\texttt{>\{a(X),<3,3>b,\$c\}, \{a(X1),M<0,0>b,\underline{b,b,b},\$c1\}}\\ \texttt{:- <-1,}\infty\texttt{>\{a(X),\$c\},\{a(X1),\$c1\}}\end{array}$$

Note that (EQ) itself allows $l\langle z_1, z_2\rangle_0 T_i$ to be unrolled more than z_1 times, resulting in the '-1' in the above example. The right number of unrolling is ensured by (*CardCond*) shown in Sect. 5.2.

5.2 Reduction Relation of QLMNtal

The reduction relation of QLMNtal is similar to that of LMNtal (Fig. 6) but with changes to handle cardinality and non-existence. While the reduction of LMNtal processes proceeds solely based on the *existence* of subprocesses that can be checked locally, checking of *non-existence* requires access to the entire process. In the terminology of programming language theory, this means we must (partly) switch from inductive small-step semantics with structural rules (w.r.t. parallel composition) to so-called *contextual semantics*. In mainstream programming languages, contextual semantics is mainly used for specifying the order of evaluation, but here we use it to access global information.

Specifically, in QLMNtal, the structural rule (R1) for parallel composition and the main rule (R6) are deleted from the rules of Fig. 6. They are replaced with the new rule (RQ) and the accompanying structural rule (R3′) below:

$$(\text{RQ}) \quad \frac{\forall l\langle z_1, z_2\rangle_0 (z_1 \leq 0 \wedge z_2 \geq 0) \ \wedge\ \forall l\langle \frown\rangle_0 \bigl(cxt(\{T, \$\gamma\})\theta \xrightarrow{neg(l,\{T,\$\gamma\}):-}{\not\longrightarrow}\bigr)}{(simp(T), \$\gamma)\theta \xrightarrow{T:-U} (simp(U), \$\gamma)\theta}$$

$$(\text{R3}') \quad \frac{R \equiv R' \quad P \xrightarrow{R} P'}{P \xrightarrow{R'} P'}$$

(RQ) consists of (i) a conclusion representing the rewriting and (ii) a premise under which the rewriting takes place. The $\$\gamma$, called a *global context*, is a special process context that represents the "context" of the head of the rule, i.e., graph elements not affected by rewriting. We could think of the existence of an *(implicit) global membrane* which contains all processes and in which all rewritings take place; $\$\gamma$ could be thought of the process context of the global membrane. Now the role of the substitution θ in (RQ) is extended to instantiate $\$\gamma$ as well as the other process contexts in the head of the rule.

We will now explain how (RQ) works step-by-step.

Rewriting of a Process. Let (RQ′) be the conclusion of (RQ), i.e.,

$$(simp(T), \$\gamma)\theta \xrightarrow{T:-U} (simp(U), \$\gamma)\theta \ .$$

The auxiliary function *simp* removes quantification templates from the head and the body of a rule, whose definition is given in Fig. 10.

$$simp(0) \stackrel{\text{def}}{=} 0 \qquad\qquad simp(p(\vec{X})) \stackrel{\text{def}}{=} p(\vec{X})$$
$$simp(\{T\}) \stackrel{\text{def}}{=} \{simp(T)\} \qquad\qquad simp(\mathcal{Q}T) \stackrel{\text{def}}{=} 0$$
$$simp((T_1, T_2)) \stackrel{\text{def}}{=} (simp(T_1), simp(T_2)) \qquad simp(\$p) \stackrel{\text{def}}{=} \$p$$

Fig. 10. *simp* used in (RQ) and (RQ′)

For example, let R = <2,2>a :- <2,2>b, which says "replacing two a's with two b's", can be transformed to R' = <0,0>a,a,a :- <0,0>b,b,b by using (EQ) twice and then be applied using (RQ′) as follows:

$$\text{a,a,a} \ \big(\equiv (simp(\text{<0,0>a,a,a}), \$\gamma)[\text{a}/\$\gamma] \ \big)$$
$$\xrightarrow{R'} \text{b,b,a} \ \big(\equiv (simp(\text{<0,0>b,b,b}), \$\gamma)[\text{a}/\$\gamma] \ \big).$$

Condition on the Outermost Cardinality Quantifier. The condition $z_1 \leq 0 \land z_2 \geq 0$ in (RQ), hereafter referred to as (*CardCond*), is a condition that all outermost quantifiers $l\langle z_1, z_2\rangle$ (the suffix '$_0$' in (RQ) standing for 'outermost') in the rule must satisfy when applying the rule. (*CardCond*) confirms that the quantified template T of $l\langle z_1, z_2\rangle T$ is replicated between z_1 and z_2 times by (EQ). For example, the rule <2,2>a :- <2,2>b described above is equivalent to <1,1>a,a :- <1,1>b,b, but this rule cannot be used for rewriting because it does not satisfy (*CardCond*).

Condition on the Outermost Non-existence Quantifier. The following (*NegCond*), the second premise of (RQ), states what all outermost non-existence quantifiers $l\langle \hat{\ }\rangle$ in the rule must satisfy when used for rewriting:

$$(NegCond) \quad \forall l\langle \hat{\ }\rangle_0 \big(cxt(\{T, \$\gamma\})\theta \xrightarrow{\ \ neg(l, \{T, \$\gamma\}):-\ \ }{\not\ \ } \big) \ ,$$

where *cxt* and *neg* are defined in Fig. 11 and Fig. 12, respectively.

$$cxt(\mathbf{0}) \stackrel{\text{def}}{=} \mathbf{0}$$
$$cxt(p(\overrightarrow{X})) \stackrel{\text{def}}{=} \mathbf{0}$$
$$cxt(\{T\}) \stackrel{\text{def}}{=} \{cxt(T)\}$$
$$cxt(\mathcal{Q}T) \stackrel{\text{def}}{=} \mathbf{0}$$
$$cxt((T_1, T_2)) \stackrel{\text{def}}{=} (cxt(T_1), cxt(T_2))$$
$$cxt(\$p) \stackrel{\text{def}}{=} \$p, p^{id}()$$

Fig. 11. cxt used in (RQ)

$$neg(l, \mathbf{0}) \stackrel{\text{def}}{=} \mathbf{0}$$
$$neg(l, p(\overrightarrow{X})) \stackrel{\text{def}}{=} \mathbf{0}$$
$$neg(l, \{T\}) \stackrel{\text{def}}{=} \{neg(l, T)\}$$
$$neg(l, \mathcal{Q}T) \stackrel{\text{def}}{=} T \qquad (\mathcal{Q} = l\langle\hat{\;}\rangle)$$
$$neg(l, \mathcal{Q}T) \stackrel{\text{def}}{=} \mathbf{0} \qquad (\mathcal{Q} \neq l\langle\hat{\;}\rangle)$$
$$neg(l, (T_1, T_2)) \stackrel{\text{def}}{=} (neg(l, T_1), neg(l, T_2))$$
$$neg(l, \$p) \stackrel{\text{def}}{=} \$p^{ct}, p^{id}()$$

Fig. 12. neg used in (RQ)

(*NegCond*) confirms that there is no process matching the non-existence quantifier in each of the "places" of the membrane hierarchy.

Intuitively, $cxt(\{T, \$\gamma\})\theta$ represents the 'target' of the non-existence check, where the target means all elements whose existence is not explicitly mentioned in the head (but is represented by process contexts including $\$\gamma$). $neg(l, \{T, \$\gamma\})$ extracts all the quantified parts of $l\langle\hat{\;}\rangle$. Pattern matching is performed between these two, $cxt(\{T, \$\gamma\})\theta$ and $neg(l, \{T, \$\gamma\})$, and its success/failure indicates that the non-existence condition prefixed with $l\langle\hat{\;}\rangle$ is unsatisfied/satisfied, respectively.

The function cxt (Fig. 11) removes all atoms and quantified templates from the process template while maintaining the original membrane hierarchy. Thus, $cxt(\{T, \$\gamma\})$ represents the extracted process contexts in the hierarchy. By instantiating those process contexts by θ, the context of the head of the rule will be extracted. It also generates a unique *id atom* in each membrane containing a process context. The purpose of the id atoms is to identify each child membrane when there are multiple child membranes within the same parent membrane.

The function neg (Fig. 12) removes all atoms and cardinality quantifications from the process template while maintaining the original membrane hierarchy, and removes the top-level non-existence quantifiers with the specified label l. Thus, $neg(l, \{T, \$\gamma\})$ represents the elements that should *not* exist in each place of the membrane structure. In addition, neg renames process contexts with fresh names (denoted as $\$p^{ct}$ in Fig. 12) (to avoid name crash with existing ones) and generates a unique id atom in each membrane containing a process context, e.g.,

$$neg(\texttt{M}, \{\{\texttt{M\^{}a(X)}, \$p\}, \texttt{M\^{}a(X)}, \$\gamma\}) = \{\{\texttt{a(X)}, \$p^{ct}, p^{id}\}, \texttt{a(X)}, \$\gamma^{ct}, \gamma^{id}\} \;.$$

For example, let

$$R = \{\texttt{M<\^{}>a(X)}, \$\texttt{p}\}, \; \texttt{M<\^{}>a(X)} \; \texttt{:-} \; \{\$\texttt{p}\}, \; \texttt{ok},$$

stating "generate an ok when there are no a's connected across any membrane."
If (RQ') is applied to R, R could perform the following rewriting, where $\theta = \big[(\texttt{a(X)},\texttt{b})/\$\texttt{p},\ (\texttt{a(X)},\texttt{c})/\$\gamma\big]$:

$$\{\texttt{a(X)},\texttt{b}\},\texttt{a(X)},\texttt{c}.\quad (\equiv (simp(\{\texttt{M<\^{}>a(X)},\$\texttt{p}\},\texttt{M<\^{}>a(X)}),\$\gamma)\theta\,)$$
$$\xrightarrow{R} \{\texttt{a(X)},\texttt{b}\},\texttt{a(X)},\texttt{c},\texttt{ok}. \quad (\equiv (simp(\{\$\texttt{p}\},\texttt{ok}),\$\gamma)\theta\,)$$

However, no such rewriting is performed by (RQ) because the outermost non-existence quantifier M<^> does not satisfy (*NegCond*):

(*NegCond*) for the M<^>

$$= cxt(\{\{\texttt{M<\^{}>a(X)},\$\texttt{p}\},\texttt{M<\^{}>a(X)},\$\ \gamma\})\theta \xrightarrow{neg(\texttt{M},\ \{\{\texttt{M<\^{}>a(X)},\$\texttt{p}\},\texttt{M<\^{}>a(X)},\$\ \gamma\})\ :\!-}$$
$$= \{\{\underline{\texttt{a(X)}},\texttt{b},\texttt{p}^{id}\},\underline{\texttt{a(X)}},\texttt{c},\gamma^{id}\} \xrightarrow{\{\{\texttt{a(X)},\$\texttt{p}^{ct},\texttt{p}^{id}\},\texttt{a(X)},\$\gamma^{ct},\gamma^{id}\}\ :\!-}$$
$$= false\ .$$

Notice that the definition of (*NegCond*) is inductive because it refers to the reduction relation of QLMNtal. This means that we only need to explicitly check that the outermost non-existence quantifier satisfies (*NegCond*); the inner quantifiers will be checked by recursion. Note also that the inductive definition of (*NegCond*) is well-founded because each application of *neg* removes the non-existence quantifier by exactly one level.

6 More Examples

6.1 Repotting the Geraniums

"Repotting the Geraniums" [15] is a problem that requires a nested quantification rule, whose description is given below and visualized in Fig. 13:

> There are several pots, each with several geranium plants. Some pots were broken because the geraniums filled the space with their roots. New pots are prepared for the broken pots with flowering geraniums and all the flowering geraniums are moved to the new pots. — [15]

The original modeling of the problem in a visual graph rewriting tool GROOVE [10] is shown in Fig. 14.

An example of QLMNtal modelling of this problem is shown below:

```
M<+>( {type(pot),flag(cracked),N<+>t(X),$p},
    N<+>{type(geranium),flag(flowering),s(X)} ) :-
M<+>( {type(pot),flag(cracked),$p}, {type(pot),N<+>t(X)},
    N<+>{type(geranium),flag(flowering),s(X)} )
```

(a) Before repotting (b) After repotting

Fig. 13. Repotting the Geraniums

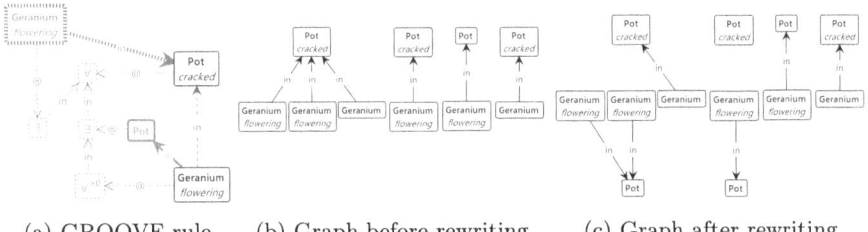

(a) GROOVE rule (b) Graph before rewriting (c) Graph after rewriting

Fig. 14. Repotting the Geraniums (screenshot from the GROOVE tool)

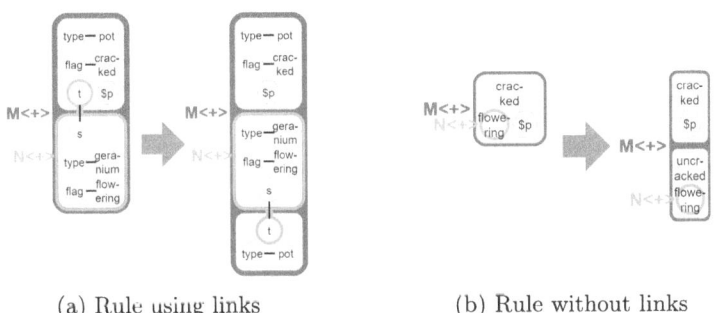

(a) Rule using links (b) Rule without links

Fig. 15. Repotting the Geraniums in QLMNtal

A visual version of this rule is given in Fig. 15(a), where the scope of the two quantifiers are shown in two different colors.

Each object is represented by a membrane, with the type (pot or geranium) represented by a unary atom connected to a type atom and the state (cracked, flowering) by a unary atom connected to a flag atom. The location of entities is represented by connecting membranes with s (source) atoms and t (target) atoms. The rewrite rule uses nested universal quantifiers, where the outer universal quantifier (labeled M) matches all cracked pots and the inner universal quantifier (labeled N) matches all flowering geraniums in each cracked pot. Because of the '<+>' quantifier, no new pots are prepared for pots without a flowering geranium, even if they are cracked.

Whereas the above rule is intended to reflect the original graph rewriting rule [15], the membrane construct of LMNtal allows more concise description of the problem that does not use any links to represent placement:

```
M<+>{cracked, N<+>flowering, $p} :-
M<+>({cracked, $p}, {uncracked, N<+>flowering}).
```

A visual version of this simplified rule is shown in Fig. 15(b).

6.2 Petri Nets

Petri Nets [4] describe discrete event systems as directed bipartite graphs. The most fundamental form, place-transition nets, consists of places (circles), transitions (rectangles), arcs and tokens (bullets). A transition fires when every input place of the transition has at least one token. When the transition fires, it removes one token from each of input places and adds one token to every output place. Figure 16 shows how t2 fires (both t1 and t2 may fire). This can be encoded concisely in QLMNtal as follows,

```
M<+>{s(A1),token,$1}, {M<+>t(A1),N<+>s(A2)}, N<+>{t(A2),$2} :-
M<+>{s(A1),$1}, {M<+>t(A1),N<+>s(A2)}, N<+>{t(A2),token,$2}
```

where the left and the right membranes in the head and the body represent places, while the middle membrane represents transition. We used labels M and N to indicate which quantifiers are interdependent and should be unrolled together. Figure 17 shows a visual representation of the QLMNtal rule.

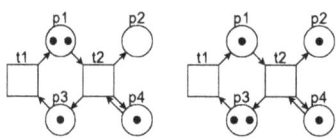

Fig. 16. Petri Net, before and after transition

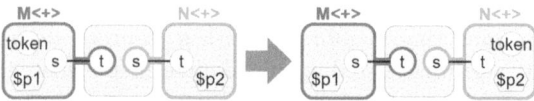

Fig. 17. Petri Net in QLMNtal

7 Related and Future Work

We first note that, while there are a number of graph rewriting tools [1,6,9,10,13,14,20], many of them are visual tools, and our work can be regarded as an attempt to design a *language with abstract syntax* and *syntax-directed operational semantics* as in other programming languages. We share this motivation with GP 2 [13], but the design principles of the two languages are very different.

Most of the graph rewriting tools provide *control structure* for rewriting, some in visual syntax (e.g., GReAT [1]) and some in textual syntax (e.g., the strategy language of PORGY [6]). In contrast to all these tools that come with separate sublanguages for execution control, (Q)LMNtal is designed as an inherently concurrent language whose execution is controlled by subgraph matching that serves as a synchronization mechanism.

While many graph rewriting tools support non-existence conditions [1,9,14,20], fewer tools feature universal quantification within a rewrite rule. The goal of QLMNtal is to enable all quantifications to be expressed within a single rewrite rule, thereby simplifying the state space.

The notion of cardinality can be found in (i) practical regular expressions, (ii) answer set programming (ASP) [8] for satisfiability and optimization problems, and (iii) some graph formalisms [1,2,5]. QLMNtal can not only specify how many copies of a subgraph should be extracted but can also rewrite all the extracted graphs in a single step.

An important feature of GROOVE [10] is that it allows combined and nested use of quantification within a single rule. Non-existence is expressed by coloring the graph, while existential and universal quantification are expressed by connecting them to special nodes representing quantifiers. In QLMNtal, the syntax-directed operational semantics allows various quantifications (including non-existence) to be combined and nested, where labeled quantifiers specify which elements of rewrite rules are jointly quantified. We believe our operational semantics can lead smoothly to logical interpretation of the proposed constructs, which is our future work.

There have been several attempts to support graph rewriting in functional languages [3,12,17]. However, in functional graph rewriting languages, handling of quantities seems (to the best of our knowledge) to be an open problem.

Future work includes extending SLIM and the LMNtal compiler. SLIM is a virtual machine for LMNtal with dedicated intermediate code. Our previous work [16] showed that a limited form of nested quantification could be checked by SLIM's instruction set. Our challenge is to support the full expressive power of QLMNtal by introducing additional control structure into SLIM that an enhanced LMNtal compiler should then support.

Acknowledgments. The authors are indebted to anonymous reviewers for their valuable comments. This work is partially supported by Grant-In-Aid for Scientific Research (23K11057), JSPS, Japan, and Waseda University Grant for Special Research Projects (2024C-432).

References

1. Balasubramanian, D., Narayanan, A., Buskirk, C., Karsai, G.: The graph rewriting and transformation language: GReAT. Electronic Communications of the EASST **1** (2006). https://doi.org/10.14279/tuj.eceasst.1.89
2. Blau, H., Immerman, N., Jensen, D.: A visual language for querying and updating graphs. Tech. rep., University of Massachusetts (2002), https://web.cs.umass.edu/publication/docs/2002/UM-CS-2002-037.pdf
3. Brus, T.H., van Eekelen, M.C.J.D., van Leer, M.O., Plasmeijer, M.J.: Clean — a language for functional graph rewriting. In: Functional Programming Languages and Computer Architecture (FPCA 1987). LNCS, vol. 274, pp. 364–384. Springer Berlin Heidelberg (1987)https://doi.org/10.1007/3-540-18317-5_20
4. Desel, J., Reisig, W.: Place/transition Petri nets. In: Reisig, W., Rozenberg, G. (eds.) Lectures on Petri Nets I: Basic Models: Advances in Petri Nets, pp. 122–173. Springer Berlin Heidelberg (1998) https://doi.org/10.1007/3-540-65306-6_15
5. Fan, W., Wu, Y., Xu, J.: Adding counting quantifiers to graph patterns. In: Proc. SIGMOD'16. pp. 1215–1230. ACM, New York, NY, USA (2016)https://doi.org/10.1145/2882903.2882937
6. Fernández, M., Kirchner, H., Namet, O.: A strategy language for graph rewriting. In: Vidal, G. (ed.) Logic-Based Program Synthesis and Transformation (LOPSTR 2011). LNCS, vol. 7225, pp. 173–188. Springer Berlin Heidelberg (2012)https://doi.org/10.1007/978-3-642-32211-2_12
7. Früwirth, T.: Theory and practice of constraint handling rules. J. Logic Program. **37**, 95–138 (1998) https://doi.org/10.1016/S0743-1066(98)10005-5
8. Gebser, M., Kaminski, R., Kaufmann, B., Shaub, T.: Answer Set Solving in Practice. Springer Cham (2013) https://doi.org/10.1007/978-3-031-01561-8
9. Geiß, R., Batz, G.V., Grund, D., Hack, S., Szalkowski, A.: GrGen: A fast SPO-based graph rewriting tool. In: Graph Transformations (ICGT 2006). LNCS, vol. 4178, pp. 383–397. Springer Berlin Heidelberg (2006).https://doi.org/10.1007/11841883_27
10. Ghamarian, A., de Mol, M., Rensink, A., Zambon, E., Zimakova, M.: Modelling and analysis using GROOVE. Int. J. Softw. Tools. Technol. Transfer **14**, 15–40 (2012) https://doi.org/10.1007/s10009-011-0186-x
11. Gocho, M., Hori, T., Ueda, K.: Evolution of the LMNtal runtime to a parallel model checker. Computer Software **28**(4), 4_137–4_157 (2011).https://doi.org/10.11309/jssst.28.4_137
12. Matsuda, K., Asada, K.: A functional reformulation of uncal graph-transformations: or, graph transformation as graph reduction. In: Proc. 2017 ACM SIGPLAN Workshop on Partial Evaluation and Program Manipulation (PEPM 2017). pp. 71–82. ACM, New York, NY, USA (2017)https://doi.org/10.1145/3018882.3018883
13. Plump, D.: The design of GP 2. Electronic Proceedings in Theoretical Computer Science **82**, 1–16 (2012) https://doi.org/10.4204/eptcs.82.1
14. Ranger, U., Weinell, E.: The graph rewriting language and environment progres. In: Applications of Graph Transformations with Industrial Relevance (AGTIVE 2007). LNCS, vol. 5088, pp. 575–576. Springer-Verlag, Berlin, Heidelberg (2008) https://doi.org/10.1007/978-3-540-89020-1_41
15. Rensink, A., Kuperus, J.H.: Repotting the geraniums: On nested graph transformation rules. Electronic Communications of the EASST **18** (2009) https://doi.org/10.14279/tuj.eceasst.18.260.262

16. Saito, R., Ueda, K.: An enhancement of negative expression and realization of universal quantification for LMNtal, a hierarchical graph rewriting language. In: Proc. 35th Conference of Japan Society for Software Science and Technology (JSSST) (2018), http://jssst.or.jp/files/user/taikai/2018/PPL/ppl5-1.pdf, (in Japanese)
17. Sano, J., Ueda, K.: Implementing the λ_{GT} language: A functional language with graphs as first-class data. In: Graph Transformation (ICGT 2023). LNCS, vol. 13961, pp. 263–277. Springer, Cham (2023).https://doi.org/10.1007/978-3-031-36709-0_14
18. Ueda, K.: LMNtal as a hierarchical logic programming language. Theoret. Comput. Sci. **410**(46), 4784–4800 (2009). https://doi.org/10.1016/j.tcs.2009.07.043
19. Ueda, K.: Logic/constraint programming and concurrency: The hard-won lessons of the fifth generation computer project. Science of Computer Programming **164**, 3–17 (2018) https://doi.org/10.1016/j.scico.2017.06.002
20. Varró, D., Balogh, A.: The model transformation language of the VIATRA2 framework. Sci. Comput. Program. **68**(3), 214–234 (2007). https://doi.org/10.1016/j.scico.2007.05.004
21. Yamamoto, N., Ueda, K.: Engineering grammar-based type checking for graph rewriting languages. IEEE Access **10**, 114612–114628 (2022). https://doi.org/10.1109/ACCESS.2022.3217913

Author Index

A
Amadini, Roberto 117
Aoto, Takahito 185

B
Barwell, Adam D. 149
Bertolissi, Clara 131
Brogi, Antonio 117
Brown, Christopher 149

C
Chong, Stephen 3

E
Erbatur, Serdar 47

F
Fernández, Maribel 131
Forti, Stefano 117
Francès de Mas, Jordina 64

G
Gazza, Simone 117
Gelderie, Marcus 166
Giallorenzo, Saverio 117

H
Hagens, Kasper 202
Hanus, Michael 27
Hu, Jingmei 3

K
Kop, Cynthia 202

L
Luff, Maximilian 166

M
Marshall, Andrew M. 47
Mishina, Haruto 220

N
Narendran, Paliath 47

P
Peltzer, Maximilian 166
Plebani, Pierluigi 117
Ponce, Francisco 117

R
Ringeissen, Christophe 47, 82

S
Sarkar, Susmit 149
Seltzer, Margo 3
Soldani, Jacopo 117
Subramani, K. 99

T
Thuraisingham, Bhavani 131

U
Ueda, Kazunori 220

V
Vigneron, Laurent 82
Vitali, Monica 117

W
Wojciechowski, Piotr 99

Z
Zavattaro, Gianluigi 117

SPRINGER NATURE

GPSR Compliance

The European Union's (EU) General Product Safety Regulation (GPSR) is a set of rules that requires consumer products to be safe and our obligations to ensure this.

If you have any concerns about our products, you can contact us on ProductSafety@springernature.com

In case Publisher is established outside the EU, the EU authorized representative is:

Springer Nature Customer Service Center GmbH
Europaplatz 3
69115 Heidelberg, Germany

The manufacturer's authorised representative in the EU is Springer Nature Customer Service Centre GmbH, Europaplatz 3, 69115 Heidelberg, Germany. If you have any concerns regarding our products, please contact ProductSafety@springernature.com

Printed and bound by CPI Group (UK) Ltd, Croydon, CR0 4YY

26/03/2026

02078962-0007